# HYDROLOGY OF THE HAWAIIAN ISLANDS

L. STEPHEN LAU AND JOHN F. MINK

# Hydrology of the Hawaiian Islands

University of Hawaiʻi Press | Honolulu

Printed in the United States of America

11 10 09 08 07 06 6 5 4 3 2 1

**Library of Congress Cataloging-in-Publication Data**
Lau, L. Stephen (Leung-Ku Stephen), 1929–
Hydrology of the Hawaiian Islands / L. Stephen Lau and John F. Mink.
p. cm.
Includes bibliographical references and index.
ISBN-13: 978-0-8248-2948-3 (hardcover : alk. paper)
ISBN-10: 0-8248-2948-4 (hardcover : alk. paper)
1. Hydrology — Hawaii. 2. Water-supply — Hawaii.
I. Mink, John F. (John Francis), 1924– II. Title.
GB832.L38 2006
551.4809969 — dc22 2006007354

University of Hawai'i Press books are printed on acid-free paper and meet the guidelines of permanence and durability of the Council on Library Resources.

Designed by April Leidig-Higgins

Frontis photo: Wailua Falls, Kaua'i, by Santos Barbasa Jr.

Printed by Maple Vail Book Manufacturing Group

To contributors to the hydrological sciences
of the humid tropics and especially the
Hawaiian Islands

CONTENTS

**ILLUSTRATIONS**

## Figures

## Plates (following page 156)

# TABLES

# PREFACE

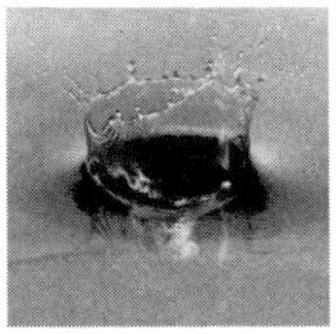

Hydrologic concepts and eventually the science of hydrology have played a vital role in the evolution of society in the islands of the Hawaiian Archipelago. Polynesians, the first settlers in Hawaiʻi, created a culture that in many ways depended on the proper application of hydrologic principles. The passage of time and the migration of peoples to the Islands from other countries brought an increase in population and the introduction of a variety of new cultures, both of which generated new demands for water and the need to expand utilization of the water resources.

In islands, resources within the confines of each island must meet water demands. This fundamental constraint encouraged the application of relatively sophisticated scientific and engineering principles to water investigations in Hawaiʻi long before this approach had become common in most other regions of the world. Besides water demand, other issues concern the protection of groundwater and coastal waters from pollution, the management of flood hazards, and the balance of stream use with protection.

Many hydrologic processes in Hawaiʻi are profoundly different from those of continents and other climatic zones. Hawaiʻi's humid tropical climate, the surrounding ocean, its volcanic earth, and high mountains govern hydrologic analysis. Management of water, land, and environments faces great uncertainties and often may be at risk of potential failure. Successful experiences in Hawaiʻi may be useful for other communities with similar environments.

A large body of literature concerning water in the Hawaiian Islands has accumulated over the last century, but the reports and documents usually relate to specific problems and their solutions. Several books that include discussions of hydrology have their primary focus on other scientific matters such as geology and volcanology. A text with its focus specifically on the fundamentals of Hawaiian hydrology is appropriate to add to the library of other texts that discuss but do not emphasize hydrologic principles.

This book comprises eight chapters, some of which discuss applications of hydrology that deal with water, land, and environmental issues in Hawaiʻi. Chapter 1, on the geological environment, presents the foundation of the water environment, providing information on rock types and their hydrologic characteristics. Chapter 2, on the hydrologic cycle, discloses a linkage of the hydrologic elements in terms of both quantity (flow cycle) and quality (transport cycle) using equations of water balance based on mass conservation. Precipitation, the starting point of the hydrologic cycle and the virtual freshwater source, is the focus of Chapter 3; and evaporation, the largest single abstractor of water, is the area of interest in Chapter 4. The subject of Chapter 5 is wetting and infiltrating the surfaces. Chapter 6 is about groundwater, the ultimate sink in the disposition of rainwater. Chapter 7 is concerned with surface runoff. Surface runoff and groundwater discharge reach the shoreline and impact the quality of coastal waters. Thus, Chapter 8, the final chapter, focuses on coastal waters.

Chapters 3 through 7 each begins with a short section on the current knowledge of the natural processes and fundamental theories involved with its specific subject. For some, this information provides background and concepts; for others, it provides a quick review. What follows are formal discussions on the distinctive characteristics of each of the hydrologic elements and their interrelations under natural and human-influenced conditions, as well as reference to modeling and database compilation, on which quantitative analysis and management are based.

Much of the book was written with the interested lay reader in mind, but some sections require familiarity with science and quantification of hydrological processes. Nevertheless, the book can serve as an introduction to the hydrology of the Hawaiian Islands as well as a reference for more advanced studies. An extensive bibliography of selected references is appended.

## ACKNOWLEDGMENTS

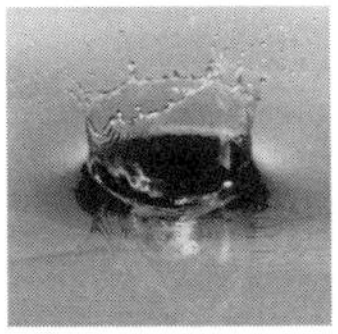

The authors are indebted to the University of Hawai'i, Water Resources Research Center's directors for their support and the Center's Publications staff for their services. The directors are James E. T. Moncur (current) and Roger S. Fujioka (former). The Publications staff involved includes Patricia Y. Hirakawa, who managed and word processed the manuscript and was responsible for the index; April W. L. Kam, who was responsible for the artwork; and Karen Y. Tanoue, who edited the initial version of the manuscript. Numerous members of the faculty and staff of various university units assisted in acquiring the literature for the References. The University of Hawai'i at Mānoa libraries provided special permission to access their book collections during the period of rehabilitation of the libraries from the devastation due to the Halloween Eve flood of 2004.

Many government agencies and nongovernment organizations, scientists, and engineers have contributed substantially to the knowledge of hydrology of the Hawaiian Islands as broadly defined in this book. They are too numerous to name individually but are partially reflected in the References. Many eminent hydrologists, including J. Bear, V. T. Chow, S. N. Davis, and D. K. Todd, and scientists in other disciplines, including P. H. McGauhey, conducted studies or offered suggestions that have contributed to the advances in the hydrology of Hawai'i.

On a personal note, the authors are grateful to their wives, Virginia M. Lau and the late Patsy T. Mink, for their understanding, encouragement, and assistance. Regrettably, coauthor John F. Mink passed away before the book was published. His contributions to the book are invaluable and appreciated.

CHAPTER ONE

# Geological Environment

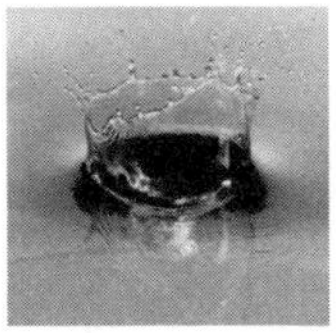

Lying in the North Pacific Ocean approximately between 19° and 28° north latitude and between 154° and 178° west longitude, the Hawaiian Archipelago is the most remote chain of islands on the planet. The arc of the Aleutian Islands is 3,700 km (2,300 mi) away; the coast of California is 3,860 km (2,400 mi) to the east; and the nearest specks of land to the west are Johnston Island (a reconstructed atoll) and Kwajalein Atoll of the Marshall Islands, 1,319 km (820 mi) and 3,931 km (2,443 mi) distant, respectively. Distances to the nearest land south of the archipelago are great also: 2,170 km (1,348 mi) to the coral reefs of Kiritimati (Christmas Island) almost due south and 7,514 km (4,670 mi) to Easter Island to the southeast. The Polynesian homelands of Samoa, the Marquesas, and Tahiti are more than 4,193 km (2,606 mi), 3,860 km (2,400 mi), and 4,410 km (2,741 mi) distant, respectively (see Figure 1.1).

The Hawaiian Islands were the last major land group to be recognized by explorers from the Western world. Not until the landfall made by Captain James Cook's expedition in 1778 was the archipelago brought to the attention of Europe and America. The Spanish may have sighted and even visited the Islands before Captain Cook, but this speculation is not supported by unequivocal historical evidence.

The Cook expedition found a thriving Polynesian society that had evolved over at least fourteen centuries beginning with the arrival of the first settlers from the Marquesas-Tahiti complex of islands south of the equator (Kirch 1985). The epic journeys northward to the remote archipelago probably took place at about the beginning of the Western calendar, A.D. 0. The original settlers landed on the barren coast of the southern tip of the island of Hawai'i. Eventually, all of the islands in the chain became populated, and a society centered chiefly on wetland agriculture blossomed.

Today nearly 1.2 million people live on seven of the eight major islands. The largest concentration, 872,000, resides on O'ahu, where the main city of Honolulu is located. The largest island, Hawai'i, has a population of 138,000, Maui has 91,000, Kaua'i has 56,000, Moloka'i has 7,000, Lāna'i has 2,000, and Ni'ihau has 250. The remaining major island, Kaho'olawe, does not have a permanent population.

The total area of the main Hawaiian Islands is 16,636 $km^2$ (6,423 $mi^2$), with the largest fraction, 10,433 $km^2$ (4,029 $mi^2$), constituting the island of Hawai'i. Second in size is Maui with 1,884 $km^2$ (727 $mi^2$), followed by O'ahu with 1,546 $km^2$ (597 $mi^2$), Kaua'i with 1,430 $km^2$ (552 $mi^2$), Moloka'i with 673 $km^2$ (260 $mi^2$), Lāna'i with 364 $km^2$ (141 $mi^2$), Ni'ihau with 180 $km^2$ (70 $mi^2$), and Kaho'olawe with 115 $km^2$ (45 $mi^2$). The highest point is on the island of Hawai'i, where Mauna Kea ascends to 4,205 m (13,796 ft) and Mauna Loa to 4,169 m (13,679 ft). Haleakalā on Maui reaches to 3,055 m (10,023 ft), Wai'ale'ale on Kaua'i to 1,569 m (5,148 ft), Kamakou on Moloka'i to 1,514 m (4,970 ft), Ka'ala on O'ahu to 1,220 m (4,003 ft), Lāna'ihale on Lāna'i to 1,026 m (3,366 ft), Pu'u Moa'ulanui on Kaho'olawe to 452 m (1,483 ft), and Pānī'au on Ni'ihau to 390 m (1,281 ft). West of Kaua'i are the small rocky and uninhabited islands of Nīhoa and Lehua, neither of which is greater than 1 $km^2$ (0.4 $mi^2$) in area. The top of Nīhoa is 275 m (903 ft) above sea level and that of Lehua is 213 m (699 ft)

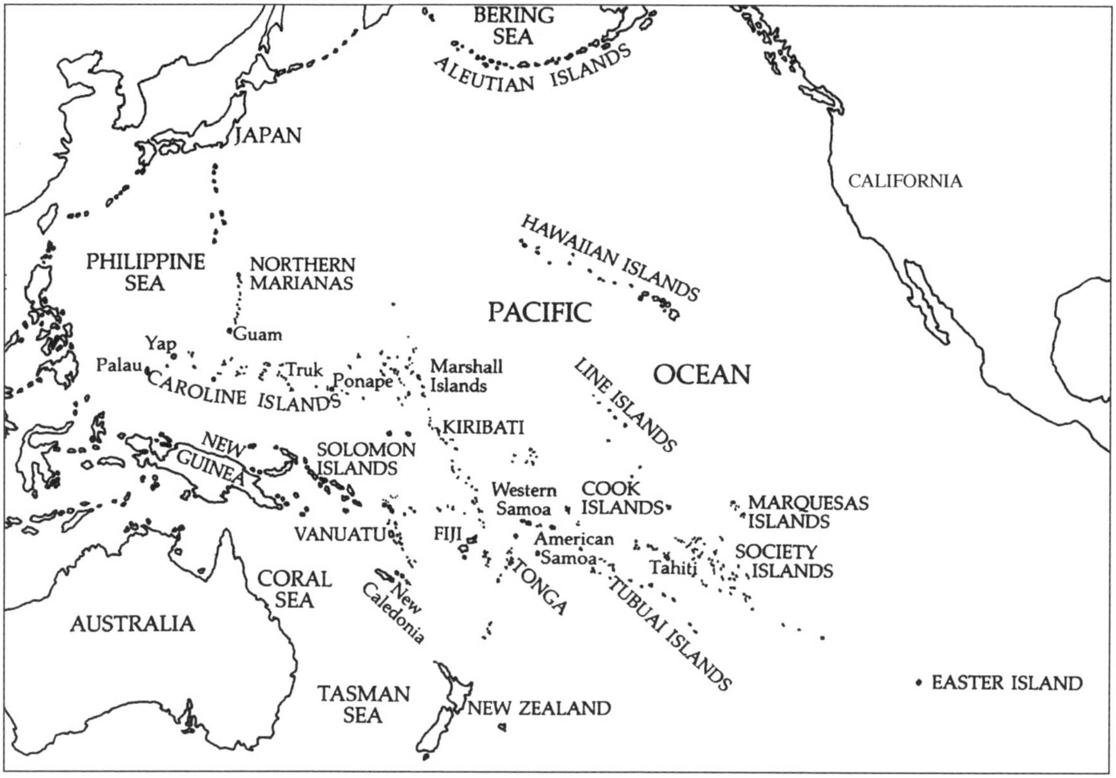


Figure 1.1. Geographic location of the Hawaiian Islands. (Adapted from Margolis 1988)

above sea level. The high topography of the islands lends them the name high islands and renders special characteristics in hydrology.

The archipelago extends northwest from Niʻihau as a string of submerged islands, of which a number surface as emerged coral reefs called the Northwestern or Leeward Islands. The farthest is Kure Atoll, a patch of exposed coral 2,034 km (1,264 mi) from Kauaʻi. Between Kure and Niʻihau are a few other fossil coral remnants named Midway Atoll, Pearl and Hermes Atoll, Lisianski Island, Laysan Island, Maro Reef, Gardner Pinnacles, French Frigate Shoals, Necker Island, and Nīhoa Island (see Figure 1.2).

Beyond Midway the submerged islands follow a linear trend that bends to a more northerly direction aimed toward the Aleutian Arc and is called the Emperor Chain. All of these submerged islands, now guyots (submarine volcanoes with a flat top), were created by the same volcanic processes that continue to generate new land on Hawaiʻi, the most southeasterly island in the Hawaiian-Emperor Chain. The entire length of the string of islands and guyots from the eastern coast of Hawaiʻi to the last identifiable submerged island nearest the Aleutians is about 4,800 km (3,000 mi). Within this extent a total of over 100 island remnants, in addition to the main Hawaiian Islands, has been counted (Clague and Dalrymple 1988).

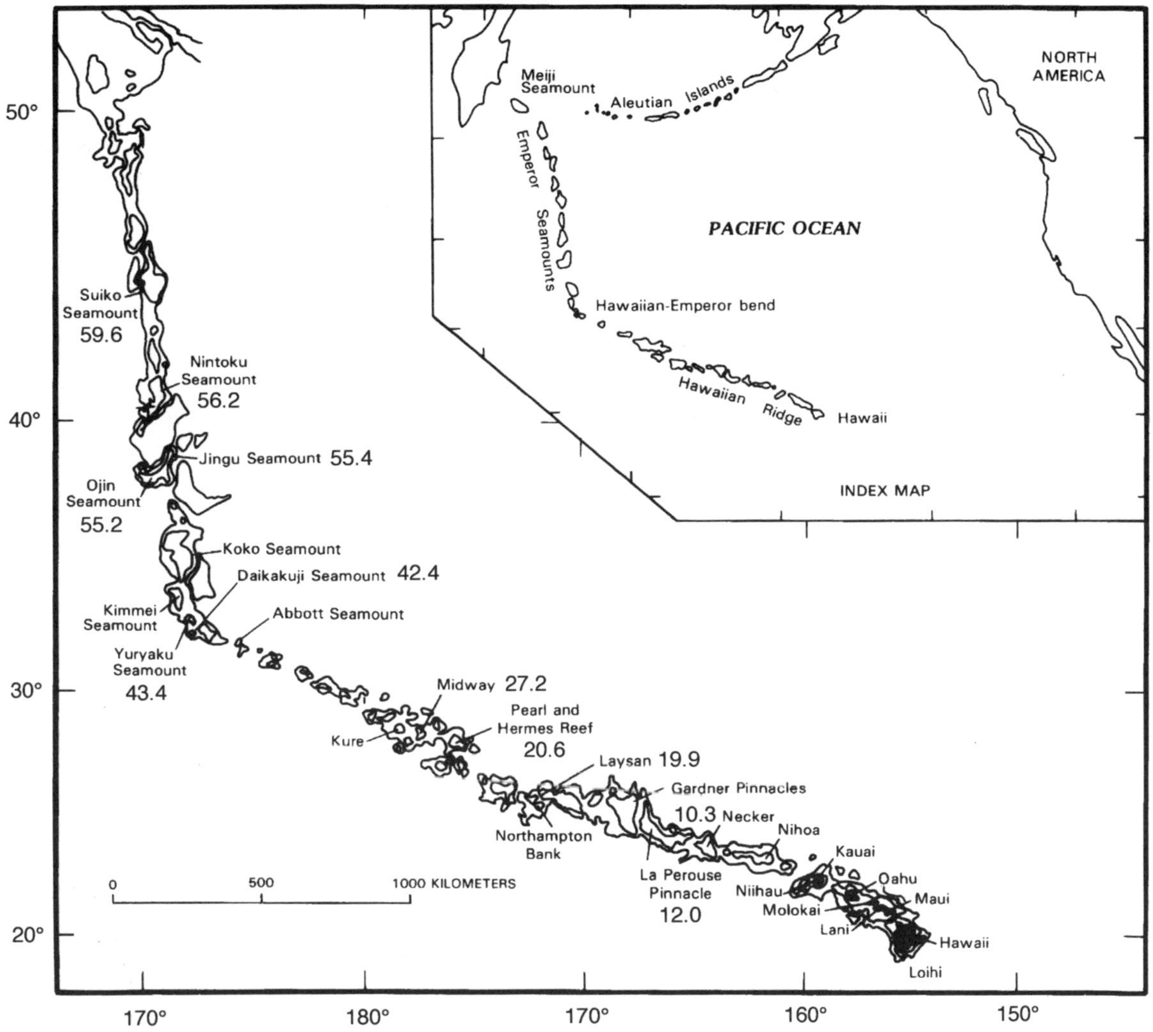


Figure 1.2. Bathymetry of the Hawaiian-Emperor volcanic chain. Contours are at 1-km and 2-km depths; numbers are age in millions of years. (Adapted from Clague and Dalrymple 1988)

## Volcanic Geology

### Origin of the Hawaiian Islands

The origin of the islands of the archipelago has been speculated upon since the arrival of the first Polynesians. As read literally, the Hawaiian *Kumulipo,* a sacred creation chant, "seems to picture the rising of the land out of the fathomless depths of the ocean" (Beckwith 1951).

The scientific rationalizations of European and American explorers and scientists before the paradigmatic shift to plate tectonics in the 1960s and 1970s generally attributed island origins to volcanic eruptions emanating from deep fractures in the Earth's crust.

The victory of plate tectonics over the diverse arguments about crustal activity that preceded it inspired new arguments about the volcanic origin of the Islands. The usually accepted explanation is that first proposed by Wilson (1963). He proposed that a "hot spot" in the mantle-crust is located about where the island of Hawai'i now rests. From this hot spot, magma ascends to the seafloor to spread out as thin layers that eventually pile up to emerge above sea level as an island. Mauna Loa and Kīlauea continue to be supplied with magma from the hot spot, and a new volcano, Lō'ihi, is building below sea level to the east of the Kīlauea coast. All of the islands, those submerged and those still above sea level, from Lō'ihi to the farthest extent of the Emperor Chain, were created with magma ascending from the hot spot.

The islands grew on the Pacific tectonic plate, whose general movement is west-northwest toward the subduction trenches that follow courses from the Aleutian Islands on the north to the Mariana Islands on the west. The slow (average 8.6 cm [2.6 in.] per year) migration of the plate carries with it the islands, but the hot spot is stationary, which accounts for the reach from the outermost fringe of the Emperor Chain to Kīlauea-Lō'ihi.

The weight of the volcanic piles created by effusions from the hot spot exceeds the strength of the crust to withstand warping. Although the volcanic masses rise higher and higher above sea level because of the frequency of eruptions, subsidence takes place simultaneously. Following the cessation of volcanic activity subsidence continues, and eventually the volcanic dome sinks below sea level to become a guyot. The fate of all volcanic islands riding on the tectonic plate is extinction, ultimately by descent into the subduction trenches that front the island arcs in the western Pacific Ocean. The successive stages in the development of a Hawaiian island are illustrated in Figure 1.3.

## Ages of the Islands

The oldest identifiable volcanic extrusions from the hot spot, which now rest at the northwestern extreme of the Emperor Chain, erupted at least 70 million years ago in Paleocene time. The youngest volcanoes, Kīlauea and Lō'ihi, as well as Mauna Loa, continue to be active. Age of individual volcanoes increases with distance from the hot spot.

Volcanic rocks of the Hawaiian Islands have been dated with the potassium-argon (KAr) method. The KAr ages were first summarized by Stearns (1966) and later by Macdonald et al. (1983). The oldest major island, Kaua'i, has rock ages ranging from 3.8 million years to 5.6 million years. Next in sequence is the Wai'anae Volcano on O'ahu at 2.7 million years to 3.4 million years, followed by the Ko'olau Volcano forming the eastern half of O'ahu at 2.2 million years to 2.5 million years. The Wai'anae Volcano probably became extinct and underwent erosion before the emergence of the larger Ko'olau dome. The West Moloka'i Volcano, whose evolution was truncated before reaching maturity, is dated at 1.8 million years, and its sister, the East Moloka'i Volcano, erupted 1.3 million years to 1.5 million years ago. Somewhat younger is the West Maui Volcano at 1.15 million years to 1.3 million years, and much younger is the East Maui Volcano at just 0.8 million years. Rocks of the Lāna'i and Kaho'olawe Volcanoes likely range from 1.0 million years to 1.5 million years. On the island of Hawai'i all of the volcanoes are less than 1 million years old. Radiometric evidence confirms the general succession of geological events and agrees approximately with the ages deduced from geological reasoning.

The voluminous fluid outpourings from the main volcanoes comprise all but a small percentage of the total rock masses exposed above sea level, but a veneer of lava and pyroclastics that erupted after the primary shields were eroded cover portions of Kaua'i, O'ahu, Moloka'i, and Maui. On Kaua'i the posterosional rocks fall in a wide age range, about several hundred thousand years to 1.5 million years old. On O'ahu the tephra edifices of Koko Head and Diamond Head are just several tens of thousands of years old. Kalaupapa Peninsula on Moloka'i is less than 100,000 years old, and late eruptions on the flanks of Haleakalā took place just over two centuries ago (1792).

In the standard geologic time scale the older islands rose above sea level in late Pliocene, the final epoch of

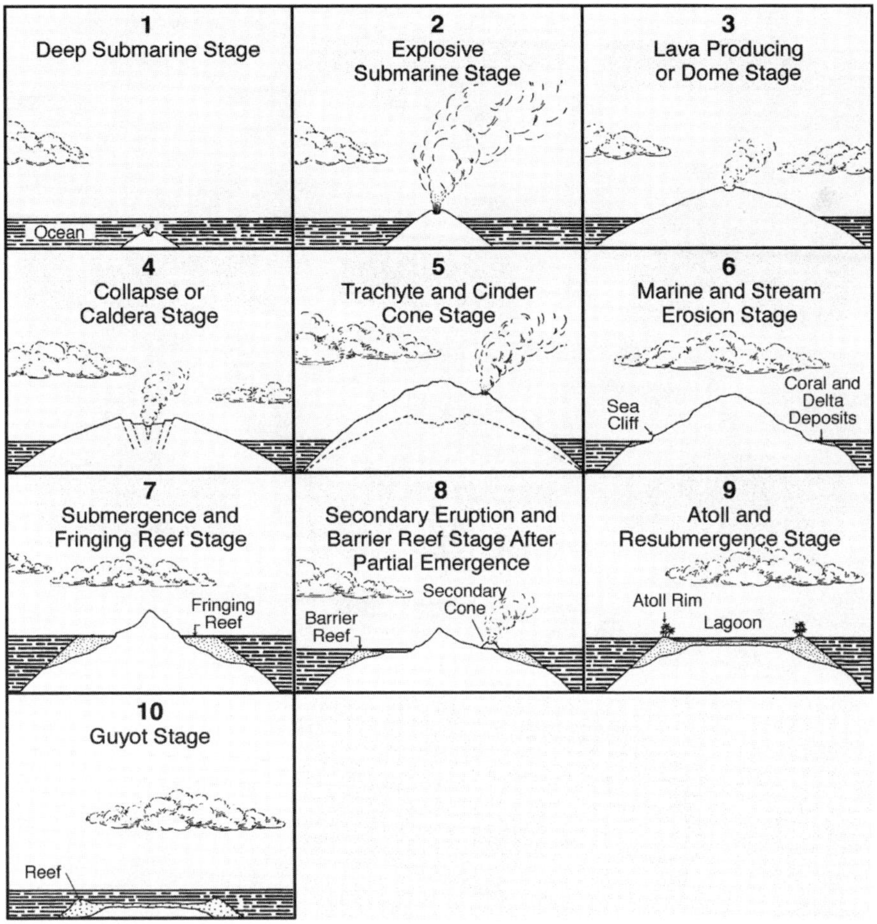


Figure 1.3. Ten stages in the geologic history of a typical volcanic island in the central Pacific. (Adapted from Stearns 1966)

the Tertiary Period. Haleakalā and the volcanoes of the island of Hawai'i emerged in the Pleistocene Epoch. The most ancient rocks on Earth have been dated at about 4 billion (4,000 million) years, and the age of the Earth is thought to be about 4.5 billion (4,500 million) years. The growth of the Hawaiian Archipelago took place over a mere 0.2 percent of the known age of the Earth.

## Volcanic Processes

The fluid lavas that reach the existing surface ascend from magma chambers, which in turn are supplied with molten rock from the hot spot. The typical initial and principal stage in the formation of a volcanic dome begins with frequent and voluminous extrusions of primitive basalt. Individual eruptions are sporadic but, viewed in the framework of geologic time, virtually continuous. Because of their fluidity, basalts can travel long distances from eruption loci. In the initial eruptive phase, basalts alter somewhat in composition to become alkalic. These basalts are more viscous than the primitive basalt and form thicker layers. The interval between eruptions widens as volcanic activity diminishes until it reaches extinction.

Each volcano is shaped like a shield and is topped with a principal caldera that connects with the magma cham-

Figure 1.4. Photographs of lavas, including *'a'ā*-clinker, *pāhoehoe,* and dikes: *(facing page, top),* *'A'ā* flows at Hālawa quarry, O'ahu, showing the dense flow interiors and interbedded unconsolidated clinker; *(facing page, bottom),* A thin *pāhoehoe* flow (shiny smooth surface) resting on the clinkery surface of an *'a'ā* flow on the south flank of Kīlauea Volcano; *(left),* Dike complex exposed in the highway cut on the H-3 freeway near Kāne'ohe Bay, O'ahu. The dikes are cutting caldera-filling lavas of the Ko'olau Volcano and average about 0.5 m thick. (Reprinted from Macdonald et al. 1983 with permission from E. Abbott and University of Hawai'i Press)

ber. Caldera diameters range from 2 to 24 km (1 to 15 mi), but most are 5 to 8 km (3 to 5 mi) wide. Associated with them are rift zones that commonly follow linear strikes from the central caldera. Lavas pour from the caldera and rift zones to create a pile of numerous thin layers known as "flanks." The layers of primitive basalt are typically less than 8 m (25 ft) thick with a lateral extent of less than 30 m (100 ft). When the extruded lavas are highly fluid, they congeal into smooth layers called "*pāhoehoe*"; when more viscous, they solidify into more massive layers called " *'a'ā.*" The upper and lower boundaries of the massive *'a'ā* core consist of "clinker," composed of rubbly and spiny fragments created from mechanical dragging (see Figure 1.4).

As the dynamics of the volcano ebbs, the caldera fills with thick and massive layers of basaltic rock along with breccia from collapsing walls. In the rift zones the lava feeder conduits solidify into thin, dense, quasi-vertical slabs called "dikes." Downslope of the caldera and rift zones are the flanks of the volcano, which are composed of thousands of thin flows piled one on another. From a distance the shape of the volcano from a rift zone to where the flank plunges into the sea or merges with another volcano follows a smooth, fairly straight line, with a slope usually of less than 10 degrees.

Embossed on the symmetry of the volcanic dome may be pyroclastic cones and irregular surfaces associated with the later phase of the primary eruptive stage and with posterosional eruptions. Within the main mass of each volcano the piling of individual primitive lava flows was too rapid for substantial weathering and erosional processes to have taken place. These flows, uninterrupted by unconformities (see Glossary), constitute the bulk of each volcano. Virtually all major aquifers in the Hawaiian Islands occur in the pile of primitive lavas on the flanks of each volcano.

During the late alkalic phase of the principal stage of volcanism when eruptive events were infrequent, minor unconformities consisting of soil beds resting on weathered zones were formed in place. These unconformities are local. In contrast, the surface that developed on the original volcanic dome as a result of the long period between cessation of primary activity and the commencement of secondary volcanism is a profound global unconformity. Except in eastern Kaua'i, posterosional volcanics are limited in area and in thickness.

## Erosion and Geomorphology

The primary volcanoes grew rapidly, too quickly to allow erosion to disfigure the symmetry of each dome. Not until the principal stage of volcanism ceased did erosion by water, wind, and the sea develop a landscape of deep valleys, narrow ridges, precipitous coastlines, and coastal plains and beaches (see Figure 1.5). The erosional processes created these features in a relatively short time — over a period of less than a million years.

Where rainfall was abundant, runoff was the most universal erosive dynamic, carving deep valleys by both physical and chemical processes. The portions of the islands facing the northeast trade winds received the highest rainfalls because of orographic lifting of moist air masses and consequently experienced the most effective down-cutting by running water. Where the crest of the mountains lies below the atmospheric (temperature) inversion layer, the lee side near the crest also receives substantial rainfall and is deeply eroded, but the drier sectors on the lee are less incised and even today retain vestiges of the original slope. The degree of landscaping by erosion correlates directly with age except for high versus low rainfall areas as just described.

In addition to surface alterations caused by surface runoff, vast landslides have been hypothesized as a cause for the precipitous mountain fronts and high cliffs along many coasts. The eastern flank of Kīlauea Volcano may be in the early stages of such a landslide. Submarine evidence of massive landslides has been found off the coasts of each of the main islands.

The deep valleys characteristic of the wet regions of each island have been cut by running water, which destabilizes the slopes by tearing away rock fragments, inducing local collapses, the debris of which remain in talus slopes or is carried downstream by floods. Both linear and arcuate deep valleys have been carved. Because high-rainfall zones are usually congruent with rift zones in which dikes are abundant, the shape of the valley depends on the orientation of the dikes. The valleys are

Figure 1.5. Photographs of initial landforms in youth stage and eroded landforms in maturity stage: *(facing page),* View northeastward toward Hōlei *Pali* and Poliokeawe *Pali,* fault scarps of the Hilina fault system on the south slope of Kīlauea, island of Hawai'i; *(above),* Ha'ikū Valley is the center of a group of three amphitheater-headed valleys that form huge scallops in the face of the Ko'olau *Pali*. Toward the camera from the ends of the steep spurs in the foreground the intervening ridges between the valleys largely have been removed by stream and wave erosion. As the face of the *pali* continues to shift westward toward Honolulu the scarp will become lower and individual peaks probably will replace the continuous cliff line. Pearl Harbor, the 'Ewa coastal plain, and the Wai'anae Range are visible in the background; *(left),* The great *pali* along the windward side of the Ko'olau Range extends from the southeastern tip of O'ahu behind the community of Waimānalo northward until it disappears from view beyond Kāne'ohe. The portion of the *pali* at the lower edge of the photograph is a sea cliff, but the rest was cut principally by stream erosion. Olomana Peak is the isolated peak in the midground. (Reprinted from Macdonald et al. 1983 with permission from E. Abbott and University of Hawai'i Press)

linear where their axes parallel the strikes of the dikes and arcuate where their direction is perpendicular to the trend of the dikes.

Linear valleys are narrow where they encounter the rift zone because dikes are ramparts against widening. On the other hand, in arcuate valleys the dikes are undercut by stream action and their collapse widens the head of the valley into an "amphitheater." The occurrence of groundwater in compartments between dikes hastens erosion where a dike is breached.

The flanks of volcanoes are less deeply eroded than the rift zones, even in high-rainfall regions. In the older islands dry areas may be appreciably eroded, but in the younger islands the contours of the original volcanic dome persist.

Wind as an erosive force has created limited but interesting landforms. The effect of winds is mostly restricted to coastal regions where large dunes were created. During sea levels that were lower than present sea level, large quantities of sand were blown considerable distances inland. Extensive areas of fossil dunes are found on the windward coasts of O'ahu, Moloka'i, and Lāna'i.

Second to running water in importance as an erosive process is the behavior of the sea. In the absence of a substantial offshore coral reef, the unimpeded surf carves a rugged coastline of exposed volcanic bedrock and sea cliffs. Where coral reefs mitigate the power of the waves, sediments accumulate and coastal plains develop. A stable sea level governs the distribution of sediments carried by streams from the erosion front inland.

The level of the sea, although stable over long intervals, has not been constant relative to the landmasses. Subsidence of the heavy volcanic load in response to isostatic adjustment has been responsible for the drowning of hundreds of feet of land surface. Sea level changes induced by waxing and waning glacial periods during the Pleistocene Epoch have been responsible for the contemporary configuration of coastal regions. The coastal plains of layered sediments are the consequence of sea level variations ranging from about 30 m (100 ft) above current sea level to 91 to 122 m (300 to 400 ft) below during the Pleistocene.

The most pronounced coastal features were caused by two separate stable sea level intervals. H. T. Stearns studied sea level changes in Hawai'i in great detail, and his nomenclature is applied to the various sea level stages. The inland reach of coastal plains is mainly the result of the Ka'ena Stand of the sea (Yarmouth Interglacial in the geological literature of North America) when the sea was 27 to 30 m (90 to 100 ft) higher than today. The large coastal plain of southern O'ahu from Barbers Point to Hawai'i Kai is the fossil remnant of this stand of the sea. The other important sea level stage, the Waimānalo (Sangamon Interglacial), stood 8 m (25 ft) higher than present sea level. It left behind fossil coral reefs and sand dunes. The consequences of these sea levels are especially evident on O'ahu.

Table 1.1 presents a history of important sea levels as identified by Stearns (1966). Those deeper than about 91 m (300 ft) below the present level were caused by stillstands during subsidence. The others are related to Pleistocene glacial and interglacial periods.

Plate 1.1 is a simplified geologic map of the eight major Hawaiian Islands.

## Rock Types: Composition and Hydrologic Character

The volcanoes of all of the Islands followed a similar early history, but the final stages of the evolution of rock types sometimes differed. The initial and most voluminous eruptive rocks were basaltic in composition and consisted of highly fluid flows that lithified into discontinuous, inhomogeneous layers averaging less than 3 m (10 ft) thick. These "primitive basalts" account for more than 95 percent of the total rock mass of the Islands and compose an even larger share of the subsurface geology in the zone of saturation.

Compositionally, these basalts are either "tholeiitic" or "alkalic." The tholeiites have relatively more silica and calcium and less sodium and potassium than the alkalic basalts, but structurally and in mode of emplacement both varieties are so similar as to be indistinguishable. Not

Table 1.1. Ancient Shorelines (Stable Sea Level) along the Island of O'ahu

| Elevation[a] m | Elevation[a] ft | Shelf or Terrace | Years before Present (approximate) | Remarks |
|---|---|---|---|---|
| −91 | −300 | Māmala | 15,000 | Maximum Wisconsin Glacial; the glaciers melted rapidly about 11,000 years ago |
| +8 | +25 | Waimānalo | 125,000 | Maximum Sangamon Interglacial |
| −18 | −60 | Waipi'o | | Probable Illinoisan Glacial; extensive dunes formed |
| +29 | +95 | Ka'ena | 475,000 | Maximum Yarmouth Interglacial |
| −366 | −1,200 | Lualualei | | Still-stand during isostatic subsidence |

SOURCE: Stearns 1966.
[a]Measurements are presented as above or below present mean sea level.

until after the primary epoch of island building, during which the essential size and shape of the volcanic shields were established, were the primitive basalts succeeded by other eruptives that differed from the basic magma.

In the early, most voluminous eruptive stage, each volcano rose as a monolithologic pile of lava units that accumulated too rapidly for substantial interflow erosion to take place. In some of the volcanoes an intermediate stage of eruption, transitional from the early stage, followed, during which the lavas formed into andesitic and trachytic rocks that were chemically and mineralogically differentiated from the basic primitive basalts. At some volcanoes a mild erosional hiatus preceded the intermediate eruptive phase, but at others the transition was unbroken. The early and succeeding stages were followed by a long period of quiescence, during which the volcanic shields were deeply eroded. Deep valleys, narrow ridges, and highly dissected terrains were formed, and large quantities of sediments were laid down in the lower reaches of valleys and along coasts. Subsidence accompanied and followed the initial erosion, and subsequent changes in relative sea level created a complicated succession of terrestrial and marine sediments at elevations below about 30 m (100 ft) above present sea level.

A final phase of eruptive activity occurred at a few of the older volcanoes in relatively recent times after erosion had reduced the shields to a mountainous landscape. The new rocks range from ordinary basalts to extremely basic nephelinitic and associated basalts. This posterosional stage of volcanism is relatively minor in overall hydrologic importance, except in eastern Kaua'i, in contrast to the earlier eruptive periods and the interval of profound erosion.

Concurrently with the main period of volcanism and for a long time afterward, the emplaced lavas in the calderas and along portions of the rift zones were affected by hydrothermal alteration — in cases so intense that the mineralogy and chemistry of the lavas were severely altered. The resulting metamorphic rocks are restricted to former zones of eruption, in particular the near vicinity of calderas, where they control hydrologic behavior.

## Hydrologic Characteristics of Volcanic Rocks

Volcanic rocks are either extrusive or intrusive. The bulk of the volume consists of the extrusive variety, which effused from fissures as molten lava onto the surface before solidifying into flow units ranging from highly fragmented piles of debris to dense, massive beds. A much smaller amount was blown out of vents as pyroclastics. Intrusive rocks solidified below the surface into more homogeneous units than the eruptive series. Although lava flows occur in all three of the principal structural features of volcanic shields (calderas, rift zones, and flanks), intrusive rocks are commonly restricted to calderas and rift zones, the spatially most-limited terrains. Pyroclastic material, also called "tephra," is most noticeable along the eruptive zones but is also found within the flank sequence—often as the result of wind dispersal of ejecta from erupting sites.

## Extrusive Rocks

The chemical composition of basic basaltic magma gives rise to fluid lava flows that travel easily down slopes for considerable distances before congealing. These lavas accumulate as thin layers to create the shield shape characteristic of Hawaiian volcanoes. The andesitic and trachytic lavas that differentiate from the basic magma are more viscous and solidify into thicker, more massive units.

By chemical composition and mineralogy, the volcanic rocks are classified as belonging to either the Tholeiitic Suite or the Alkalic Suite (Macdonald and Abbott 1970; Clague and Dalrymple 1988). The Tholeiitic Suite is dominated by tholeiitic basalt and accounts for all but a small percentage of the total volcanic mass. These basalts, characterized by a chemical composition of about 10 percent CaO and 49 percent $SiO_2$ and mineralogy dominated by calcic plagioclase and pyroxene, are dark and fine-grained and often contain phenocrysts of olivine. Basalts having slightly more $SiO_2$ but about the same percentage of CaO are called alkalic basalt of the Tholeiitic Suite. They are similar to tholeiitic basalt in mineralogy and texture.

The Alkalic Suite is gradational from the Tholeiitic Suite and comprises a series of types from andesitic rocks to trachyte. In this series the CaO decreases and the $SiO_2$ increases. The molten lavas of the Alkalic Suite are more viscous than are tholeiites. The eruptive succession of primitive basalts is frequent enough to forestall the generation of substantial interflow weathering and erosion, but the much more sporadic frequency of Alkalic Suite eruptions often allows local weathering and erosion to take place.

In Hawai'i, andesitic rocks are classified as hawaiite or mugearite. Hawaiite is nearest to tholeiites in composition; mugearite is farther along the differentiation sequence. Hawaiite is widespread in East Maui and in the Wai'anae Range of O'ahu, and mugearite blankets much of the Kohala mountains on the island of Hawai'i. The position of these rocks as a thin cover on the tholeiite shield limits their opportunities to be primary aquifers.

The end member of the Alkalic Suite are trachytes, which volumetrically are very small, but where they occur these rocks form dramatic landscapes. Trachytes are the most viscous of extrusive rocks. Rather than constituting a succession of irregular layers, they extrude as bulbous masses that congeal near the locus of eruption. Their principal occurrences are in West Maui, where they were named "bulbous domes" by Stearns and Macdonald (1942), and at Pu'u Wa'awa'a on the island of Hawai'i. Nowhere are they developable as aquifers.

The extrusives of the Alkalic Suite are the final product of the volcano's constructional stage. Most of the shield was built over a period of about 1 million years by the tholeiitic outpourings, and the final cover of andesitic rocks accounts for several hundred thousand years of activity. Dormancy was followed by a long period of erosion, during which the shape of the present landscape was created. Then in portions of a few of the Islands a new but less voluminous phase of volcanism took place. Originally referred to as the "post-erosional" stage (Stearns and Vaksvik 1935) but since renamed the "rejuvenation" stage (Macdonald et al. 1983), the active vents produced unusual basalts containing nephelinite in place of calcic plagioclase. In eastern Kaua'i much of the terrain is covered by nephelinitic basalts, and in southeastern O'ahu the most famous landmarks, such as Diamond Head and Punchbowl Crater, were derived from pyroclastics (fragments that are thrown into the air and then fall back to the ground) of these late volcanics. In Kaua'i the rejuvenation-stage volcanics play an important role in hydrology. In southeastern O'ahu the formations are less extensive but locally are hydrologically significant.

The final structure of a lava unit is largely determined by the grade down which the original fluid magma flowed. In the Hawaiian Islands practically all of the observable lavas have been subaerially extruded; few submarine lavas are exposed or have been encountered in borings. The flanks of the volcanic shields have slopes usually greater than 3 degrees and less than 10 degrees, a grade permitting rapid flow and subsequent cooling into heterogeneous piles of *pāhoehoe* and *ʻaʻā*, the principal extrusive rock forms. These rocks differ in structure but are chemically identical. In calderas, in pit craters of rift zones, and in topographic depressions from which effusions could not escape, layers of dense and massive lavas have accumulated. However, the layered flow units along the rift zones are mainly of the flank type.

*Pāhoehoe* refers to flows having smooth, hummocky, glassy surfaces surrounding highly vesicular interiors. A single *pāhoehoe* unit may consist of many individual toes, pods, and other ovoidal shapes. Although highly porous, *pāhoehoe* has little inherent permeability because the vesicles may not be connected, but en masse it may be highly permeable because of secondary structural features. This permeability arises as a result of flow units that are unconformably matched, leaving openings between the units; lava tubes that were originally conduits from which lava drained before cooling; and cooling joints that are normal to flow surfaces. Lava tubes are the largest single permeability element, having diameters ranging from 0.3 to 6.1 m (1 to 20 ft). They are not, however, as common as other permeability elements. *Pāhoehoe* is more characteristic of primitive basalts than of basaltic differentiates.

In contrast to the smoothness of *pāhoehoe*, the *ʻaʻā*-clinker association consists of a dense, massive, and discontinuous central phase, the *ʻaʻā* core, bounded by spiny, fragmented lava breccia called "clinker." Vesicles in *ʻaʻā* tend to be relatively few, large, and irregular in shape. As in *pāhoehoe*, practically all permeability of the *ʻaʻā*-clinker association results from structural features created in the course of emplacement and cooling of the fluid lavas. The openings in clinker beds are the most effective of the common permeability elements of the extrusive rocks. In the massive *ʻaʻā* phase, cooling yields vertical joints, also important permeability elements. This vertical component of permeability is enhanced by frequent bridging of clinker across flows. Andesitic lavas fall mostly into the *ʻaʻā*-clinker association.

*Pāhoehoe* and *ʻaʻā*-clinker cannot be treated as separate rock masses with respect to hydrological characteristics. The unit is usually limited laterally to an order of hundreds of feet and vertically to an order of tens of feet. A *pāhoehoe* unit may be succeeded by an *ʻaʻā* unit, which in turn may be covered by another *pāhoehoe* flow. A representative hydrogeologic volume includes both varieties of emplacement.

*Pāhoehoe* is more common near zones of eruption because of its fluidity, whereas *ʻaʻā* is more common toward the downslope margins of the shield. In leeward Oʻahu at a distance of 8 km (5 mi) from the Koʻolau rift zone, about 75 percent of the rock consists of *ʻaʻā*, of which clinker composes up to 45 percent (Wentworth and Macdonald 1953). Electric logs taken in Kaimukī, also 8 km leeward of the Koʻolau crest, were interpreted as showing 45 percent clinker (Lao et al. 1969). Transects up Kīpapa Valley in the central Koʻolau Range, 3 to 5 km (2 to 3 mi) leeward of the rift zone, show 50 to 70 percent of the rock consisting of *ʻaʻā*-clinker. Few soil or pyroclastic layers break the monotony of the primitive extrusive sequence.

Pyroclastic rocks account for only a small fraction of the total extruded mass — no more than 5 percent and probably less than 1 percent. On an areal scale the dominant pyroclastic forms are ash and tuff. Ash, which consists of particles having a diameter of 4 mm (0.016 in.) or less, consolidates and compacts into tuff, but tuff may also include similarly compacted larger ejecta. Constituents, particularly glass, in the ash and tuff alter at ordinary temperatures into complex hydrates called palagonite. Palagonitized ash and tuff display extremely low permeability. Cinder and larger ejecta retain a higher fraction of their original permeability than the finer-grained material.

The most widespread exposure of pyroclastics occurs on Hawaiʻi Island. Pāhala Ash covers 518 to 777 $km^2$ (200 to 300 $mi^2$) of the northern and eastern sectors of the island to a maximum thickness of about 15 m (50 ft). Pyroclastics of the rejuvenated-stage Honolulu Volcanic series cover an appreciable area of southeastern Oʻahu.

Only in a large-scale sense can statistically describable aquifer parameters be assigned to the volcanic rocks. Aquifer hydraulics of the extrusive rocks are controlled by regional values of hydraulic conductivity and effective porosity, but local characteristics may control the behavior of individual wells and other engineering constructions. Also, a tremendous difference exists between essentially unweathered lava flows and in situ weathered units. All of the great aquifers (see Glossary) of Hawai'i are composed of extrusive rocks, chiefly of tholeiitic primitive basalts and olivine basalts.

Laboratory determinations of hydraulic conductivity and porosity need to be interpreted cautiously because only a tiny piece of the heterogeneous rock matrix can be tested at one time. Consequently, few laboratory measurements have been attempted. Wentworth (1951) found a range of porosity of 5.2 to 51.4 percent in drill cores of the Ko'olau basalt and 1 to 16 percent in cores of the posterosional Honolulu series. He also reported laboratory measurements of hydraulic conductivity ranging from $6 \times 10^{-4}$ to 713 m/d (0.002 to 2,340 ft/d) for ash and $8 \times 10^{-2}$ to 30 m/d (0.27 to 98 ft/d) for cinder from the Honolulu area. Ishizaki et al. (1967) determined the porosity of dense, hard *'a'ā* ("blue rock") as 7.7 to 10.4 percent and hydraulic conductivity as $1.1 \times 10^{-5}$ m/d ($3.5 \times 10^{-5}$ ft/d)—exceedingly small indeed and not at all representative of the extrusive rock assemblage.

Gravity surveys in shafts and tunnels in O'ahu, Maui, and Hawai'i showed the regional porosity of unweathered lavas to be about 20 percent (Huber and Adams 1971). The effective porosity is considerably less, however.

The regional hydraulic properties of extrusive volcanic rocks are commonly deduced from analysis of pumping test results or by estimates of total groundwater flow derived from hydrologic budget accounts. Pumping tests give transmissivity from which hydraulic conductivity is inferred from the relationship $K = T/b$, where $K$ is hydraulic conductivity, $T$ is transmissivity, and $b$ is depth of flow to the pumping well (refer to Chapter 6). It is impossible to stipulate categorically the depth of flow in a thick basal groundwater that is not confined below and which wells and galleries are only partially penetrating. Usually, depth of flow in an unconfined lens is taken as coincident with the theoretical thickness of the static thickness of the freshwater—calculated as equal to 41 times the freshwater head in the ideal buoyancy case. This assumption leads to a minimum value for hydraulic conductivity (Chapter 6).

Wentworth (1938) conducted the first regional aquifer test in the Hawaiian Islands and computed the hydraulic conductivity of the Ko'olau basalt as lying in the range of 554 to 1,072 m/d (1,818 to 3,516 ft/d). Many similar tests have been conducted since, and a summary of the most informative ones are given in Williams and Soroos (1973). Further summarizing their work, a total of 51 pumping tests on wells in O'ahu and Maui show an average transmissivity of $1 \times 10^5$ m²/d ($1.1 \times 10^6$ ft²/d)—equivalent to a hydraulic conductivity of 365 m/d (1,200 ft/d) if the theoretical thickness of the lens is taken as depth of flow.

A regional aquifer analysis of south central O'ahu was performed by Mink (1980) employing data derived from the aquifer response to the instantaneous shutdown of pumping due to a labor strike in the sugar plantations in 1957. At that time virtually all pumpage in the region was done for sugarcane irrigation. Analysis of recovery curves following the cessation of pumping yielded a hydraulic conductivity of 457 m/d (1,500 ft/d) and an effective porosity of 5 percent. Employing the same data obtained on resumption of pumping, Souza and Voss (1987) calculated hydraulic conductivity as 457 m/d and specific yield (effective porosity) as 4 percent.

Although not definable as a precise, unvarying number for any of the aquifers, the regional hydraulic conductivity of the major aquifers consisting of unweathered primitive basalts and olivine basalts that were laid down as flank flows dipping between 3 and 10 degrees lies in the range of 305 to 1,524 m/d (1,000 to 5,000 ft/d), with the most probable values centering around 457 m/d (1,500 ft/d). On the island of Hawai'i the newer lavas are even more permeable. A value of 2,167 m/d (7,110 ft/d) was calculated from data of a field experiment returning ocean water at a high rate in a trench in the Hualālai Volcanics (Lau and El-Kadi 1995).

The intermediate extrusive rocks could be expected to have a somewhat lower overall hydraulic conductivity because of the greater thickness of the massive parts of the flows. The posterosional lavas congealed frequently on gentler slopes, resulting in dense and massive rocks for which a considerably lower average hydraulic conductivity would apply.

Manifestly, the regional hydraulic conductivity of extrusive rocks is extremely high. On casual observation the layered sequencing of lava units suggests that the horizontal component of permeability should be greater than the vertical component. However, a quantitative test of this hypothesis would be extremely difficult to make, and none has been accomplished in Hawai‘i. Nevertheless, in analyses involving aquifer behavior, judgments are made that the horizontal component is considerably larger than the vertical component. A value of 200:1 was suggested by the 1957 poststrike drawdown data (Souza and Voss 1987). Whether such an assumption is justified or not remains a question.

Extrusive rocks that have been weathered, especially those underlying a blanket of older alluvium and marine sediments, exhibit hydraulic properties grossly inferior to those of unaltered rocks. The differences are so great that weathered rocks are a hydrologic regime distinct from the deeper aquifer. The effects of weathering are most pronounced in valleys and in coastal plains where a column of sediments overlies the original basalt. On interstream ridges and slopes situated above valley sediments, the effects of weathering, although evident, are far less effective than in valleys.

Typically, the weathered section (called saprolite) overlying fresh basalt is 15 to 46 m (50 to 150 ft). Its hydraulic conductivity is several magnitudes less than that of unaltered basalt. In the weathered material the original permeability elements are clogged by in situ chemical alteration and by clays and colloids that precipitate from percolating solutions. Wentworth's (1938) laboratory determinations of hydraulic conductivity for four samples of weathered Ko‘olau basalt showed a range of 0.025 to 0.039 m/d (0.083 to 0.128 ft/d). A field test for permeability of the weathered Ko‘olau section underlying older alluvium in lower Waiawa Valley in southern O‘ahu yielded an average value of only $8.5 \times 10^{-8}$ m/d ($2.8 \times 10^{-7}$ ft/d) — virtually impermeable (Towill Corporation 1978). Laboratory testing of drill cores in the weathered zone (saprolite) taken for contamination studies in central O‘ahu showed hydraulic conductivities over a range of 0.13 to 3.65 m/d (0.42 to 12.0 ft/d) (Miller et al. 1988). Values of hydraulic conductivity for seven laboratory-tested small cores of saprolites from Schofield Barracks, O‘ahu, ranged from $1.0 \times 10^{-5}$ to $2.1 \times 10^{-2}$ m/d ($3.4 \times 10^{-5}$ to $6.8 \times 10^{-2}$ ft/d). However, an in situ value obtained from an infiltration test in a large (20.9 $m^2$ [225 $ft^2$]) instrumented basin at the same location was much higher, about 0.9 m/d (3 ft/d), probably because of possible fractures in the saprolites (Harding Lawson Associates 1996). A pump test in saprolite at Kunia, O‘ahu, gave a maximum hydraulic conductivity of 0.31 m/d (1 ft/d).

## Intrusive Rocks

Intrusive rocks of Hawaiian volcanoes consist almost entirely of magma congealed in the structures through which it was moving on its way to effusion at the surface. There are no large plutonic masses in the Islands. Individual intrusive units are small-scale features, but because they are concentrated along rift zones and in the caldera region, they play a fundamental role in the hydrology of each island.

Usually, the extrusive zones of a Hawaiian volcano comprise a caldera and several narrow rift zones radiating from it. Calderas are usually less than 11.3 km (7 mi) in diameter and contain an intrusive assemblage of dikes, stocks, and sills mixed with collapsed breccia and pyroclastics. Rift zones are usually less than 4.8 km (3 mi) wide but extend for tens of miles along linear trends. Their dominant intrusive rocks are dikes. Sills, mostly as the horizontal expression of dikes, are frequent but limited in dimensions. Stocks, as subjacent intrusive bodies, and other small intrusive bodies are rare and hydrologically insignificant.

Dikes of the rift zones are the most widespread intrusive rocks. Volumetrically, they account for only a tiny proportion of the volcanic masses, but their hydrologic significance is immensely greater in degree. The dike bands behave as very low permeability barriers to the flow of groundwater, in contrast to the extraordinary flow characteristics of the layered flank lavas.

Single dikes tend to have a vertical attitude and a thickness of less than 3.1 m (10 ft). In the main part of the Koʻolau Range of Oʻahu the average dike is 1.5 to 1.8 m (5 to 6 ft) thick. Multiple dikes with practically no extrusive rocks between them are common. Between most dikes lie compartments of ordinary, layered lavas that often form small aquifers.

Where dikes are most frequent, typically along the axis of a rift, the zone is called the "dike complex." Here the dikes account for more than 10 percent of the total rock mass, and the flow lavas cut by the dikes are too small in volume to behave as unit aquifers. The remainder of the rift consists of the "marginal dike zone," where dikes make up less than 5 percent of the rock mass. In the marginal zone interdike aquifers of considerable dimensions are often found.

Most dikes have such low permeability that they serve, individually or in combination, as nearly impermeable vertical boundaries. They are composed of dense basalt with a thin coat of glassy selvage at their contact with the extrusive lavas.

### Metamorphic Rocks

Metamorphic rocks in Hawaiʻi are associated with caldera and accompanying hydrothermal activity, usually deep in the rift zones. Surface exposures are limited to the vicinity of a caldera and do not play an important role in regional hydrological processes.

Basalts altered by volcanic gases and hydrothermal solutions become metamorphic rocks. Olivine converts to serpentine and talc, and pyroxenes to chlorite (Macdonald and Abbott 1970). Precipitation from circulating solutions yields quartz, opal, calcite, and zeolites that fill vesicles and clog permeability elements, greatly reducing the porosity and hydraulic conductivity of the rock mass. Metamorphic rocks are too poorly permeable to behave as aquifers.

## Hydrologic Characteristics of Sedimentary Rocks

For the island of Oʻahu, Wentworth (1951) divided the sedimentary rocks into older, intermediate, and recent alluvial and marine sediments. This classification is most applicable to the older islands of Kauaʻi, Oʻahu, Molokaʻi, and Maui but less so to Lānaʻi and Hawaiʻi. Nevertheless, it is convenient for descriptive purposes.

The older terrestrial alluvium consists of the detritus generated during the initial period of profound erosion. It fills the deeply carved valleys and spreads out as deltaic platforms along coastal margins. The older alluvium is composed of various-sized particles of volcanic rock from silt to boulders, weakly cemented in a clay matrix in which the individual particles retain their apparent original shape but have been altered by weathering and compaction into a coherent mass.

The older alluvium lies immediately above weathered volcanic bedrock in the lower valley reaches and beneath coastal plains. This alluvium, which is referred to as "valley fill," is thickest below an elevation of approximately 30.5 m (100 ft), but it may extend inland as a narrow tongue to an elevation of several hundred feet. Below the coastal plain and in lower valleys it is hundreds of feet thick. The older alluvium is a major component of the caprock that rims portions of the older islands. The maximum depth of sedimentary aquitards overlying basal aquifers is undetermined but is at least 457 m (1,500 ft) below sea level on Oʻahu (Stearns and Vaksvik 1935).

Wentworth (1938) measured parameters of the older alluvium in the laboratory. Eight samples gave a range in porosity of 46.4 to 62.4 percent and in hydraulic conductivity of 0.006 to 0.113 m/s (0.019 to 0.37 ft/d) — approximately 10,000 times less than the hydraulic conductivity of fresh basalt but of about the same magnitude as that of the saprolite of weathered basalt. The most effective confining layering in the caprock consists of weathered basalt overlain by older alluvium.

The older marine sediments interfinger with and overlie the older alluvium. The basal part of the marine section consists predominantly of estuarine and lagoonal mud, silt, and sand. Fossil coral reefs and associated detritus appear in the middle and upper portions of the section. The older marine sediments grade without noticeable unconformity into the intermediate marine sediments, which reach to an elevation of 30.5 m (100 ft) above sea

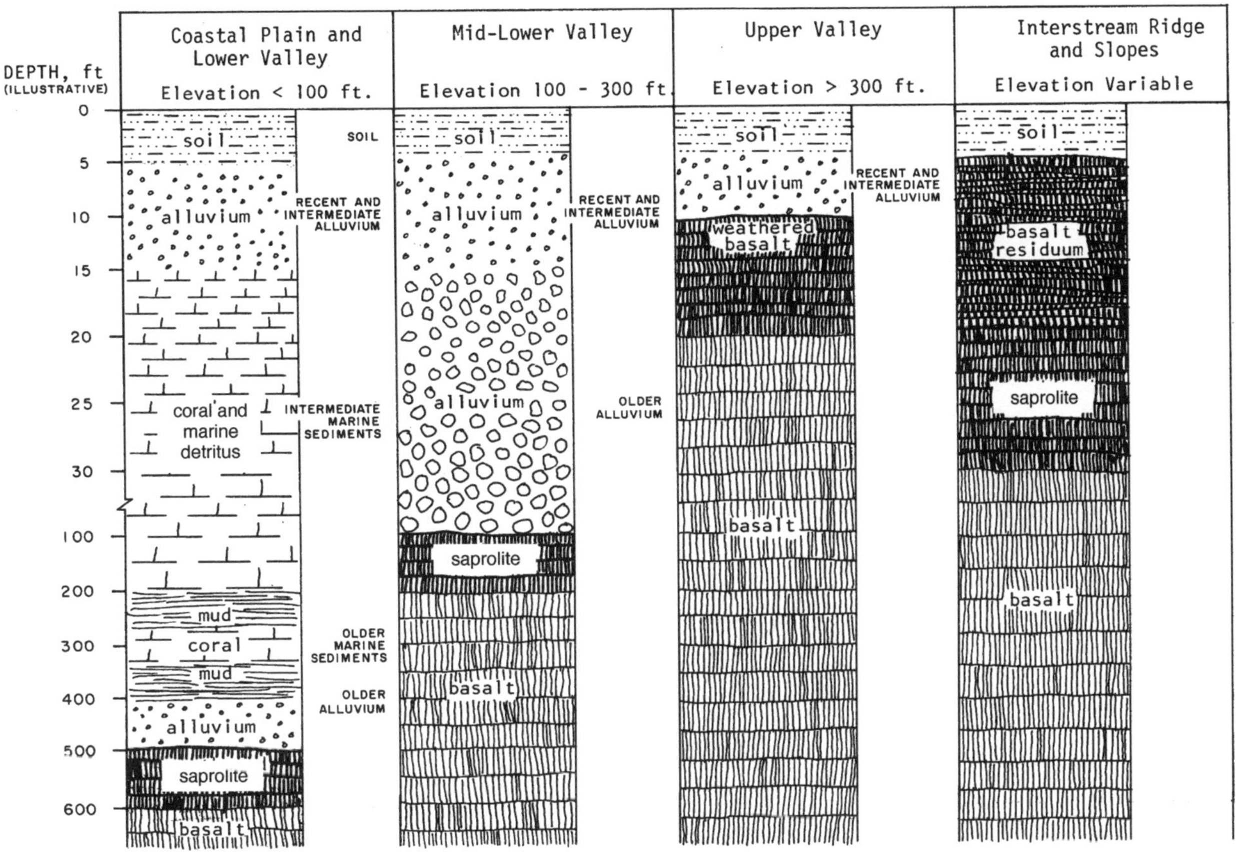


Figure 1.6. Typical rock sequences in older Hawaiian islands. (Adapted from Mink and Lau 1980)

level. Coral reef deposits are more common in the intermediate sequence. In the ʻEwa coastal plain of southern Oʻahu the top 30.5 to 61.0 m (100 to 200 ft) of the caprock consists chiefly of limestone of the intermediate series.

The clays, muds, and silts of the marine sediments are as poorly permeable as the older alluvium, but fossil coral reefs often are more permeable than unweathered basalt. The whole of the caprock is saturated with salty to brackish water, with the least salty water restricted to the upper limestone section. An examination of the logs of borings in the Honolulu coastal plain by H. S. Palmer (Wentworth 1951) showed the distribution of materials in the caprock as 50 percent clay, 40 percent coral, 7 percent gravel and sand, and 3 percent ash and tuff.

The intermediate alluvium embraces the detritus that was formed subsequent to the major erosional stage but before contemporary activity. It includes colluvium, talluvium, and, in general, the slope wash mantle overlying bedrock in the mountain areas. Neither as voluminous nor as stable as the older alluvium, it serves as aquifers on only a small scale and in unusual circumstances.

Recent alluvium is even less important hydrologically than intermediate alluvium. It consists of the clays, sands, and boulders deposited since the last high stand of the sea, the Waimānalo Stand (+8 m [25 ft]) of about 125,000 years ago. Except in unique topographic situations, its accumulation is less than 3.05 m (10 ft) thick. Similarly, the recent marine sediments comprise the detritus of coral reefs laid down on and along the coasts since the Waimānalo Stand of the sea. Like the recent alluvium, the recent marine sediments are unimportant in groundwater hydrology.

Within the sedimentary rock assemblages, aquifers occur, yet the sedimentary column is also significant hydrologically for the role it plays as an aquitard overlying volcanic rock aquifers. Figure 1.6 illustrates the typical rock sequence from ground surface down to the basalt aquifer in regions covered by sediments and weathered zones.

CHAPTER TWO

# Hydrologic Cycle

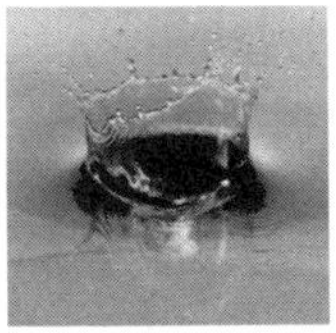

Water exists in three different phases or states: liquid as water, gas as water vapor, and solid as snow and ice. A change in phase is accompanied with transfer of heat, as is manifested in evaporation and condensation. Water and water vapor are of most interest in the Hawaiian Islands, where the lowland climate is classified as that of the humid tropics and where snowfall occurs only occasionally during winter storms above 2,000 m (6,562 ft) elevation on Hawai'i's three highest mountains (Mauna Loa, Mauna Kea, and Haleakalā), but there is no perennial snow cover.

Freshwater in Hawai'i occurs as atmospheric water vapor, rainwater, surface waters, and groundwater. Rainfall is episodic, surface waters are commonly concentrated in perennial streams and lakes, and groundwater is virtually ubiquitous but usually invisible. Water vapor is everywhere but visible only when it appears as clouds.

## Hydrology

In its broadest sense, hydrology is the science of the occurrence, movement, and distribution of water. It also embraces water quality and reactions of water in the living environment (U.S. Federal Council for Science and Technology 1962). Hydrology is a natural science that deals with water appearing in the atmosphere, the earth, and the oceans (National Research Council 1991a). Commonly, hydrology is restricted to the science of freshwater, and it is in that sense to which the following discussions refer.

The quantity of water is measured in terms of volume, depth, and surface area. Movement of water is measured in terms of velocity (displacement per unit of time) and discharge or flow rate (volume per unit of time). Many other physical variables—including mass, energy, power, force, pressure, and temperature—are also important in hydrologic studies.

The freshest water in nature is not chemically pure. It always contains some dissolved and suspended impurities that are physical, chemical, and microbiological in nature. These impurities characterize the quality of water. The concentrations of impurities determine the fitness of the water for use by humans, animals, and vegetation.

Like water quantity, water quality is measurable. Until the 1800s, human senses—taste, sight, smell, and feel—served as crude measuring instruments. Chemical water-quality parameters are measured as concentration (i.e., weight per unit volume of water). For example, saltiness of water is assessed by chloride ($Cl^{-1}$) in milligrams per liter (mg/l). The accepted chloride concentration standards are 250 mg/l for drinking water and 1,000 mg/l for irrigation water. Seawater contains approximately 18,980 mg/l of chloride.

Water quality is an intrinsic dimension of water and is inseparable from water quantity. A source of water requires assessment of both for whatever purpose or use it may be intended.

# Hydrologic Cycle

## Flow Cycle

Heated by solar energy, the ocean water is evaporated in extremely large amounts — six times greater than the amount from land surfaces. Energized by the sun, the atmosphere in which water vapor is a very small part is set into motion by thermodynamic processes. Thus, the water in the environment begins a journey known as the hydrologic cycle.

The cycle is endless but may start with rain that results from condensation of water vapor. Rain initially wets the various surfaces of the land, and as the rain continues, it may pond on the land surfaces and flow overland, draining into stream channels or into low-lying land and water bodies. These events are common and visible, but what is not obvious is evaporation, which removes water from the wetted surfaces, and infiltration into the ground. Rain may last only a few hours, but evaporation continues for many days afterward. Percolation of the infiltrated rainwater usually follows a circuitous path before reaching the saturated zone to become groundwater. Groundwater discharges slowly into wetlands, streams, lakes, and ultimately oceans.

Circulation of hydrologic elements is measured by residence time, which is computed by dividing the volume of water in storage by the rate of flow. The atmosphere, although it supports a global average precipitation of about 1.13 m (44 in.) per year, contains only $12.9 \times 10^3$ km$^3$ or about 0.03 percent of the Earth's freshwater at any moment. However, atmospheric water storage is replenished once every 8.2 days on the average through evaporation from global surfaces ($5.1 \times 10^8$ km$^2$). The residence time for slowly circulating waters such as global groundwater is on the order of tens to hundreds of years (Dooge 1984).

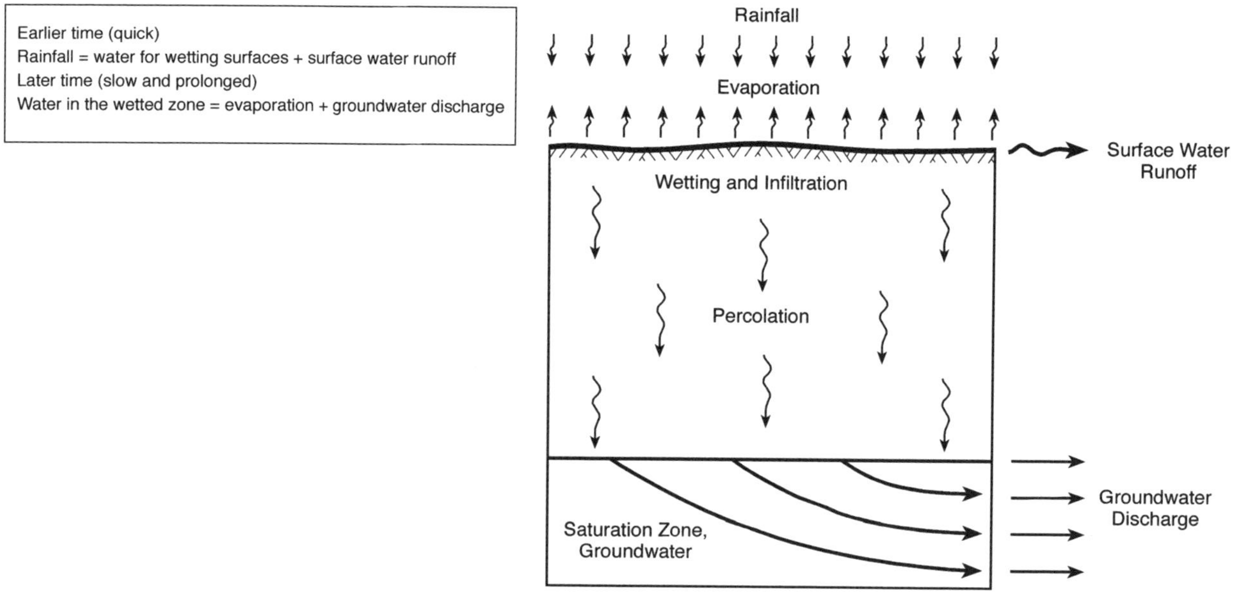


Figure 2.1. Disposition of rainfall: A simplified schematic vertical section.

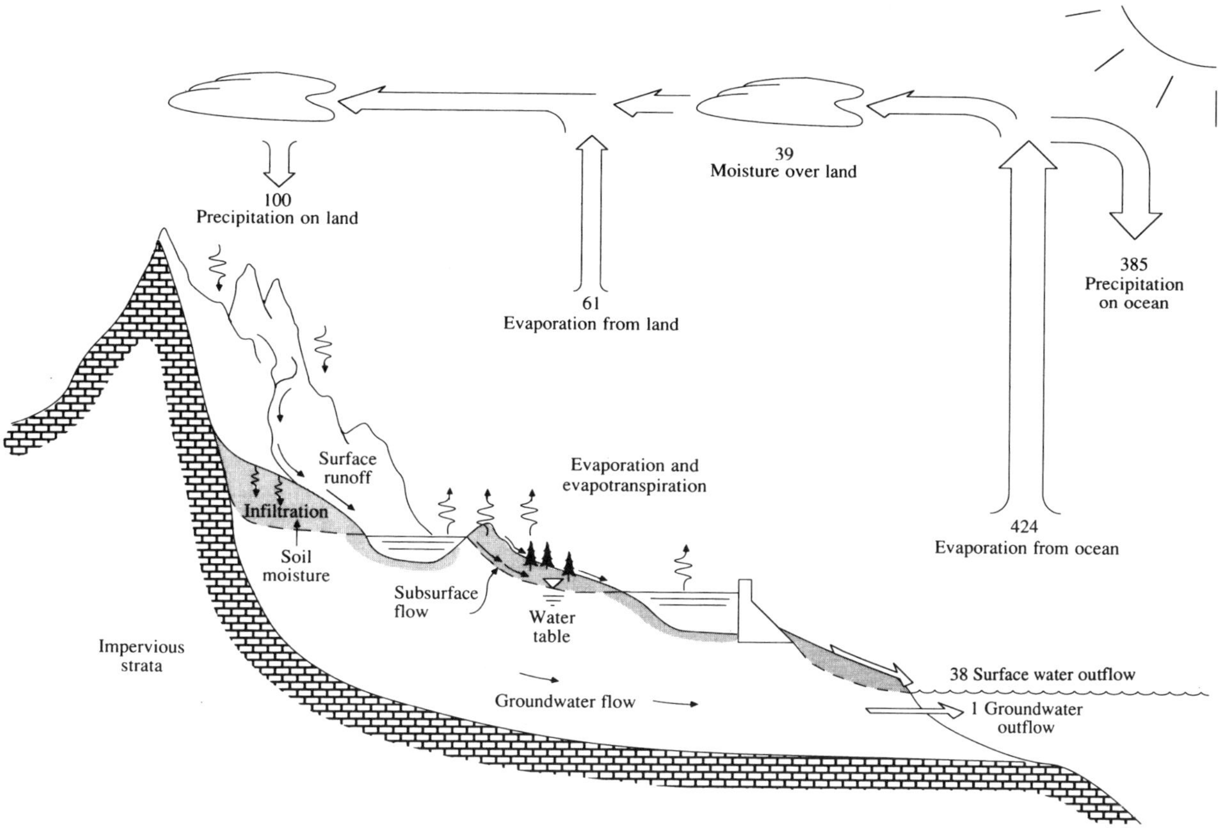


Figure 2.2. Hydrologic cycle and global average annual water balance. Numbers are annual volume of major components in units relative to that of land precipitation of 100. (Reprinted from Chow et al. 1988 with permission from McGraw-Hill Companies)

Groundwater typically moves at an average velocity on the order of only 1 m (3.28 ft) or less per day, whereas dry-weather streamflow rushes at a rate of about 1 m per second — a difference of five orders of magnitude.

Disposition of rainwater falling on land is schematically presented in Figure 2.1. The real world is far more complex, as will be discussed later. This exercise serves the purpose of showing only the major dispositions and the two very different time periods during which some of the hydrologic elements dominate over the others. A common schematic representation of the hydrologic cycle is shown in Figure 2.2.

## Transport Cycle

Water quality is altered continuously as water journeys through the natural environment. The transport cycle describes the cycle of creation, transport, and disposition or fate of the water-quality parameters that are associated with the different hydrologic elements of the water cycle. The traits of water quality can augment the flow data in identifying the origin, circulation, and distribution of water in the environment.

A pioneering study relating geochemical processes to the elements of the water cycle was conducted in San Joaquin Valley, California (Davis et al. 1959). These processes include incorporation of a small amount of windblown

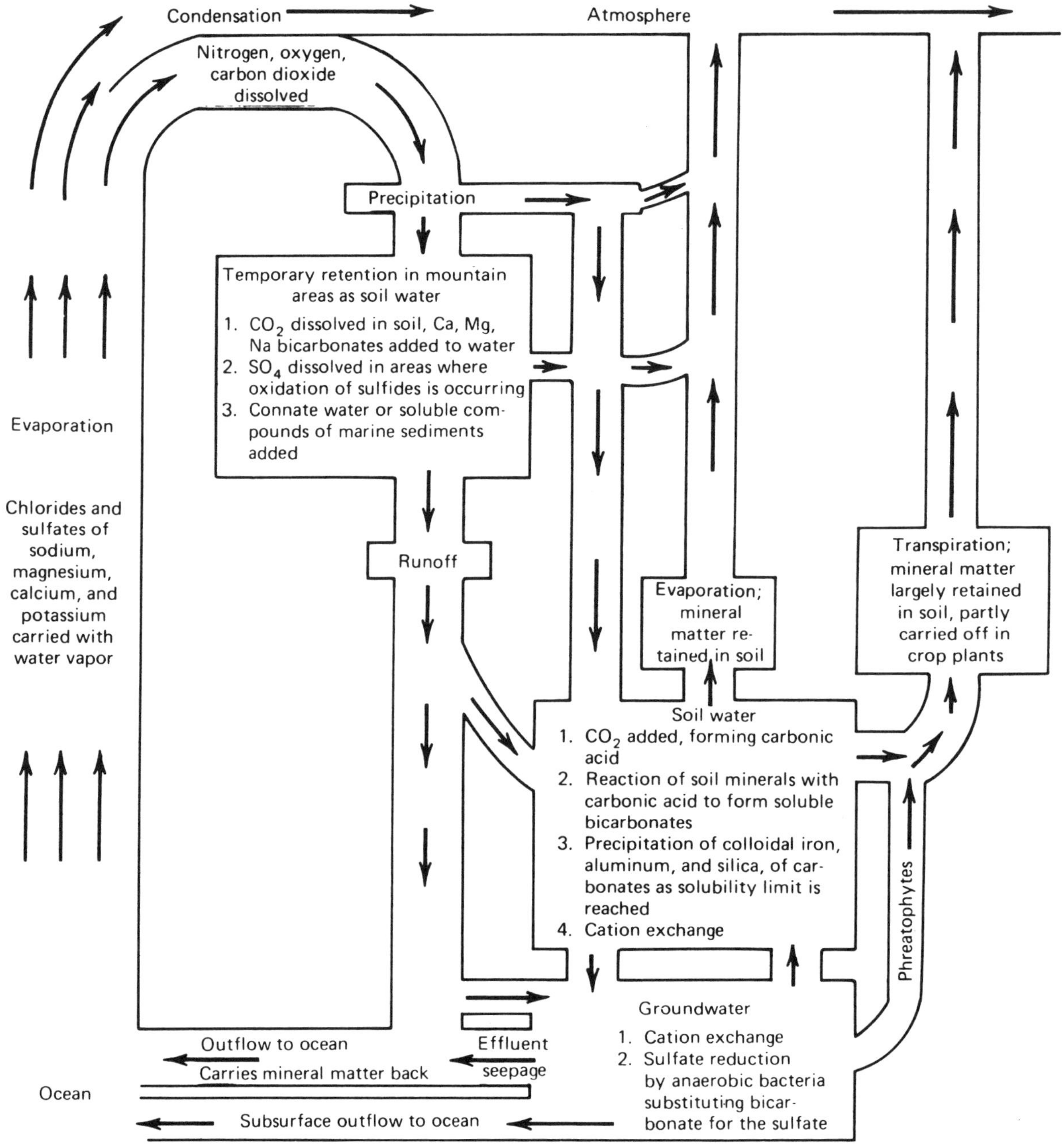


Figure 2.3. Geochemical cycle of surface water and groundwater in San Joaquin Valley, California. (Based on Davis et al. 1959; reprinted from Todd 1980 with permission from John Wiley and Sons, Inc.)

dissolved minerals in ocean evaporation; solution of atmospheric gases as water vapor condenses as precipitation; dissolution of carbon dioxide gases (originated from decomposed product of organic matter) as rainwater infiltrates into the soils; solution of mineral fragments to form bicarbonates and carbonates and salts in marine sediments; chemical reactions including chemical precipitation, ion exchange, and reduction of organic matter in the subsurface; and evaporation and transpiration of water, which leave the minerals behind in the soils. Finally, much of the mineral material is transported by streamflow or groundwater flow toward the ocean (Figure 2.3). Chemical geohydrology was subsequently introduced as a science to deal with the chemical characteristics of water in the natural environment (Back and Hanshaw 1965).

Other natural processes are also involved in the transport cycle. In the growth and decay of organic matter, cycling of nitrogen, phosphorus, and sulfur takes place, involving carbon and bacteria under aerobic (presence of dissolved oxygen) or anaerobic (absence of dissolved oxygen) conditions. The resulting water-quality parameters are added to the transport cycle (McGauhey 1968). Also transported in the cycle are microorganisms—including bacteria, fungi, algae, protozoans, plant and animal viruses—and isotopes of both radioactive and stable natures. Many of these substances originate from anthropomorphic uses of land and natural resources.

## Hydrologic Balance

### Phenomena, Models, and Applications

Water is treated as a continuum, as an aggregate of water molecules rather than individual molecules. This is permissible when the length dimension of the problem is much larger than the distances between molecules (Shapiro 1961). Most hydrologic phenomena fit this conceptual model. The real fluid is replaced with a model of continuous matter having continuum properties so defined as to ensure that on a macroscopic scale the behavior of the model duplicates the behavior of the real fluid. The detailed molecular structure of the real fluid is ignored. Examples of continuum properties are density, velocity, viscosity, internal stresses, and porosity.

Water and water vapor are set into motion by acquired energies and applied forces. The motion is described as velocity and can be treated as momentum. Momentum equals mass times velocity and is related to the applied forces according to Newton's law of momentum. The Eulerian view describes the motion at every fixed point in space. The other but less common approach is Lagrangian, in which a fixed mass of water is followed or traced as it moves. To properly account for mass and momentum, it is necessary to adopt a framework—literally a box—known as a control volume or a system with its position fixed in space. Water and water-transported substances enter and leave through the surfaces of the box as input and output, respectively, or are stored or transformed within the box.

According to Chow (1964a), hydrologic models are mathematical formulations used to simulate natural hydrologic phenomena, which can be processes or systems. The model is deterministic if the phenomena are assumed to follow only laws of definite certainty and randomness is ignored. With these assumptions, a given input always produces the same output. But when chances of occurrence are taken into account, the concept of probability must be introduced in formulating the model. The phenomenon and its model are described as stochastic or probabilistic. Stochastic differs from probabilistic. The process is stochastic if the sequence of occurrences is relevant. In reality, all hydrologic processes are more or less stochastic. Some hydrologic processes may be close to deterministic, such as evaporation, and others close to stochastic, such as precipitation, the other extreme. They are assumed differently only to simplify the analysis (Chow 1964a).

All hydrologic processes vary with time. For example, the water level in a stream and the water level in a water well continuously rise and fall with time. This is known as the unsteady or transient state. However, if after a long period of time without perturbation, the average water level remains more or less constant and the degree of fluctuation is relatively small, approach to the steady state has been achieved.

Another aspect of the analysis is how to treat spatial distribution. The lumped approach ignores spatial distribution. Some water problems may be treated by the lumped approach because they are not appreciably affected by spatial distribution. For example, the water surface in a large reservoir in which the surface slope is small may be approximated as level rather than gently sloping, if the primary concern is the change in water volume stored in the reservoir. On the other hand, flow in an open channel during flooding should be treated as a distributed problem. The lumped approach simplifies the analysis, but the results are limited by the assumptions.

The end product of the conservation analysis of mass and momentum is a set of mathematical equations, usually partial differential equations, with certain descriptors of flow and water quality as dependent variables and space and time as independent variables. These equations are mathematical models used to simulate natural phenomena. Solution of the equations for initial and boundary conditions provides prediction of the behavior of the descriptors. The equations that follow are written as one-dimensional for simplicity, but they can be expanded into two- or three-dimensional equations. Derivations of the equations are omitted, but they are readily available in the references cited.

Modeling protocol is reasonably established. Calibration of a model is usually performed first. In this procedure, the values of model parameters are adjusted and new model parameters may be created to obtain a best fit to the historical field data. A calibrated model is said to be verified if it can simulate a second independent data set. However, because of data unavailability, this step is usually omitted. If modeling involves the use of a computer, the computer program (code) needs to be checked (also known as verified) by comparing the numerical solution with the analytical solution of the mathematical formulation. A code is a computer program that is used to solve numerically a set of mathematical equations. A code is generic and is written only once, but a model is designed for each application (Anderson et al. 1993; Anderson 1995). Verification is often called validation, adding to the confusion (Oreskes et al. 1994).

Because a model is a simplified version of the phenomena, there exists no unique model of the phenomena. Different sets of simplifying assumptions will result in different models, each approximating the phenomenon in a different way (Bear and Verruijt 1987).

Many hydrologic processes are complex and difficult to analyze and quantify. Heterogeneity in nature is the rule, and uncertainty makes predictions of water behavior difficult. In applications of hydrology, reasonable assumptions are essential and more than one method of solution is possible. The results, however, must make common and scientific sense.

A statement of purpose is essential in the application of hydrology to solve water resource problems. A water resource proposal of economic and environmental importance calls for marshaling relevant facts, science and technology, and experience to formulate alternative plans that are practical, economical, and environmentally and socially acceptable.

## Balance in Flow

The fundamental descriptors of the flow process for surface water and groundwater are usually water volume, water-surface elevation, velocity, and discharge.

### Surface-Water Flow

Typically, surface-water motion takes place as overland flow over areally extensive and relatively flat surfaces or as channel flow in relatively narrow channels. Surface water interacts with subsurface water through the process of seepage (groundwater emerging on the surface) or the process of infiltration (surface water infiltrating into the subsurface). In mass conservation these interactions may have to be considered when the magnitude of seepage and infiltration is appreciable when compared with the surface-water motion.

The principle of mass conservation states that the excess of mass inflow entering the control volume over the outflow leaving the box during a specified time interval must be equal to the change of mass inside the box.

### *Lumped flow*

Open-channel flow without lateral inflow is used as an example of mass conservation:

For unsteady (transient) flow

$$I(t) - Q(t) = \frac{d}{dt}S(t) \tag{2.1}$$

where *I(t)* is the inflow, *Q(t)* is the outflow, and *S(t)* is the volume of water in storage between the upstream and downstream locations. Besides this equation, storage is related to *I* and/or *Q* by a storage function (not shown) that reflects the hydraulics involved, the geometry of the storage space, and any controlling condition at the downstream end.

For steady-state flow (i.e., $dS/dt = 0$ and neither *I* nor *Q* varies with time)

$$I = Q \tag{2.2}$$

This is a familiar concept—what goes in must come out under steady-flow conditions.

### *Distributed flow*

Distributed flow varies with space as well as time. Common situations call for this treatment when lateral flow distribution cannot be ignored and the flow conditions are desired at several intermediate locations between the inlet and outlet of the channel.

This flow problem was addressed in 1871 by Barre de Saint-Venant in his study of floods and tides in rivers in France. The analysis is rigorous and addresses both mass conservation and momentum conservation (Chow et al. 1988). But analytical solution is not possible and numerical solution with a high-speed computer had not been attempted until the 1990s.

The equation of mass conservation is

$$\frac{\partial(Au)}{\partial x} + \frac{\partial A}{\partial t} - q_l = 0 \tag{2.3}$$

and that for momentum conservation is

$$\frac{\partial u}{\partial t} + u\frac{\partial u}{\partial x} + g\left(\frac{\partial h}{\partial x} + S_f\right) = 0 \tag{2.4}$$

where $x$ is the distance along the longitudinal axis of the watercourse, $A$ is the cross-sectional area, $u$ is the velocity, $q_l$ is the lateral inflow or outflow, $h$ is the water-surface elevation above a datum, and $S_f$ is the friction slope, which may be evaluated using an empirical equation for uniform steady flow such as Chezy's or Manning's equation. The descriptors solved for are velocity $u$ and water depth $h$.

Even though these equations appear comprehensive, they are based on many assumptions, including mildly sloping water surface, hydrostatic vertical pressure distribution, constant water density, fixed and immobile channel bed and banks, and relatively small bottom slope (<15 percent).

## Groundwater Flow

Groundwater flow is diffused and extremely slow by five orders of magnitude compared with overland flow. Unlike surface water, groundwater typically flows in a network of extremely small, variably sized, interconnected, and indescribably complex pore space. However, the continuum approach bypasses the impossibility of describing this network and of solving the Navier–Stokes equation, which is the momentum equation for viscous fluids at the microscopic level in porous material (Bear 1979). The geologic medium is replaced by a porous medium that consists of interconnected voids in a solid matrix within a representative elementary volume. The continuum properties enable definition of porosity and other parameters of the porous medium at a mathematical point.

The mass conservation equations for unsteady flow are as follows:

$$\frac{\partial q_x}{\partial x} = S_s\frac{\partial h}{\partial t} \tag{2.5}$$

The left-side term is the excess of mass inflow over mass outflow, and the right-side term is the change of water volume in storage with time. In equation (2.5), $q_x$ is the specific discharge, $h$ is the hydraulic head, and $S_s$ is the specific storativity. According to Darcy's Law, specific discharge equals hydraulic conductivity $K$ times hydraulic gradient $J = \partial h/\partial x$. The average groundwater velocity $V$ equals $q_x$ divided by porosity $n$.

For steady flow (i.e., $\partial h/\partial t = 0$), equation (2.5) becomes

$$q_x = \text{constant} \tag{2.6}$$

(i.e., inflow = outflow).

These equations are similar in mathematical form to those for surface water (lumped flow). In the event of additions (sources) such as recharge by infiltration, terms are added to the left side of equation (2.5). In groundwater problems, momentum conservation is not explicitly expressed as in surface-water problems. Groundwater momentum is negligibly small, and all dynamic effects of groundwater flow are already embodied in Darcy's Law (Hubbert 1940).

For unsaturated flow in the subsurface, an additional variable is needed: volumetric water content θ, which is water volume per bulk volume of the porous medium. A common situation is redistribution of infiltrated water in the surface soils. The basic mass conservation equation is as follows:

$$\frac{\partial q_z}{\partial z} = \frac{\partial \theta}{\partial t} \tag{2.7}$$

where $q_z$ is the specific discharge in the vertical direction. The hydraulic conductivity $K$ is complicated, being dependent on the water content θ.

Mass conservation is also applicable to atmospheric water, but more than conservation of mass is involved in air flow. Because air flow is rapid, momentum of the moist air must be considered. Conservation of heat is also important because heat transfer is effective in air flow. The independent variables are temperature, pressure, density, and mixing ratio of the water vapor. Mixing ratio is the ratio of water-vapor mass to dry-air mass in grams per kilogram, with values typically at 20 g/kg in the tropics. It is approximately equal to specific humidity, which is the ratio of water-vapor mass to total mass. A three-dimensional numerical model employed for air-flow studies on the island of Hawai'i is based on conservation of mass, momentum, and heat (Clark 1979).

## Balance in Transport

Air flow in the atmosphere transports water vapor and other substances of hydrologic interest. Radioactive isotopes such as tritium and radiocarbon that were spiked into the atmosphere by high-altitude testing of nuclear bombs in the 1950s have been circling the Earth as a part of the atmosphere. They are dissolved in rainwater, and their precision in decay with time provides a distinct signal for tracing and dating meteoric water after it enters the hydrologic cycle (Hufen et al. 1972). Air flow also transports acid rain resulting from airborne industrial wastes (sulfur dioxide and oxides of nitrogen) dissolved in water vapor. Transport may occur over long distances.

Water transports solutes, microbes, sediments, and heat. Solutes are dissolved chemical compounds. Many solutes and some microbes (bacteria, viruses, protozoans) that exceed water-quality criteria and standards are considered pollutants or contaminants. Although solutes and microbes are invisible, sediments consisting of insoluble solid particles vary in size from very fine clay (240 nm or $9.5 \times 10^{-6}$ in.) to very large boulders (4,000 mm or 160 in.). Sediments are transported as suspended load and bed load in flowing streams. A major concern about the origin of sediments is the erosion of productive soils. Deposition of sediments reduces the volume available for storage of water in reservoirs and, in Hawai'i and other coastal areas, smothers coral reefs. The deposited sediments may be considered as pollutants if they contain adsorbed toxic chemicals. The transport of heat is of special importance in creating and sustaining climate regimes. Heat is also important in overall water quality, and it impacts chemical reactions and biological growth rates in water.

Solutes are transported by advection, dispersion, and diffusion. Advection is the movement of a mass expressed by the velocity of water in the general direction of flow.

Dispersion is due to microscopic velocity variation in a porous medium. Diffusion is also known as turbulent diffusion in large surface-water bodies where turbulence prevails (Brooks 1960). Diffusion in groundwater is called molecular diffusion and becomes effective only in very slow–velocity regimes. Many attenuative processes affect solutes in water. Examples include adsorption (adhesion of solute to a solid surface), volatilization (evaporation of solute into the atmosphere), biodegradation (microbial breakdown of the solute), and radioactive decay.

Microbes respond to most of the processes that affect solutes. This is partly due to their extremely small sizes (viruses 10 to 300 nm [0.4 to $11.8 \times 10^{-6}$ in.]). Colloidal particles are also small (1 to 1,000 nm) and electrically charged. The transport of viruses and colloids is not well understood but is assumed to be similar to that of solutes as a first approximation. Some of the attenuation processes such as sorption also apply. Bacteria may either die off or multiply in the water environment, but animal viruses usually cannot survive after leaving their warm-blooded hosts and are inactivated in hostile (hot, dry, sunny) environments.

Sediments behave differently from solutes and microbes. The weight and size of the solid particles are special factors in transport processes. For example, the suspended load is, in part, the result of balancing the force of turbulent motion of the streamflow with the submerged weight of the solid particles and the drag force exerted on the particle surfaces. Sediments are, by and large, conservative and do not diminish.

The complexity of the natural processes governing sediment transport in surface water makes mathematical formulation quite difficult in comparison with that of solute transport in groundwater. However, the principle of mass conservation applies to sediment transport as well. Bed-load transport in streams is the rate of movement of the sediment particles that roll, slide, and hop along the streambed. Under equilibrium conditions — the bed elevation remains unchanged over a period of time — the streamflow will erode a certain amount of sediment particles from the streambed, but an equal amount of sediment particles will be deposited from the flow onto the streambed (Einstein 1950). Verified by experiments, this statement is essentially the principle of conservation of mass. The resulting equations Einstein developed for bed load (not shown) laid the foundation for most subsequent studies (Shen and Julien 1993).

The most common descriptor of transport is concentration, C, which is the weight of the solute or substance being transported in milligrams per unit volume of water in 1 liter (l). One mg/l virtually equals the weight ratio of 1 mg/kg or 1 part per million because water weighs 1 mg/ml under standard conditions. The milliequivalent per liter (meq/l) is often used, especially for analytical purposes. Equivalent weight is the atomic or molecular weight divided by its valence. For bacteria, viruses, and protozoans, the unit is counts per volume or density. For virus, it is plaque-forming unit; for *Giardia,* which is a protozoan, it is cysts per 100 liters; and for *Cryptosporidium,* also a protozoan, it is oocysts per 100 liters. Sediment concentration is usually expressed as weight of sediments to total sediment + water weight, given in parts per million, or as weight of sediment to total sediment + water volume, in kilograms per cubic meter (kg/m$^3$) or $10^3$ mg/l.

In hydrology, it is necessary to recognize regional differences in the value of parameters and variables of natural processes. For example, in the Yellow River in northern China, named for the color of its water, the suspended solid concentration annually averages 37,600 mg/l. This value is four orders of magnitude higher than that for Kahana Stream, Oʻahu, which remains in its near-pristine state.

The basic methodology for doing a mass balance of a transported contaminant is similar to that for the flow of water. First, a control volume is selected. The excess of the contaminant inflow over the contaminant outflow (mathematically known as negative of divergence of a flux vector) is equated to the increase of the contaminant in the control volume. The inflows and outflows are due to advection, dispersion, or diffusion, whichever is applicable. The attenuation processes may be accompanied by injection or extraction of water containing the contaminant. The additional terms are combined mathematically with the basic equation that describes the contaminant mass increase or decrease in the control box. Solute transport in surface water and groundwater are used in the following sections to demonstrate the mass balance equations.

### Solute Balance in Surface-Water Flow

The mass conservation equation is expressed as follows (Fischer et al. 1979):

$$\frac{\partial}{\partial x}\left(E_x \frac{\partial C}{\partial x}\right) - u\frac{\partial C}{\partial x} = \frac{\partial C}{\partial t} \tag{2.8}$$

where $E_x$ is the turbulent diffusivity. In equation (2.8), the diffusive transport and advective transport (the first and second terms on the left side or the negative divergence of these two fluxes) equal the change in contaminant mass in the control volume with time. A single term is commonly added to the left side to account for all of the contaminant loss due to sedimentation, photolysis, volatilization, sorption, and biodegradation. The term is usually a first-order kinetics in the form of $-kC$ where $k$ is the first-order rate coefficient. Complex formulation calls for second-order kinetics (Sawyer et al. 1994). A classical predecessor of transport in surface water is the depletion of dissolved oxygen in the water of a stream due to discharge of sewage, as formulated by Streeter and Phelps (1925). Sufficient dissolved oxygen is essential not only for the survival of fish and other fauna but also for the maintenance of acceptable aesthetic conditions.

### Solute Balance in Groundwater Flow

The basic mass conservation equation is as follows (Bear 1979):

$$D_x\frac{\partial^2 C}{\partial x^2} - V\frac{\partial C}{\partial x} = \frac{\partial C}{\partial t} \tag{2.9}$$

where $D_x$ is the coefficient of dispersion and $V$ is the average groundwater velocity. The attenuation processes in groundwater transport are commonly treated individually in contrast with those of surface-water transport. Some of the surface-water attenuation processes such as photolysis and sedimentation are not active in the subsurface. Processes such as adsorption are especially effective in groundwater transport because of long residence time, large contact surfaces, and sorptive organic matter. Other effective processes are hydrolysis for organic chemicals and radioactive decay for radioisotopes, both being treated as first-order kinetics with a half-life parameter.

For unsaturated flow, the balance equation for advection-dispersion is complicated (see Chapter 5). When coupled with vegetal growth, the unsaturated zone provides the most effective natural processes for the protection of groundwater quality.

## Global Water Balance

Global water quantities can only be estimated because of the magnitude of their values and the complexity of their pathways. Several selected statistics are cited here to indicate their order of magnitude. An early but still acceptable set of global water estimates is that by Wolman (1962). Although not totally unexpected, the distribution of global water provides some surprises. Ocean water accounts for the bulk of the world's water (97 percent), leaving only 3 percent as freshwater. This 3 percent represents approximately $41 \times 10^6$ km$^3$ ($11 \times 10^{18}$ gal). About three-quarters of the freshwater is frozen as ice and glaciers. Groundwater accounts for nearly all of the other one-quarter of freshwater. The amount held in lakes is extremely small (only 0.3 percent of freshwater) and in rivers and streams even smaller (0.03 percent of freshwater). Atmospheric water, which drives the hydrologic cycle, is also minute (0.03 percent of freshwater). The global average annual precipitation on land surfaces is approximately 860 mm (34 in.).

Statistics of transport have not been adequately established because transport phenomena have been subject to detailed assessment only in recent years. Current estimates of waterborne sediment and dissolved species fluxes from the Earth's continents are available (Dooge 1984; National Research Council 1991a). For instance, in North America, the annual sediment yield is estimated at $76.4 \times 10^3$ kg/km$^2$ (218 tons/mi$^2$) and the annual dissolved

Table 2.1. Disposition of Precipitation

| | North America | | North America | Island of O'ahu | All Hawaiian Islands |
|---|---|---|---|---|---|
| | Annual Amount | | Annual Amount (as % of precipitation) | | |
| Parameter | (mm) | (in.) | | | |
| Precipitation | 670 | 26.4 | 100 | 100 | 100 |
| Water for wetting surfaces | 467 | 18.4 | 70 | 76 | 69 |
| Surface-water runoff | 203 | 8.0 | 30 | 24 | 31 |
| Evaporation | 383 | 15.1 | 57 | 40 | 39 |
| Groundwater discharge | 84 | 3.3 | 13 | 36 | 30 |

SOURCES: Data for North America from Wolman 1962; data for Hawai'i from Hawai'i Department of Land and Natural Resources 1977.

NOTE: Precipitation = water for wetting surfaces + surface-water runoff; water in the wetted zone = evaporation + groundwater discharge.

load at $40.8 \times 10^3$ kg/km$^2$ (117 tons/mi$^2$). The global dissolved solid concentration is estimated at 142 mg/l.

The global water balance is illustrated by a simplified model for the disposition of rainfall in Figure 2.2 (Chow et al. 1988). The North America statistics are compared with the statistics for the island of O'ahu and all Hawaiian Islands (Hawai'i Department of Land and Natural Resources 1977) in Table 2.1.

Groundwater recharge (or discharge) in the Hawaiian Islands accounts for about one-third of the rainfall and is substantially higher than that for North America. These statistics underscore the distinction of Hawaiian Island hydrology and the importance of groundwater as the premier water resource in Hawai'i.

# Appendix 2.1. A Note on Development of Hydrology

The historical development of hydrology as flow phenomena has been extensively documented (Meinzer 1942; Chow 1964b; Biswas 1972; National Research Council 1991a). Especially insightful is Chow's analysis of eight eras of hydrology and Biswas' book on the history of hydrology. The early years cover speculation (ancient to 1400), observations (1400 to 1600), measurements (1600 to 1700), experiments (1700 to 1800), modernization (1800 to 1900), and empiricism (1900 to 1930). In the sixteenth century, Palissy of France and da Vinci of Italy achieved a reasonably correct understanding of the hydrologic cycle based on observations. But it was not until the seventeenth century that the separate and sequential works of three scientists — Perrault of France (1608–1680), Marriotte of France (1620–1684), and Halley of England (1646–1742) — firmly established the concept of the hydrologic cycle. Their conclusions were based on experimental investigations concerning measurements of rainfall, evaporation, and streamflow and proved that the water evaporated from the seas came down as rainfall in ample amounts to sustain the flow of rivers and springs.

During a period of merely 20 years (1930 to 1950), which Chow called the period of rationalization, rational analyses instead of empiricism were initiated to solve hydrologic problems. Among the contributions were those of Sherman (unit hydrograph), Horton (infiltration), Theis (transient hydraulics in wells), Gumbel (frequency analysis), Bernard (hydrometeorology), and Einstein (bed-load transport). The next era, the era of theorization, has been dominated by numerical modeling, which gained popularity in the 1980s. Advanced technologies such as remote sensing (Engman and Gurney 1991) have also been phased in since the 1980s.

An important nongovernmental organization at the international level is the International Association of Hydrological Sciences (IAHS), which was instituted in 1922. Within the United Nations, the U.N. Educational, Scientific and Cultural Organization (UNESCO) has been the prime mover of water science with programs, notably the International Hydrological Program, which includes a program for the humid tropics. The World Meteorological Organization has focused on hydrological operational practices. In the United States, hydrology was added as a section of the American Geophysical Union (AGU) in 1930.

The premier research periodical in the United States is *Water Resources Research*, published by the AGU; its counterpart in Europe is the *Journal of Hydrology*. IAHS publishes the *Hydrologic Science Journal*. In addition, journals are published by organizations of several major constituent scientific disciplines, including engineering and geology. Examples are *Hydrological Engineering* and *Water Resources Planning and Management* by the American Society of Civil Engineers and the *Ground Water Journal* by the Association of Ground Water Scientists and Engineers.

An early hydrology book is *Hydrology*, edited by Meinzer (1942). It treats hydrology as a multidisciplinary science, and its contributors include Sherman (engineering), Bernard (meteorology), Baver (soil science), and Meinzer (geology).

Before 1980, only a few significant hydrology textbooks were printed, including Linsley, Kohler, and Paulhus' *Hydrology for Engineers* (1958, latest edition 1982); Todd's *Groundwater Hydrology* (1959, 2nd edition 1980); Chow's *Open-Channel Hydraulics* (1959); Eagleson's *Dynamic Hydrology* (1970); Bear's *Hydraulics of Groundwater* (1979); and Freeze and Cherry's *Groundwater* (1979). Chow's (ed.) *Handbook of Applied Hydrology* (1964b) is an encyclopedia of facts and methods of analyses and investigations. Since the 1980s and especially in the 1990s, hydrology books have proliferated. Some are specialized, such as Stephens' *Vadose Zone Hydrology* (1996) and Dingman's *Physical Hydrology* (1994). Maidment, a student of Chow, edited the most recent *Handbook of Hydrology* (1993).

Universities provide formal education and award academic degrees in hydrology and closely related fields, including civil and environmental engineering, and geosciences, yet the number of degrees in hydrology is still relatively few. In the United States many universities, including the University of Hawai'i, are members of a national network named the Universities Council on Water Resources (created in 1964). At the University of Hawai'i, hydrology courses were first offered in the civil engineering department. Subsequently, the geology, soil sciences, geography, and meteorology departments have broadened the spectrum needed for graduate education in hydrology.

Research and graduate education are inseparable. Universities have been the premier institutions fostering basic hydrological research. In 1964, a national network of water resources research institutes or centers was created, one at a land-grant university in each state, including the University of Hawai'i for the state of Hawai'i. Hydrological investigations are conducted by many governmental agencies, including the U.S. Geological Survey, the U.S. Agricultural Research Service, the U.S. Environmental Protection Agency, the National Oceanic and Atmospheric Administration, and the National Weather Service. The Illinois Water Survey is a notable institution with a long history at the state level.

In Hawai'i the Commission on Water Resource Management is in the process of increasing its investigative activities. The Honolulu Board of Water Supply has a long and illustrious history of studies on the hydrology of O'ahu.

# CHAPTER THREE

# Precipitation

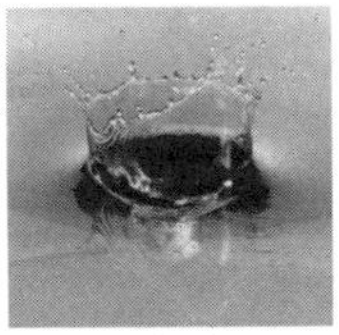

The ultimate source of the freshwater in an oceanic island is precipitation of meteoric origin. This statement is so well accepted today that it is sometimes difficult to appreciate the long history and arduous evolution of hydrological sciences. Aristotle (385–322 B.C.), who wrote the first book on meteorology, was only partially correct in explaining the process of precipitation. He recognized condensation and evaporation but did not clearly distinguish water vapor from air (Meinzer 1942; Biswas 1972). Today's knowledge about precipitation is still far from adequate to explain fully climate and weather, but great strides are being made with the aid of satellites, radar and other observational advances, and computer modeling.

## Atmospheric Environment and Precipitation

Both weather and climate describe the state of the atmosphere, but they are not synonymous. Weather refers to the atmospheric condition at a particular time. Climate deals with the atmospheric condition averaged over a number of years. Climate has been called average weather; it defines the long-term state of the atmosphere, encompassing the total effect of weather phenomena.

### Atmospheric Water

In air, water exists in a gaseous state as water vapor. Much of the time, about 60 percent on a global basis, water vapor becomes visible as clouds, which manifest saturation of water vapor in the air. Infrequently, the vapor condenses into liquid as rain, and even less frequently it becomes solid forms of precipitation such as snow and ice.

Water vapor is a very small fraction, about 1 percent by volume, of the atmosphere. The amount of water vapor is reflected by the vapor pressure, $e$: the more water vapor in a given volume, the higher the value of $e$. Saturation is the state in which the maximum amount of water vapor exists at a given temperature, $T$. The saturation vapor pressure, $e_s$, increases with temperature (see Figure 3.1). Air can become saturated by adding water vapor, by cooling, or by some combination of both. Relative humidity, $R_h = e/e_s$, as a percentage, expresses the degree of saturation. By cooling the air alone, the temperature at saturation is called dew-point temperature, $T_d$.

Vapor pressure is conveniently determined by computation employing the measured values of the dry-bulb temperature, $T$; wet-bulb temperature, $T_w$ (temperature registered on a wet muslin–covered mercury bulb); and atmospheric pressure, $P_a$. Tables for simplifying computations are presented in numerous textbooks (Linsley et al. 1982). The daytime atmosphere on the Honolulu coastal plain on a typical October day may be represented by the following readings: temperature (high), 32°C (90°F); relative humidity, 71 percent; dew-point temperature, 26°C (79°F); vapor pressure, 34.2 mbar; saturated vapor pressure, 48.1 mbar; and atmospheric pressure, 1,016 mbar.

Virtually all water vapor is contained in the troposphere, which is the lowest layer of the atmosphere. It extends to an altitude of about 11.6 km (38,000 ft) generally and, at its highest point, to about 16 km (52,800 ft) near the equator. Within the troposphere, the air tempera-

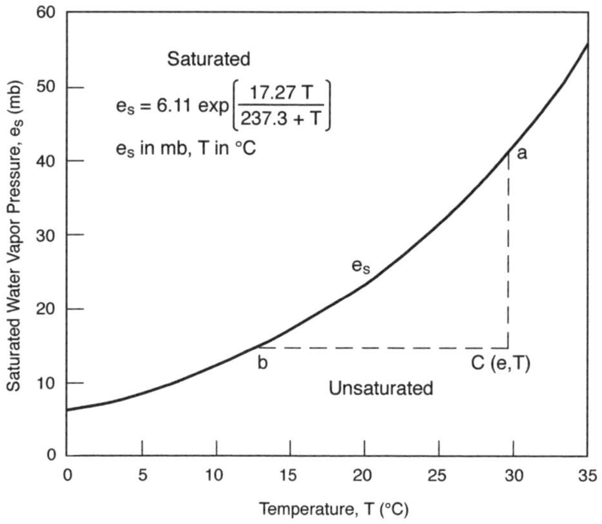


Figure 3.1. Temperature dependence of saturated water vapor pressure (1 mbar = 100 Pa = 14.5 psi). Point a is saturation vapor pressure and point b is dew-point temperature for the air C (e,T).

ture decreases with altitude at an average rate of about 6.5°C per 1 km (3.6°F per 1,000 ft), and the water vapor decreases sharply. The top of the troposphere is called the tropopause, where the temperature is, at a minimum, about −60°C (−76°F). The tropopause, which varies with latitude, acts as a "lid," keeping beneath it any rising convective activities. Atmospheric pressure decreases at high altitudes: at the tropopause it is about 100 to 300 mbar.

Temperature inversion may occur regionally as a layer in the lower troposphere. Within the inversion, temperature increases instead of decreases with altitude. The trade-wind inversion, which profoundly affects the climate and weather of the Hawaiian Islands, is an example.

The troposphere is full of various air masses. An air mass is a large body of air whose properties — especially temperature, moisture, and lapse rate (temperature change with altitude) — are relatively uniform horizontally for hundreds of kilometers (Barry and Chorley 1992). The contact between two air masses is called a front. An air mass is identified by its source and temperature, such as maritime or continental, tropic or arctic.

## Condensation

Moist air must be cooled to its dew point to begin the formation of precipitation. Among the several cooling processes in the natural environment, only vertical uplift of the air mass into higher altitudes where the environmental temperature is cooler causes a rate of cooling high enough to produce substantial precipitation.

Condensation may occur in clean air but with great difficulty. Water vapor must find a surface on which to condense. In free air the surface is provided by hygroscopic nuclei, which are wettable microscopic particles such as dust, smoke, salts, and chemical compounds.

Cloud is a visible air mass containing tiny water droplets or ice particles formed by condensation of water vapor around condensation nuclei and held in suspension. The water droplets may grow in size from a few to a few tens of micrometers to become raindrops that can be as large as 1,000 μm (1,000 μm = 1 mm = 0.039 in.). Theories differ greatly on the growth of droplets. A cloud physics study in Hawai'i indicates growth by recirculation of raindrops within the clouds (Takahashi 1988). The kinetic energy released by large raindrops on the ground is sufficient to break the bonds between soil particles to initiate surface erosion (Ekern 1950).

Clouds are classified by altitude and appearance (Figure 3.2). For example, cumulus, usually scattered low-altitude fluffy clouds, is typical of fair weather. Clouds are associated with impending weather changes. Cumulus may develop into cumulonimbus, which is towering and may generate thunderstorms. Stratocumulus is associated with warm fronts.

## Precipitation Types

Atmospheric disturbances are treated meteorologically as synoptic scale and mesoscale. Synoptic scale covers a few thousand kilometers and a few days to a week for the life span of the disturbances. Mesoscale is smaller, covering a few hundred kilometers and a time range of hours to several days. The Hawaiian Islands experience precipitation on both scales.

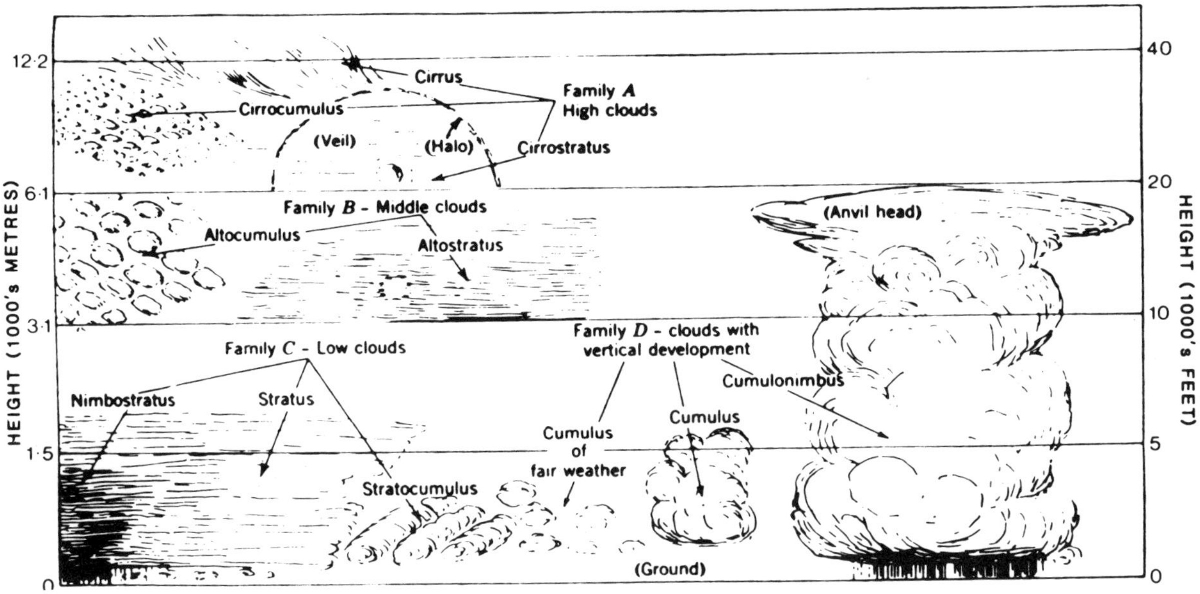


Figure 3.2. Ten basic cloud groups classified according to height and form. (Reprinted from Barry and Chorley 1992 with permission from Routledge/Taylor and Francis Group, LLC)

Three types of precipitation are named after the mechanisms that can effectively elevate air masses: (1) cyclonic, (2) orographic, and (3) convective. Hydrologists are concerned with precipitation in terms of amount, intensity, and duration.

Cyclonic precipitation is the result of systems with low atmospheric pressure (cyclones) on a synoptic scale. Convergence of various air masses creates ascent of air and eventual occlusion of warm and cold air (Figure 3.3). Warm fronts result when an advancing warm (lighter) air mass overtakes and glides slowly over a retreating cold (heavier) air mass. The resulting precipitation extends over a long distance as drizzles that grade into steady rain. A cold front is formed when fast-advancing cold air wedges into a warm air mass, forcing the warm air to rise rapidly, producing showers and perhaps thunderstorms with high rain intensity. The passage of a front generally takes one to a few days. Warm fronts last longer than cold fronts.

Orographic precipitation results from blocking, thereby lifting airflow up the windward slope of a mountain range. The height, size, and alignment of the barrier greatly influence the precipitation volume and distribution. Further, a layer of temperature inversion, if present, can limit the height reached by the airflow and alter its path. If the inversion is positioned above but close to the ridge, the airflow may accelerate and squeeze through the gap formed between the inversion and the ridge and carry with it some precipitation to the lee slopes. For a case in which the mountain is much higher than the inversion, the airflow becomes trapped below the inversion, depositing precipitation on the windward slopes and following a path in a direction parallel to the mountain range.

Convective precipitation is typically associated with towering cumulus and cumulonimbus clouds that are produced from the strong heating of a surface. The precipitation may be intense but brief and may be accompanied by thunder and lightning that affect areas of 20 to 50 km$^2$ (7.7 to 19.3 mi$^2$). However, cumulonimbus cells may organize themselves and produce very heavy rainfall, affecting much larger areas.

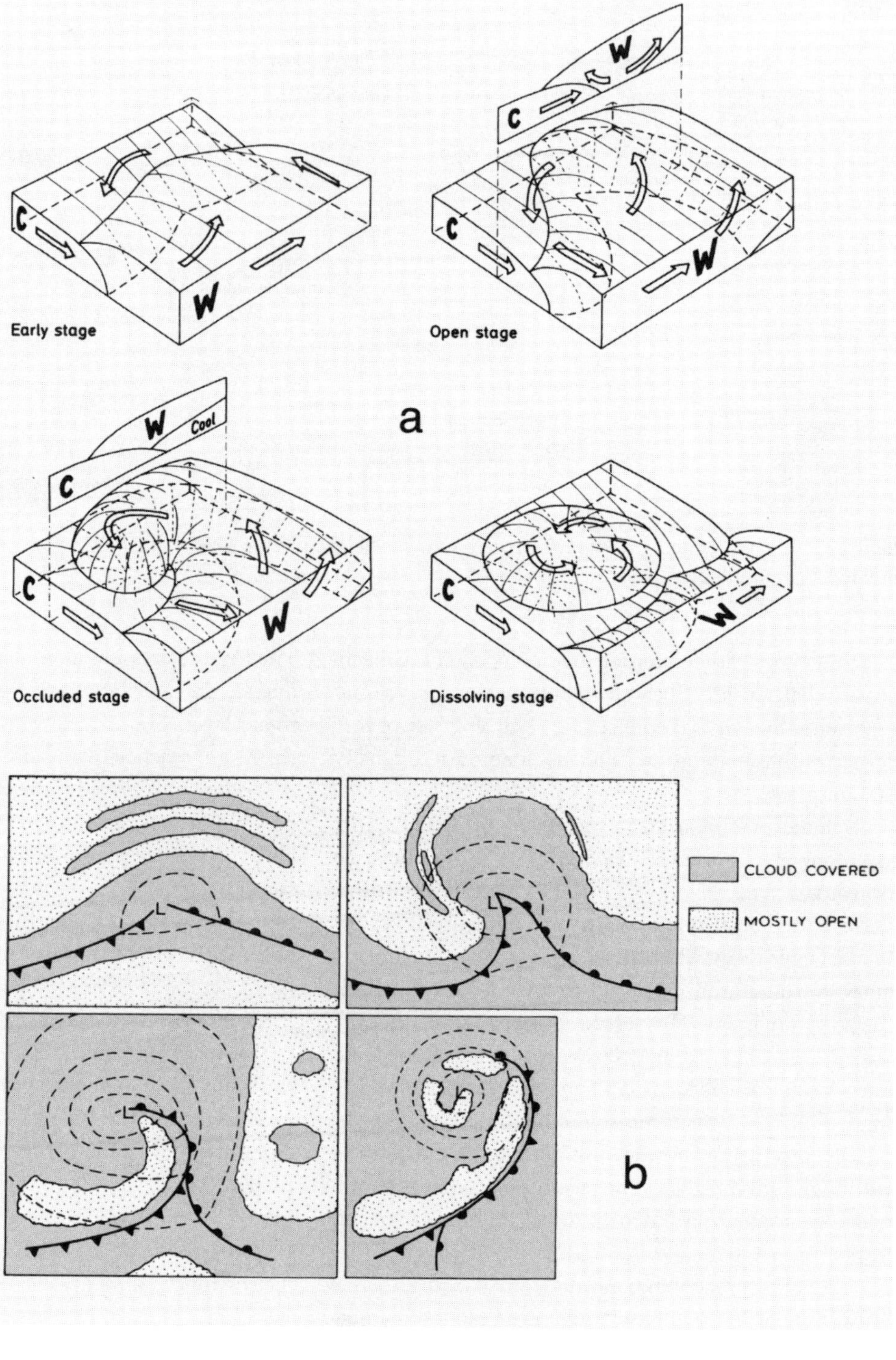


Figure 3.3. *a,* Four stages in the typical development of a mid-latitude depression; C, cold air; W, warm air. (Mostly after Strahler 1965, modified after Beckinsale); *b,* Satellite view of the cloud systems corresponding to these stages (After Boucher and Newcomb 1962). (Reprinted from Barry and Chorley 1992 with permission from Routledge/Taylor and Francis Group, LLC)

## Current Topics: Mesoscale Convective Systems and El Niño–Southern Oscillation

Knowledge of precipitation has been advanced as a result of new observational capabilities, including satellite remote sensing, weather radars, and in-situ measurements with recording rain gages (Hudlow 1988). Computer capabilities have become sufficiently sophisticated to facilitate numerical solution of atmospheric general circulation models.

Mesoscale processes are medium in scale between synoptic scale and individual cells. Convective rain cells can organize themselves into mesoscale structures of nearly circular clusters or linear squall lines of thunderstorm cells extending hundreds of kilometers (Barry and Chorley 1992). In other mesoscale convective systems (MCS) airflows that are lifted by convection can interact with topography and get anchored or "locked in." The process becomes nearly stationary over a region for a relatively long duration, causing heavy downpours and flash floods. An example is the disaster in Big Thompson Canyon, Colorado, in 1976. The deluge brought by the MCS dumped 305 mm (12 in.) of rainfall in less than 6 hours, resulting in a flash flood that caused 139 deaths and property damage totaling more than $50 million.

Conceptual and numerical models of the structure, propagation, and life cycle of an MCS are available (National Research Council 1991a). In Hawai'i, the disastrous New Year's Eve flood of 1988 on O'ahu was analyzed as a mesoscale event (National Research Council 1991b). Since then Geostationary Operational Environmental Satellite and Doppler radars have become available to help forecast and warn of flash floods.

El Niño (EN) and the Southern Oscillation (SO) are two closely associated atmospheric-oceanic events with global consequences. EN concerns the warming of ocean water in the equatorial eastern Pacific Ocean and especially off the coast of Peru. SO refers to the reversal of the usual atmospheric pressure at the east and west ends of the southern Pacific Ocean. Their association was firmly established after the 1982–1983 ENSO event (Wyrtki 1982; Rasmusson and Wallace 1983). ENSO is a prime example of the teleconnection among large-scale atmospheric and oceanic interactions. The Hawaiian Islands are profoundly affected by ENSO, even though they are located at its northern fringe. Weather patterns resulting from ENSO have been linked to floods and droughts that have caused havoc in and beyond the Pacific region.

El Niño is a large-scale, 1° to 4°C warming of sea-surface water in the equatorial eastern Pacific Ocean. A warm-water current moves eastward from the Indo-Australian region and displaces the usually northward ocean currents off Peru and the upwelling of cold nutrient water. The phenomenon occurs usually every December. But at irregular intervals (2 to 10 years), the effects of the warm-water body become much more extensive and severe, halting upwelling totally and causing profound atmospheric disturbances and hydrologic extremes (see Figure 3.4). The major features of an El Niño event may take about 18 months to pass. The Southern Oscillation is manifested when there is a reversal of the usual sea-level atmosphere pressure gradient. Usually, the pressure is low in Indonesia and northern Australia (Darwin) and high in the eastern South Pacific (Tahiti) (see Figure 3.5). During non-ENSO years, the trade winds push ocean surface water westward and offshore from South America, allowing upwelling of cold water and causing sea level to rise in the western South Pacific.

During an ENSO episode, the warm pool of ocean water drifts eastward because of weakened trade winds. This shift upsets the Hadley cell, which is a large-scale thermal cell of air circulation, in such a way that the synoptic systems, including Kona and cold-front storms that normally produce winter rainfall in the Hawaiian Islands, are displaced away from the Islands. As a result, dry conditions prevail during an ENSO episode (Schroeder 1993a; Chu and He 1994).

The Southern Oscillation is a part of the global circulation called the Walker circulation. Walker apparently was unaware of the associated episodes of warm sea-surface temperature, but he discovered the coherent pattern of the air pressure at sea level.

As El Niño warm water recedes, cool water replaces it, setting the stage for a possible La Niña, or cold phase,

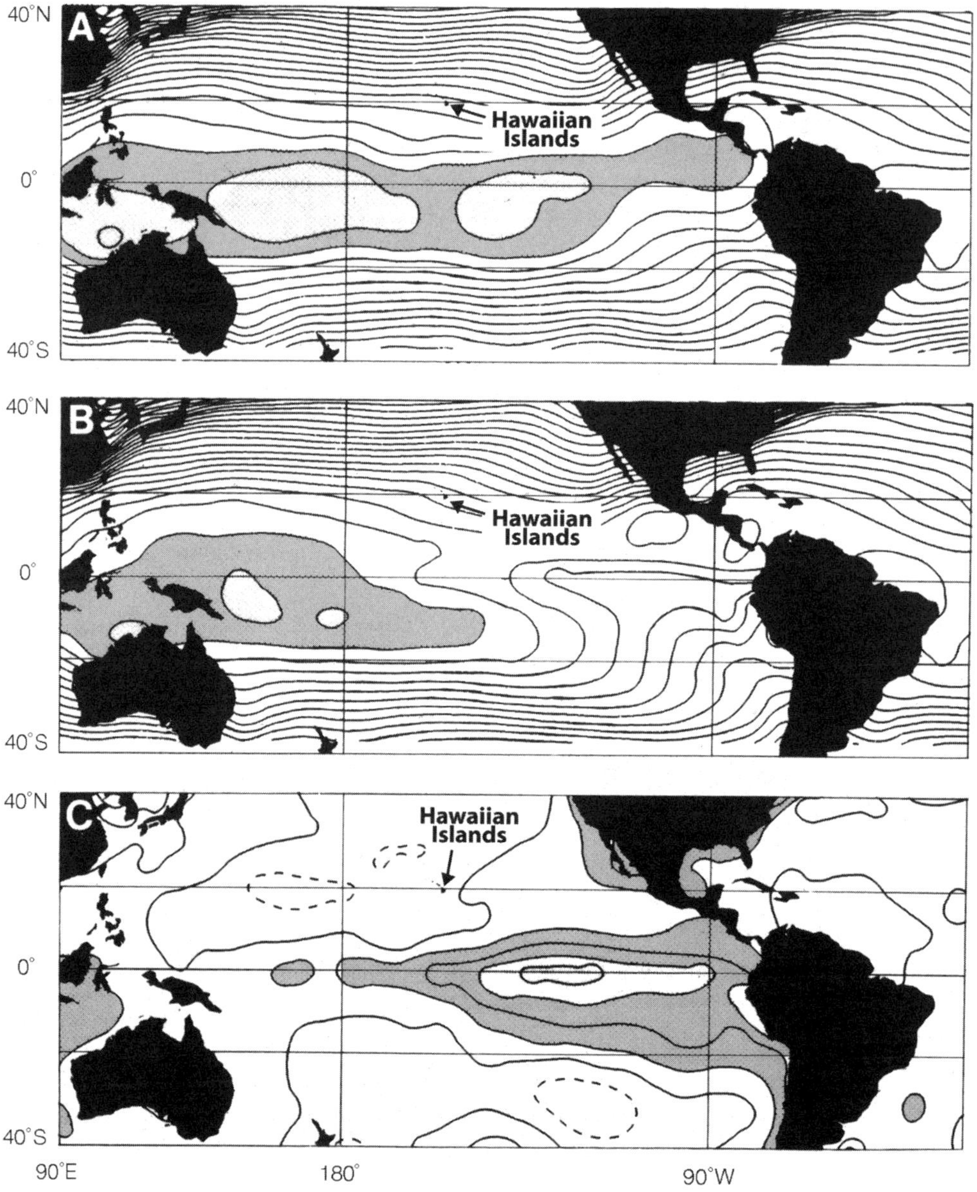


Figure 3.4. Sea-surface temperature patterns during the 1982–1983 ENSO episode. Contour interval, 1°C (1.8°F). *A*, December 1982 to February 1983; *B*, normal December–February, the heaviest shading is for sea-surface temperatures >29°C (84.2°F); *C*, anomalies for December 1982 to February 1983. The heaviest shading is >3°C (>5.4°F). (Adapted from Rasmusson and Wallace 1983)

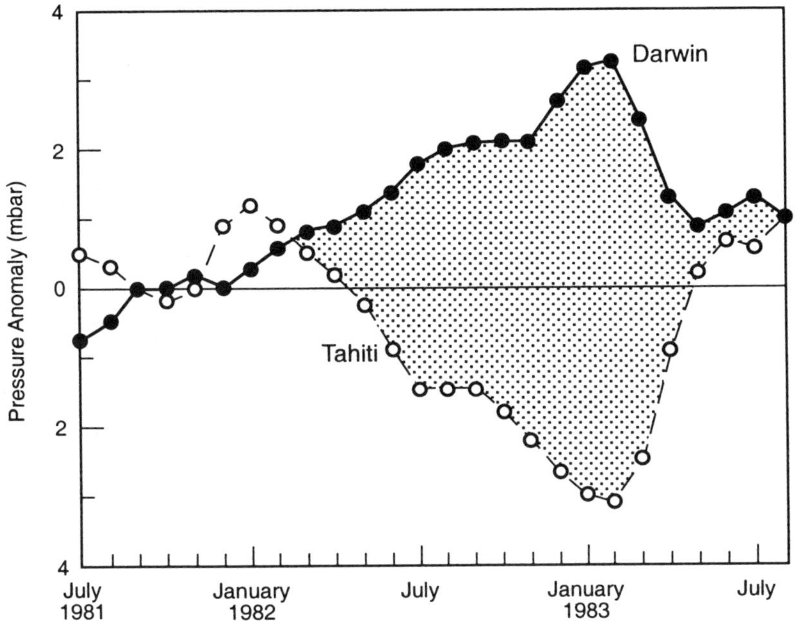


Figure 3.5. Sea-level pressure anomalies at Tahiti (17° S, 150° W) and Darwin, Australia (12° S, 131° E), during the 1982–1983 ENSO episode. (Adapted from Rasmusson and Wallace 1983)

event. La Niña's effect on rainfall is much weaker than El Niño's. Following El Niño in 1997–1998, La Niña began in the summer of 1998. A wet rainy winter season was forecast for 1998–1999, but rainfall records did not confirm the forecast. The lee of the entire Hawaiian Island chain actually suffered a prolonged drought (Chu 1999).

## Traits of Hawai'i Precipitation

The rain climate in Hawai'i is highly variable, even though the general impression is that it is warm, sunny, or partly cloudy, with gentle trade winds and occasional *mauka* (mountain) showers. Mt. Wai'ale'ale on the island of Kaua'i is one of the world's wettest spots. Mean annual precipitation there is 11,267 mm (443.5 in.). Only 25 km (15 mi) away to the lee of the mountain is Kekaha, a parched coastal town with only 543 mm (21.3 in.) of mean annual rainfall. The high islands of Hawai'i are able to recruit a great deal of rainwater, many times more than that which occurs over the surrounding ocean. As a whole, the Islands receive 1,905 mm (75 in.) of rain, averaged annually, or $1 \times 10^3$ m$^3$/s ($3.5 \times 10^4$ ft$^3$/s or $22.8 \times 10^9$ gallons per day). Virtually all precipitation is rain. Fog occurs in a band at moderate altitudes (1,200 to 1,800 m [3,900 to 5,900 ft]), and snow occurs only at the highest altitudes.

### Climate Controls

In the case of precipitation, three primary controls are paramount: the Hadley cell, the oceanic position of the major Hawaiian Islands (154° to 160° west longitude, 19° to 22° north latitude), and the high mountains. These controls interact to create profound variations in the rain climate of Hawai'i.

The Hadley cell is a model of the general circulation of the atmosphere—the large-scale patterns of wind and pressure that persist throughout the year and that transport heat and momentum. In the Northern Hemisphere, the Hadley cell operates at latitudes approximately between the equator and 30° north. In a simplified Hadley cell, which is a thermal direct cell, warm air near the equator rises and generates a low-level flow toward the equator. The Earth's rotation deflects the air currents,

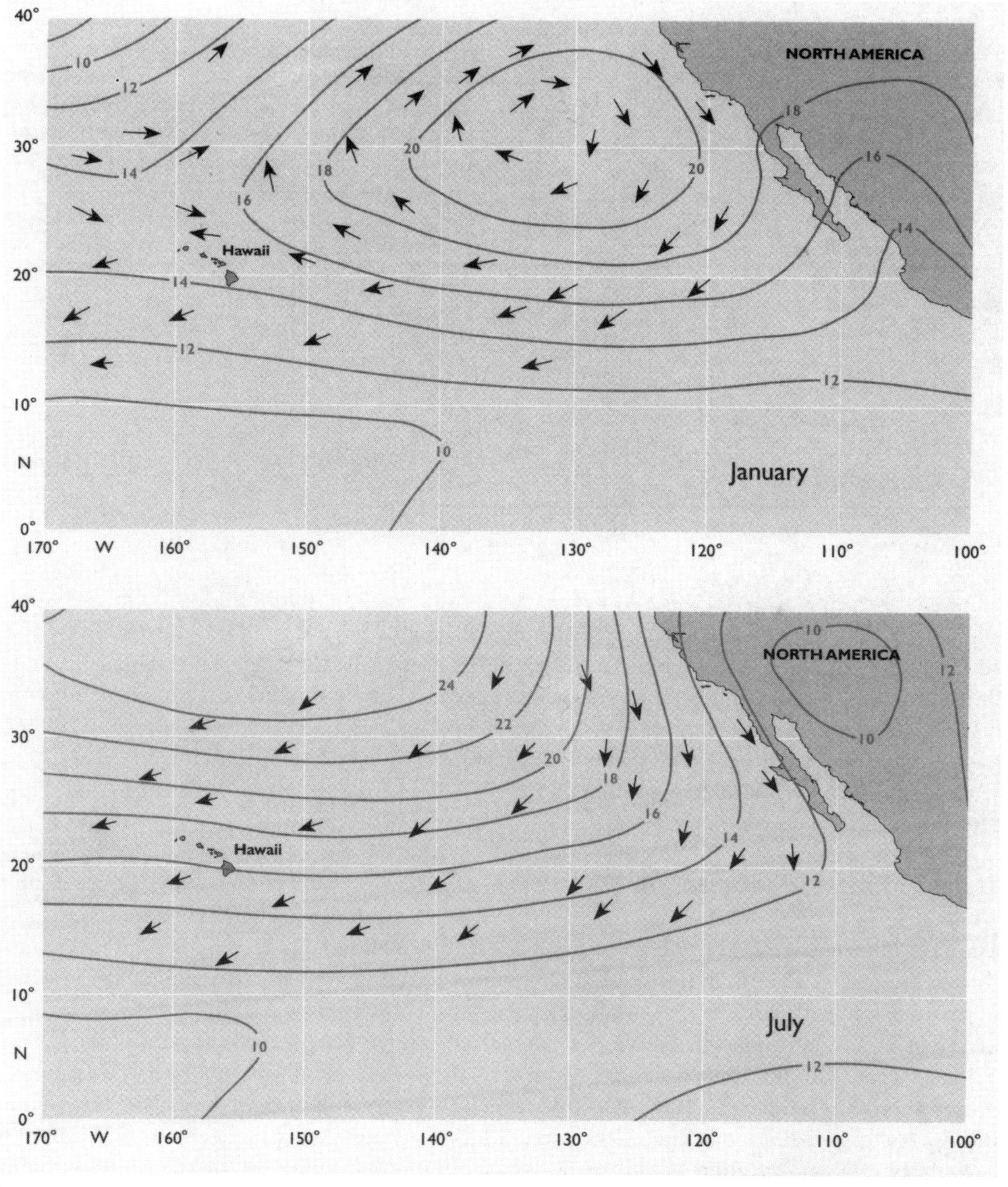


Figure 3.6. Sea-level pressure and surface winds indicating the Pacific Subtropical Anticyclone in the northeastern Pacific Ocean. Units of pressure are millibars (mbar) above 1,000 mbar. (Reprinted from Schroeder 1993a with permission from University of Hawai‘i Press)

thus forming the northeasterly trade winds. The cell is completed by the aloft poleward counter-current, which is cooled by radiation and sinks at about 30° north latitude as the Pacific Subtropical Anticyclone. Although the reality is more complicated than stated, the Hadley cell has commonly served as the starting model and has not yet been replaced by any other model (Barry and Chorley 1992).

Trade wind is the direct result of the Pacific Subtropical Anticyclone. Established by observations, the anticyclone is a permanent high-pressure cell with a seasonal shift (see Figure 3.6). The anticyclone is extensive in July, with the center located about 37° north and 150° west; it shrinks and shifts southeast to about 30° north and 130° west in January. Geographically, the Hawaiian Islands are situated within or on the fringe of the anticyclone; therefore, this shift brings about the seasonal variations of trade winds in Hawai'i. The ocean moderates the climate on coastal land by differential heat absorption and advection of heat.

A layer of temperature inversion separates the moist surface air from the dry air sinking from aloft, which warms due to compression. Within the inversion layer the temperature gradient reverses, increasing instead of decreasing with altitude (see Figure 3.7). The relative humidity of the air below the inversion is high, ranging from 60 to 80 percent, with the highest values in the windward areas. Above the inversion, the air is dry, with relative humidity ranging from 40 to even 5 percent. The trade-wind inversion reveals itself by the presence of nearly flat cloud tops and by the sharp change in climate along the slope of the high mountains that rise above the inversion (Schroeder 1993a).

The clear effect of the high and massive mountains on the major Hawaiian Islands is the creation of orographic precipitation. The trade-wind rains along the Ko'olau Mountain Range of O'ahu is a classical example. The effectiveness is moderated by the alignment of the mountain relative to the trade wind, the shape of the mountain, and the height of the mountain relative to the altitude of the inversion layer. The diurnal rain patterns are attributed to the orographic effect (Schroeder et al. 1977a; Chen and Nash 1994).

Less obvious is the fact that mountains also enhance cyclonic and convective precipitation. The central mass of Mt. Wai'ale'ale is shaped in such a way that the orographic effects are robust for winter cyclonic storms and other frontal activities (Schroeder 1993a). Anchor effects have been identified with intense convective systems that have resulted in flash floods in Hawai'i (Schroeder 1976). The anchor is defined as a discontinuity in surface roughness, such as a coastline or a mountain range, encountered in the path of storms. Ramage (1971) concluded that intense tropical rainstorms are usually anchored by a discontinuity.

## Climate Zones

### Classifications

The humid tropics can be defined in different ways, depending on the objectives. In 1993, Chang and Lau briefly summarized the three best-known approaches: Köppen, Thornwaite, and Gernier (Chang and Lau 1993). The approaches are used in a large context for the classification of climate. For instance, the island of Hawai'i, which has the extremes of climate in Hawai'i, is classified according to Köppen (Juvik et al. 1978). In addition to the expected humid tropics, three other climates are also identified and delineated as arid and semiarid, temperate, and ice (Figure 3.8).

The current humid tropical classification is reported and applied to the world (Figure 3.9) by the United Nations Educational, Scientific, and Cultural Organization (UNESCO) (Chang and Lau 1993). The classification considers vegetal growth, requires minimal data, and is practical to use. It is based on thermal criterion and hydrologic growth season (length of wet season). Tropics are defined as zones with a mean monthly temperature above 18°C (64.4°F) for the coldest month and with a wet season. The value of temperature is selected because it prevails over 95 percent of the lowland between the north

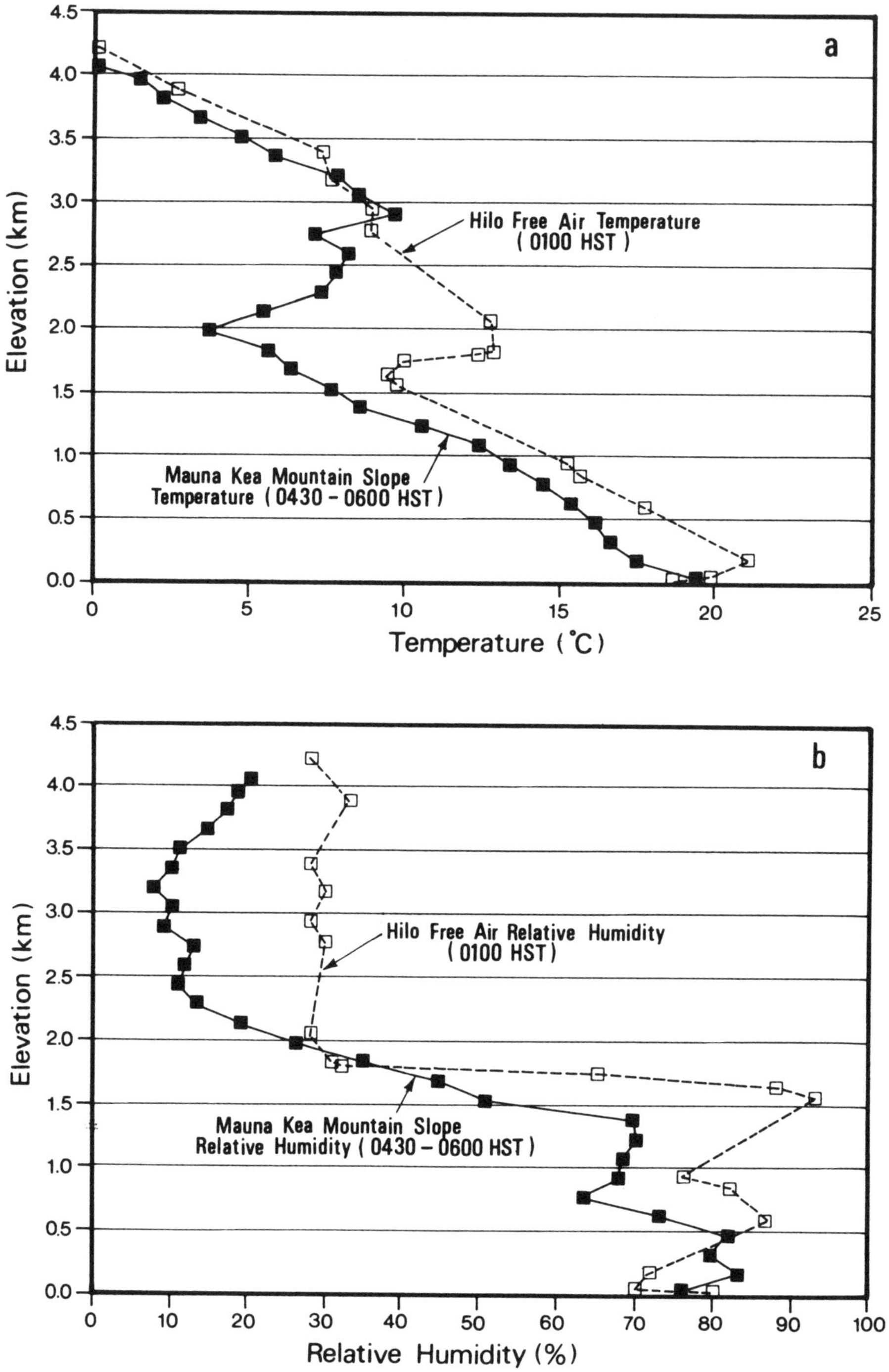


Figure 3.7. Atmospheric profiles of (*a*) air temperature and (*b*) relative humidity under typical trade-wind inversion condition. Data from both ground-based measurements and free atmosphere rawinsonde, February 27, 1993, Mauna Kea, Hawaiʻi. 1 km = 3,281 ft. (Reprinted from Nullet and Juvik 1994 with permission from Blackwell Publishing Ltd.)

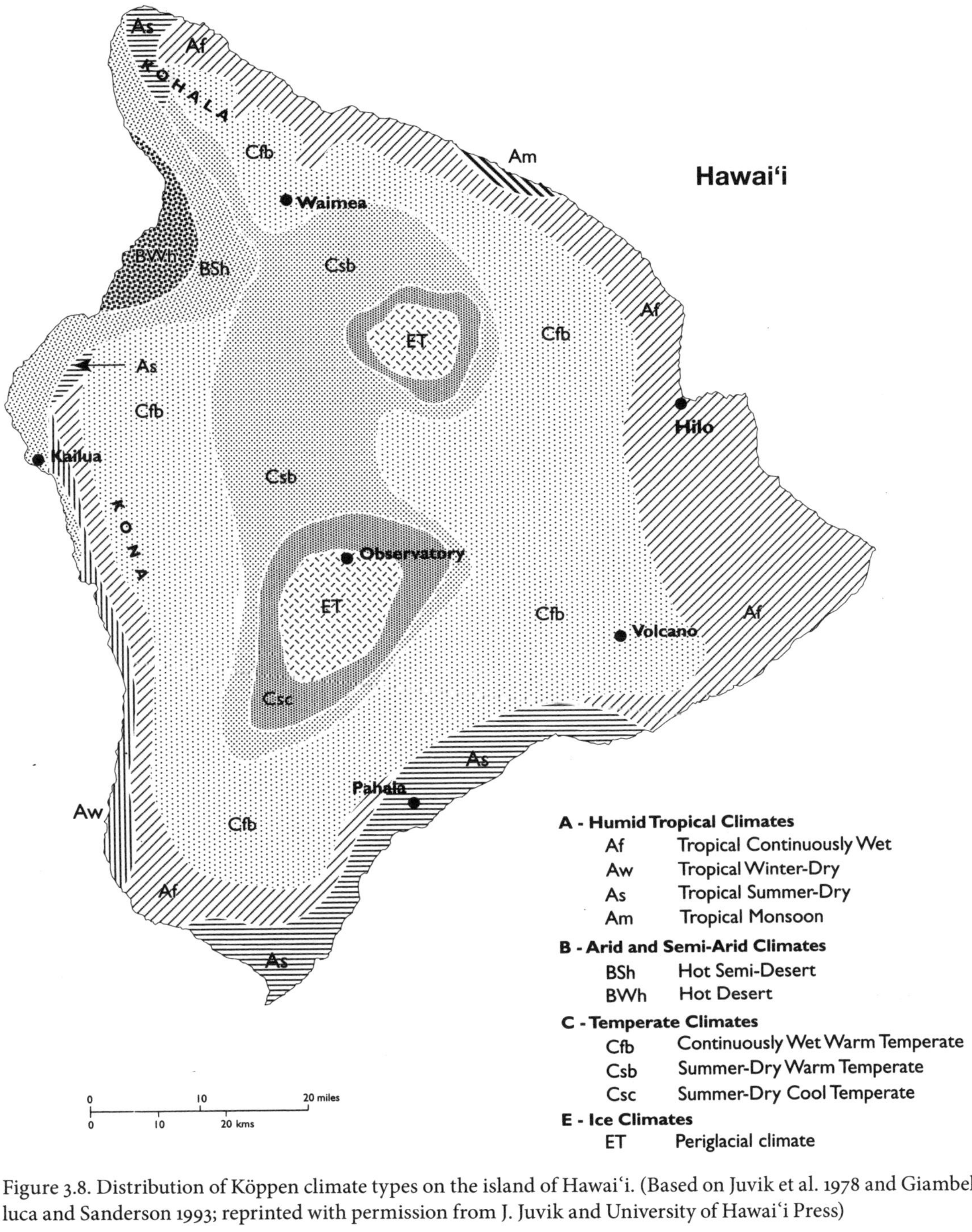


Figure 3.8. Distribution of Köppen climate types on the island of Hawai'i. (Based on Juvik et al. 1978 and Giambelluca and Sanderson 1993; reprinted with permission from J. Juvik and University of Hawai'i Press)

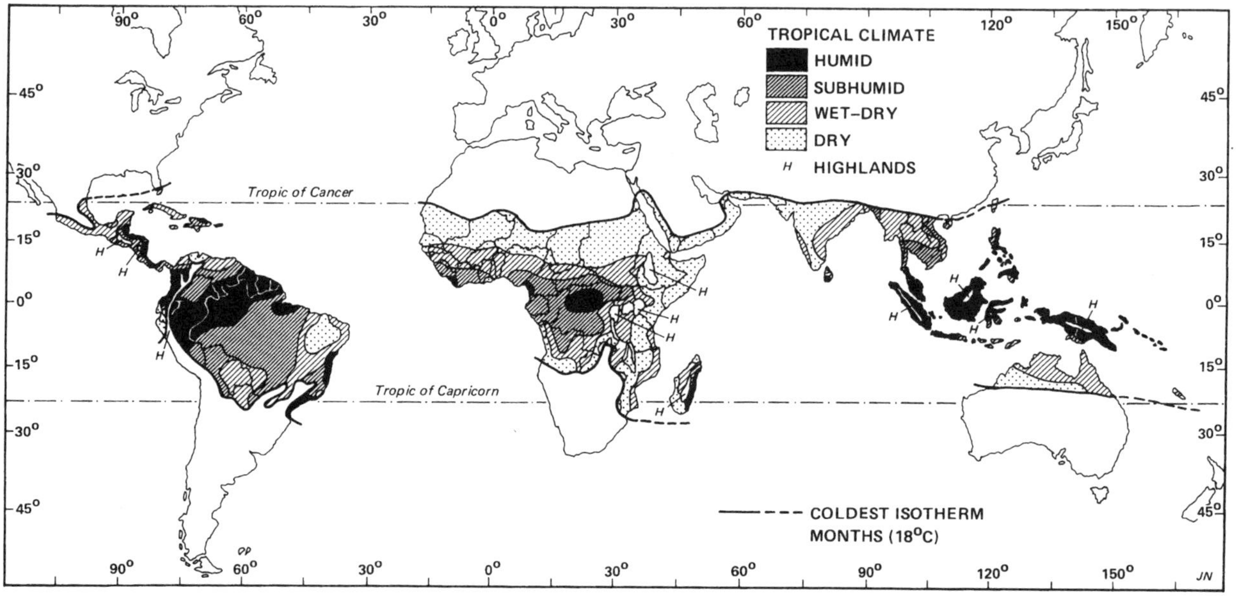


Figure 3.9. World map showing distribution of the four climatic subtypes (humid, subhumid, wet-dry, dry) of the tropics: 9.5 – 12 wet months, humid; 7 – 9.5 wet months, subhumid; 4.5 – 7 wet months, wet-dry; <4.5 wet months, dry. 1 wet month = 100 mm monthly rain, 0.5 wet month = 60 – 100 mm monthly rain; 100 mm = 3.9 in. (Reprinted from Chang and Lau 1993 with permission from Cambridge University Press)

and south 23° 30' parallels, the conventional definition of tropics. A wet month is defined as one in which rainfall exceeds evapotranspiration for agricultural and vegetal growth. According to this definition and supporting data, a wet month is a month having more than 100 mm (3.9 in.) of rainfall. A month with rainfall between 60 mm (2.4 in.) and 100 mm is credited as one-half wet month. Tropics are divided into four subzones: dry (less than 4.5 wet months), wet-dry (4.5 to 7 wet months), subhumid (7 to 9.5 wet months), and humid (9.5 to 12 wet months). Except for the highlands and a few other sites, Hawai'i falls within the classification of humid tropics.

## Climate Parameters

Solar radiation, air temperature, and winds are among the key parameters for determining precipitation. The sun radiates short wave energy, providing the external heat source of the Earth. Earth radiates its own energy in long waves with much less intensity. The intensity and the daytime hours of solar radiation are the parameters for assessing plant growth and conversion to electricity.

Upon striking a surface, radiation is either reflected or absorbed. The reflected fraction is called albedo. The lower the albedo, the greater the amount of heat absorbed. Water bodies are the greatest absorbers of radiation, with average albedo values of only 0.06 to 0.10. The albedo value depends on the angle of incidence and the type of vegetation — about 0.15 for tall crops and trees and about 0.25 for short crops such as grass (Barry and Chorley 1992).

Net radiation accounts for the net gain or loss of all the radiative energies — short and long, incoming and outgoing. Net all-wave radiation is, therefore, the single most useful parameter. In Hawai'i, evaporation is far more dependent on net radiation than on temperature. The average annual net radiation in Hawai'i varies between 121 and 145 $J/m^2/s$ (250 and 300 $cal/cm^2/d$).

Advected heat energy is the net horizontally air-transported energy and is distinct from radiation. It must be separately accounted for in heat balance and evaporation budgets, especially in land–sea interfacing environments.

The water-absorbed radiant heat evaporates water; the reflected radiant heat warms the air by thermal conduction and thermal convection. The heat involved in warming the air is known as sensible heat. Air temperature at sea level and in lowlands in Hawaiʻi is considered warm and pleasant. Its range of variation — day to day and month to month — is small. The mean daily temperature at Honolulu is 27.4°C (81.4°F) in August, the warmest month, and 22.4°C (72.4°F) in January, the coolest month. In the highlands, the annual temperature cools at about 6.5°C/1,000 m (3.6°F/1,000 ft) and at a lesser lapse rate, 4°C/1,000 m (3°F/1,000 ft), above 1,200 m (3,900 ft), the approximate altitude of the inversion layer. The mean annual temperature is a chilly 11.3°C (52.4°F) at the top of Haleakalā (station elevation 2,144 m [7,100 ft]) and 7.0°C (44.6°F) at the top of Mauna Loa (station elevation 3,400 m [11,200 ft]). The highest recorded temperature is 38°C (100°F) at Pāhala, island of Hawaiʻi, but that was a rare event (Nullet and Sanderson 1993).

Wind manifests air pressure difference and atmospheric circulation. Hydrologic interests in winds stem from rainfall and evaporation. Trade winds are most important; other winds include sea breeze, land breeze, mountain wind, valley wind, and strong winds associated with hurricanes.

Trade-wind rainfall prevails nearly everywhere in all Hawaiian high islands. In an analysis of the daily synoptic maps of Oʻahu over a 25-year record (1916 to 1940), over 90 percent of the July rainfall is attributed to trade winds (Yeh et al. 1951b). Trade-wind rainfall is substantial even in the three months of December–January–February. Trade winds, which prevail only 43 percent of the time during these months, account for 45 to 65 percent of the rainfall, with the higher values occurring in mountain locations.

Orographic rains in Hawaiʻi are due to trade winds, with few exceptions. "Kona" rains inland of the Kona coast, island of Hawaiʻi, are due to sea breezes that are thermally induced by rapid heating over land. Convective rain showers are blown upslope, creating a summer rainfall maximum that is unique in Hawaiʻi. *Nāulu* is the Hawaiian word for a sudden shower of high intensity and short duration. The *nāulu* that occur offshore over the southern leeward coast of Molokaʻi in the summer result from convergence of two air currents: the cloud and rain formed by sea breezes and blown on land in the morning and blown offshore in the early afternoon by the typical trade winds (Leopold 1949). On the island of Hawaiʻi, diurnal rainfall in general and the early morning rainband offshore of the windward slopes in particular are influenced by a complicated wind regime. The phenomenon was the subject of a scientific controversy, in part because of insufficient observational data to validate theories, until the Hawaiian Rainband Project (HaRP) (Chen and Nash 1994). This project yielded comprehensive observations of wind patterns on the island of Hawaiʻi during a 42-day summer trade-wind period. The distinct diurnal variation of convergences is similar for both strong- and weak-wind days (Figure 3.10).

Additional interests in winds concern human comfort and wind energy. During periods of weak trade winds (less than 6.25 m/s [14 mph]), people are apt to sense the high humidity. Trade winds prevail 80 to 95 percent of the time during the five months from May to September but decrease to 50 to 80 percent of the time in the seven months from October to April (Blumenstock and Price 1967). High wind-energy sites are the corners of the islands, crests of low mountain ridges, and saddles between large mountains. Examples of these areas are, respectively, Kahuku, the north corner of Oʻahu; Nuʻuanu Pali of the Koʻolau Range on Oʻahu; and the isthmus on Maui. Annual wind energy has been evaluated for all major Hawaiian Islands (Schroeder 1993b).

## Rainfall Patterns and Trends

### Rainfall Patterns

Spatial distribution or patterns are important traits of rainfall in Hawaiʻi. Isohyets are lines of equal rainfall

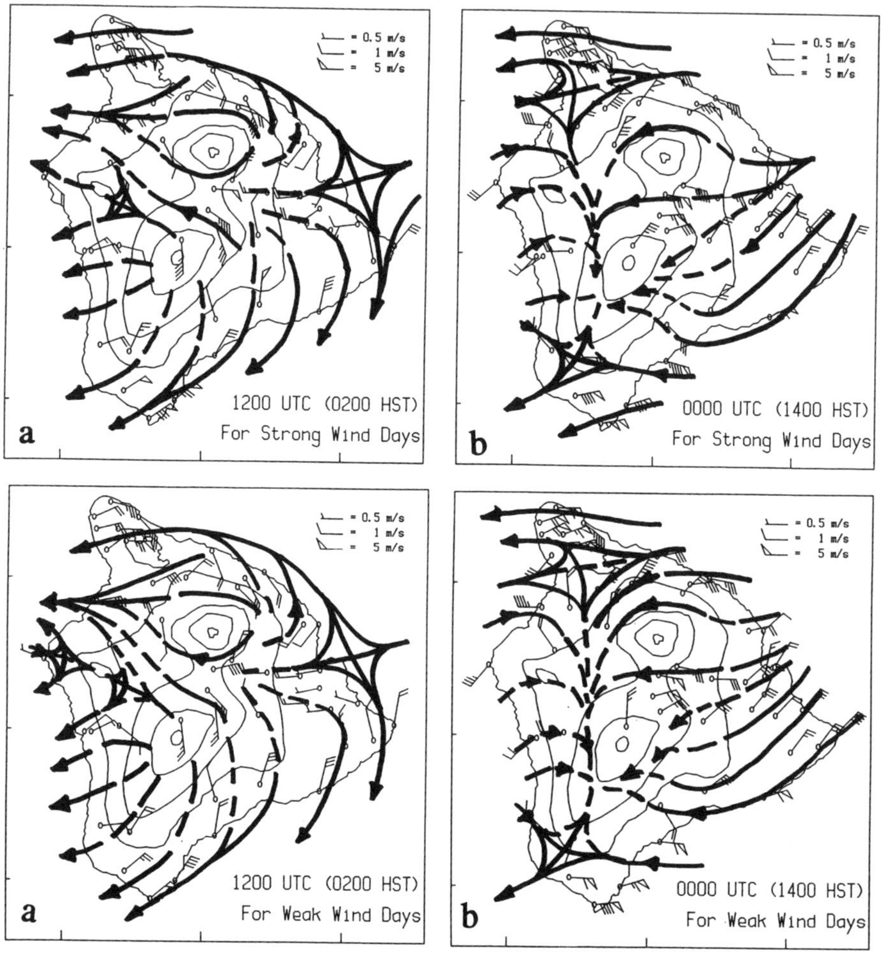


Figure 3.10. Surface-streamline analysis for strong and weak trade-wind days: *a,* 0200 hours HST and *b,* 1400 hours HST, July 11–August 24, 1990, summer trade-wind period, island of Hawai‘i. The solid stream lines indicate regions where air at the surface stays at the surface for the entire path as it moves upslope-downslope. The broken stream lines indicate regions where air at the surface does not stay at the surface for the entire path. (Reprinted from Chen and Nash 1994 with permission from American Meteorological Society)

drawn from rain-gage data and are used to suggest rainfall patterns. The Hawai‘i isohyets parallel the topographic ground contours with two exceptions. The resemblance is due to orographic effects, and the exceptions are due to atmospheric temperature inversion and strong sea breezes. Analyses of Hawai‘i rainfall patterns have included annual, monthly, and shorter durations. Maps showing the mean and median annual rainfall are presented in Figures 3.11 to 3.15 (Giambelluca et al. 1986).

To emphasize the mountain and trade-wind influences, an analysis of the annual patterns is indexed to the rain that falls into the ocean surrounding the Hawaiian Islands. The referenced value is 600 mm (23.6 in.), which is adopted from two similar (close) estimates of ocean rainfall by Elliott and Reed (1984) and Dorman and Bourke (1979). Their studies are based on marine (shipboard) observations. In their isohyet maps for the ocean, the Hawaiian Islands are located in the dry part of the Pacific Ocean.

The annual rainfall on windward coasts is much higher than the annual rainfall on the ocean. The windward coasts of three great mountains rising above the inversion—Haleakalā, Mauna Kea, and Mauna Loa—register as much as five times the rainfall on the ocean.

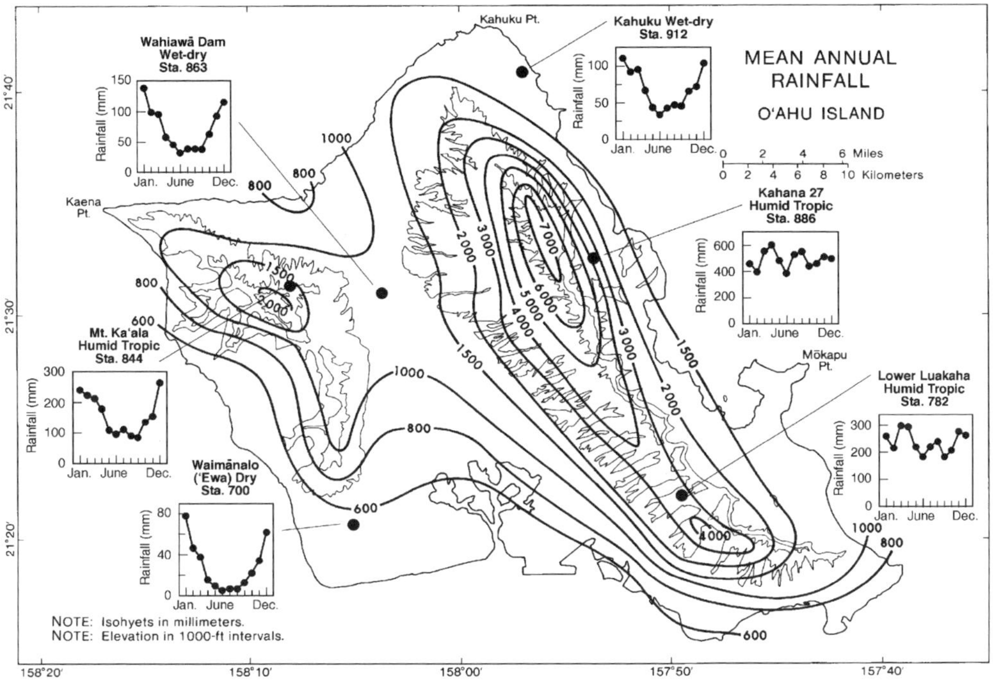


Figure 3.11. Mean annual rainfall and annual rainfall cycles of selected stations, O'ahu, Hawai'i. Adjusted median monthly rainfalls; 1,000 mm = 39.4 in. (Adapted from Giambelluca et al. 1986)

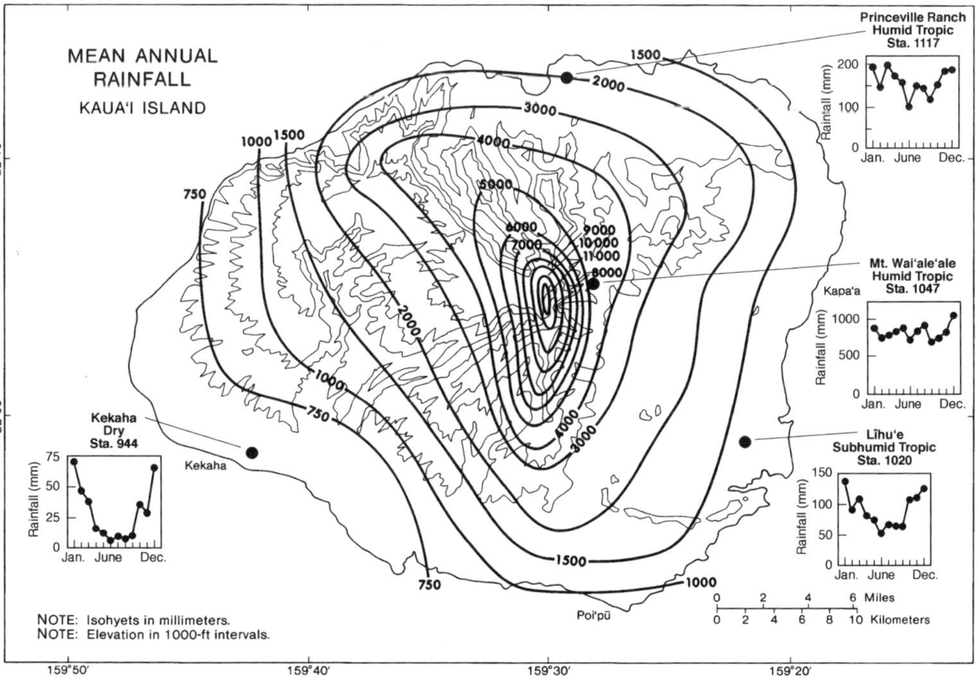


Figure 3.12. Mean annual rainfall and annual rainfall cycles of selected stations, Kaua'i, Hawai'i. Adjusted median monthly rainfalls; 1,000 mm = 39.4 in. (Adapted from Giambelluca et al. 1986)

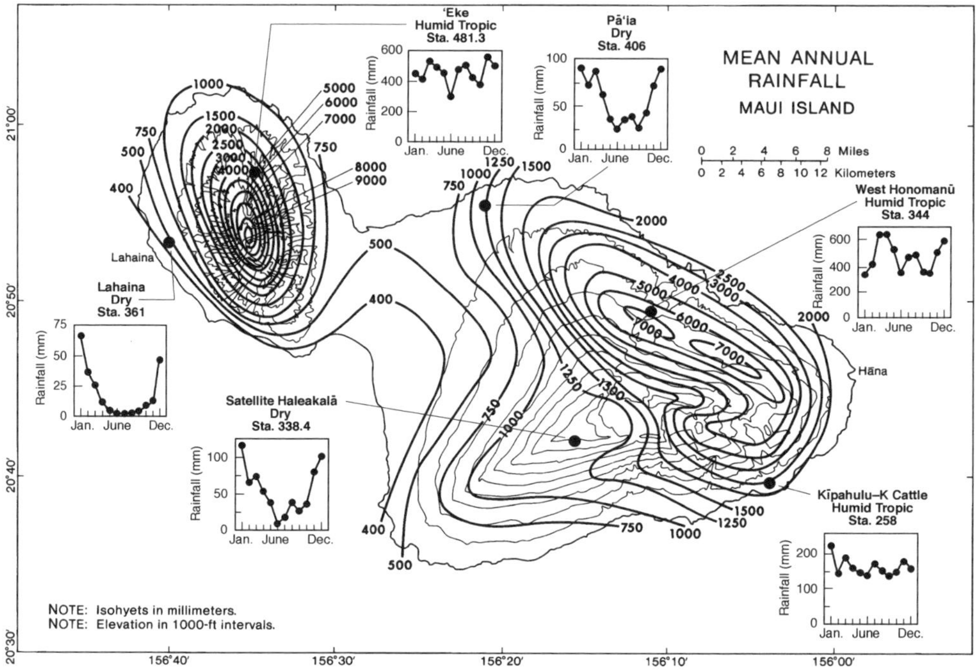


Figure 3.13. Mean annual rainfall and annual rainfall cycles of selected stations, Maui, Hawai'i. Adjusted median monthly rainfalls; 1,000 mm = 39.4 in. (Adapted from Giambelluca et al. 1986)

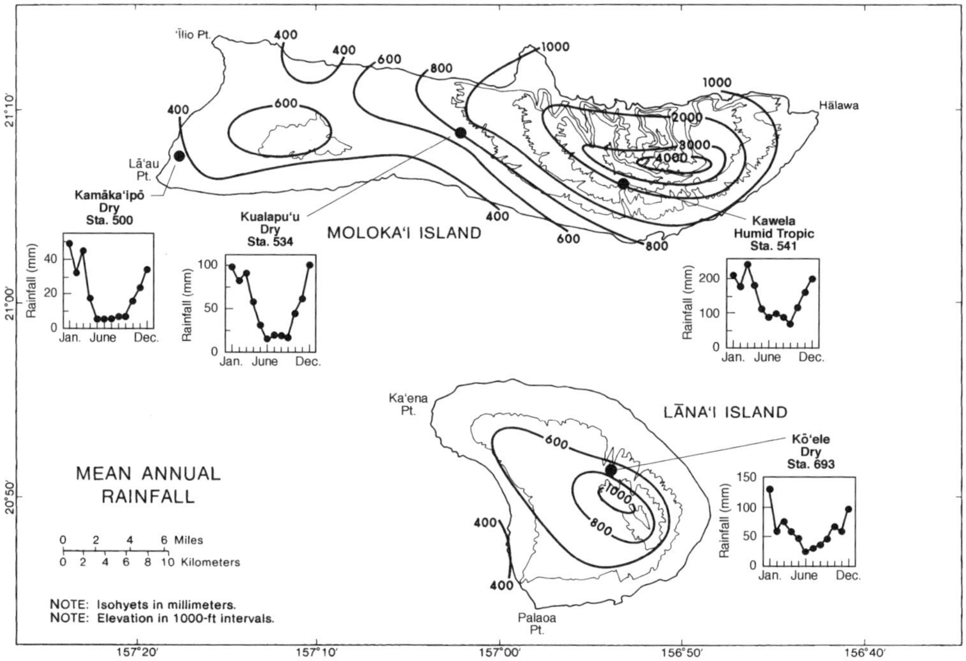


Figure 3.14. Mean annual rainfall and annual rainfall cycles of selected stations, Moloka'i and Lāna'i, Hawai'i. Adjusted median monthly rainfalls; 1,000 mm = 39.4 in. (Adapted from Giambelluca et al. 1986)

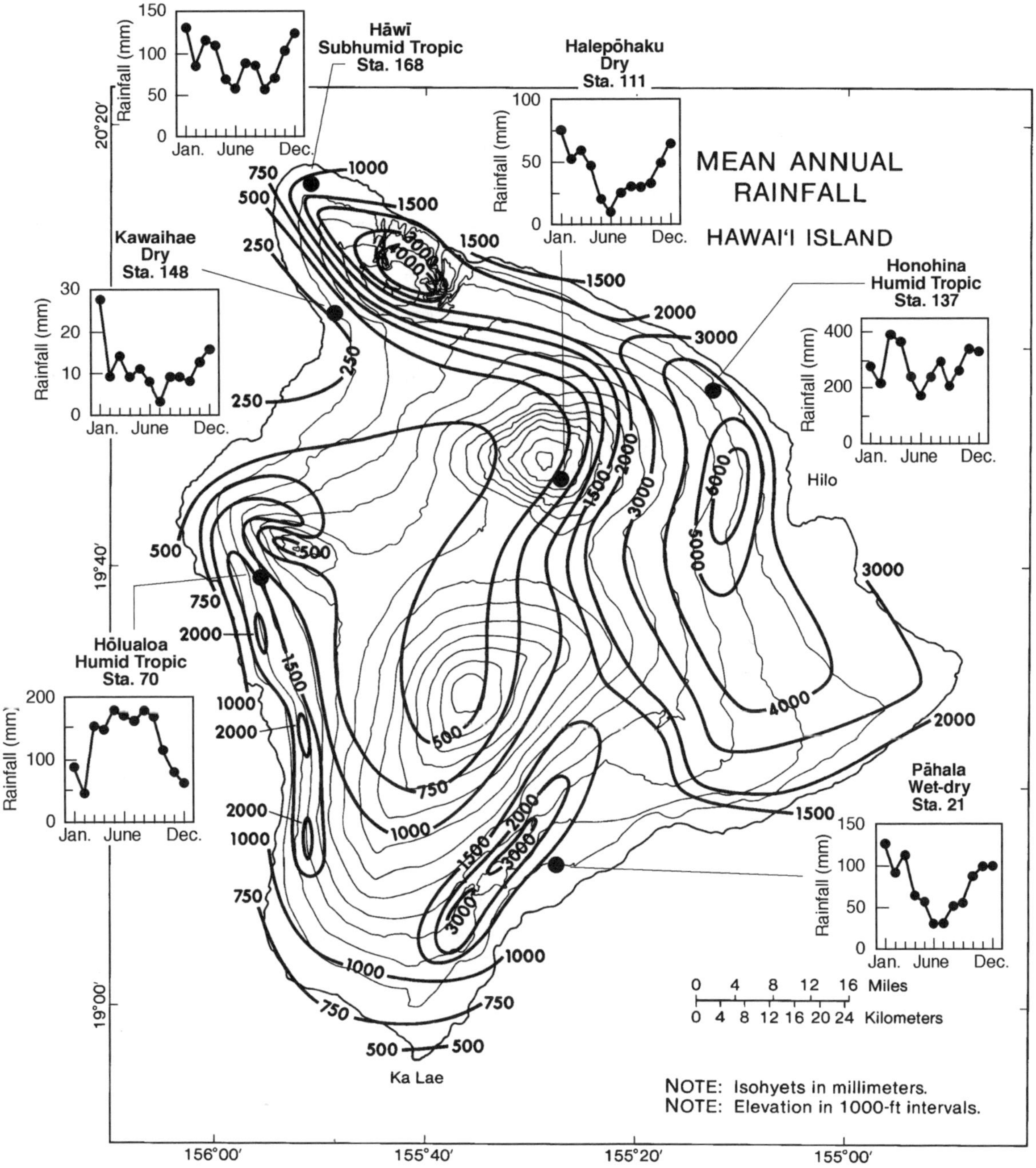


Figure 3.15. Mean annual rainfall and annual rainfall cycles of selected stations, Hawai'i Island, Hawai'i. Adjusted median monthly rainfalls; 1,000 mm = 39.4 in. (Adapted from Giambelluca et al. 1986)

Elevated only slightly above sea level, the Maui isthmus receives only ocean rainfall. Located in the rain shadow of West Maui and East Moloka'i, the whole island of Lāna'i experiences low rainfalls except for Lāna'ihale, which is situated at the highest elevation (1,026 m [3,366 ft]) of the island.

As the trade winds move up the mountain slopes, rainfall increases sharply. For mountains situated beneath the cloud base, the rainfall maximum occurs just leeward of the summit of the ridge. Examples are West Maui (fifteen times ocean rainfall), O'ahu's Ko'olau Range (twelve times), East Moloka'i, and the Kohala mountains, Hawai'i (both almost seven times). For those mountains rising above the cloud base, the maximum rainfall occurs on the slopes at an elevation below but close to the cloud base: almost twelve times ocean rainfall for Haleakalā and ten times for Mauna Kea and Mauna Loa.

Where summits and slopes emerge above the inversion clouds, rainfall drops off sharply and the climate is totally different from that of the humid tropics. The higher the summits are, the lesser the rainfall recorded (Haleakalā about two times ocean rainfall; Mauna Kea and Mauna Loa less than ocean rainfall). Some precipitation on these high mountains occurs in the form of snow.

The corners of the islands are transitional from a windward rain climate to a leeward rain climate. The corners are the two ends of an island exposed to the trade winds. When relief at the corner is relatively low and loses its airflow blocking and lifting effects, the amount of rainfall drops close to that of ocean rainfall (only 30 to 60 percent higher). But when relief at the corner is high, the rainfall can be several times the ocean rainfall. Hāna, Maui, and the downwind coast from Cape Kumukahi, Hawai'i, are examples of effective orographic rain sites that produce more than three times ocean rainfall.

With one exception, all leeward slopes lose rainfall at lower elevations. A very steep rainfall gradient of 640 mm/km (41 in./mi) over a 16-km (10-mi) span occurs in Waimea, Kaua'i. An even sharper one occurs in West Maui at 1,075 mm/km (68 in./mi) over an 8-km (5-mi) distance. The exception is the reach from Kailua-Kona to Pali Kaholo, Kona, Hawai'i, where sea breezes create a high-rainfall belt amounting to two to three times ocean rainfall between the elevations of 610 and 915 m (2,000 and 3,000 ft). A gentle rain gradient occurs over the leeward coastal plain of Honolulu (350 mm/km [22 in./mi]) over a 4-km (2.5-mi) distance. All leeward coasts are dry, with most experiencing about the same amount of rainfall as ocean rainfall. Lower rates occur at leeward coasts on Maui and Moloka'i, where rainfall is slightly less than ocean rainfall. South Kohala has the driest leeward coast, with only about 40 percent of ocean rainfall.

A rain-shadow effect occurs over the entire Wai'anae Mountain Range, which is located downwind of the Ko'olau Mountain Range on O'ahu. Even though the Wai'anae summit is higher in elevation, rainfall there is only about three times ocean rainfall as compared with twelve times for the Ko'olau Range. The moist airflow becomes drier after losing much of its moisture to the Ko'olau Range. The island of Kaho'olawe also experiences a rain-shadow effect, which in this case is exerted by the island of Maui. The West Moloka'i mountain is not high enough to generate orographic rainfall; the rainfall there approximates ocean rainfall.

Average monthly isohyet maps for the six largest islands for each of the 12 months exhibit patterns that are only slightly different from patterns on annual maps (Giambelluca et al. 1986). Trade-wind rains alone produce a pattern that is remarkably similar to the total rainfall pattern, reflecting the mesoscale interaction with the mountains. The best example is the observations over a 42-day duration (July–August 1990) of the Hawaiian Rainband Project, which confirmed the familiar windward high and leeward low maxima and minima (Figure 3.16).

These long-term averages of annual and monthly rainfalls are essential to water-supply purposes directly or indirectly. Deficiency in rainfall relative to evapotranspiration will necessitate the use of irrigation water for crop growth.

## Rainfall Trends

Trend analyses of rainfall deal with variations with time. They can be studies of wet and dry seasons, calendar

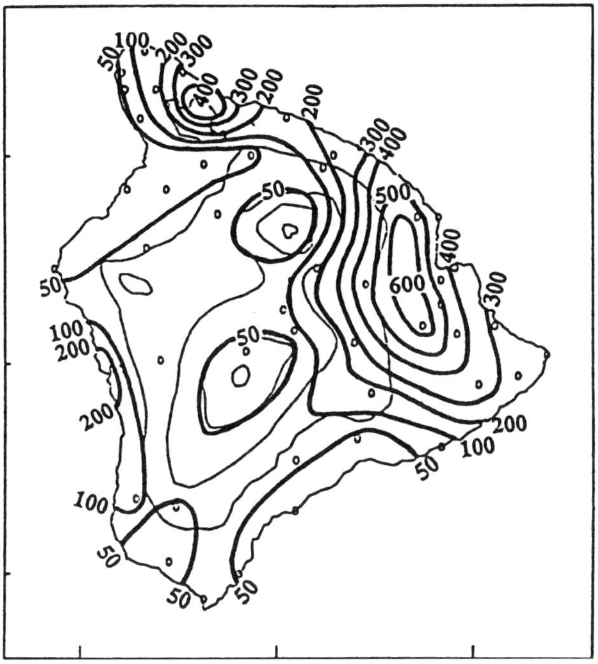


Figure 3.16. Total trade-wind rainfall (mm) accumulated over 42 days (July 11 – August 24, 1990) during the Hawaiian Rainband Project. 100 mm = 3.9 in. (Reprinted from Chen and Nash 1994 with permission from American Meteorological Society)

months, as well as year-by-year and even hour-to-hour variations. All of these have been studied in Hawai'i in part because of their practical applications, including the consequences of meteorological drought. Finally, second-to-second variations have been measured for many showers in the Hilo area employing ingenious rain-intensity gages.

The ancient Hawaiians recognized two 6-month seasons: *kau* (May–October) and *ho'oilo* (November–April). Modern analysis of climatic records shows the soundness of this Hawaiian system of seasons. However, it is more accurate to recognize a summer season of only five months (May–September) and a winter season of seven months (October–April) (Blumenstock and Price 1967). In this classification, the summer season is definitely warmer, dominated by trade winds, and rarely experiences area-wide rainstorms. The Hawai'i Department of Health (2000) has adopted the two 6-month seasons to discriminate between dry and wet seasons in promulgating surface and coastal water-quality standards.

Several trends of monthly rainfalls in a yearly cycle are apparent from an examination of the median values for selected stations (Figures 3.11 to 3.15). Reflecting the seasonal variation of trade-wind rains, the trend of dry summer and wet winter is pronounced in leeward lowlands. The trend is less evident in high-rainfall areas where the average monthly rainfall exceeds 200 mm (7.9 in.). For highlands above the atmospheric inversion, the trend of very dry summers and mildly wet winters persists because only winter cyclonic rains can reach such high-elevation stations as Halepōhaku, Hawai'i, and Satellite Haleakalā, Maui. An exceptional trend of wet summer and less-wet winter occurs at Hōlualoa, Hawai'i, which represents the high-rain belt along the leeward slopes from Kailua-Kona to Pali Kaholo. It reflects the dominance of sea-breeze rains in the summer and few winter cyclonic rains.

Cursory examination of the statistics for annual rainfall indicates that means differ only slightly from medians. Grouping thirty-five stations according to their geography and topography, Meisner (1979) noted that in windward lowlands and rainy mountain slopes the mean is consistently larger than the median. The frequency distribution of the annual rainfall has drawn little attention since Landsberg's analysis (1951) of twenty-two O'ahu stations, which suggested a nearly normal distribution. A record of about 30 years is required for the mean value to stabilize at O'ahu stations (Meisner 1979).

An early study of diurnal rainfall in Honolulu indicates a relatively higher chance of rain at night and through the early morning (8 P.M. to 8 A.M.) than during the rest of the day (Loveridge 1924). This diurnal phenomenon, known as the nocturnal maximum (Figure 3.17), has been attributed to rapid radiative cooling aloft, which causes convective overturning of the air during those hours. However, this trend does not occur in many parts of Hawai'i. Diurnal rainfall was studied for the six largest islands, with hourly rainfall data for the period 1962–1973 (Schroeder et al. 1977a, 1977b). The afternoon maximum in Kailua-Kona (Figure 3.18) is attributed to the blocking

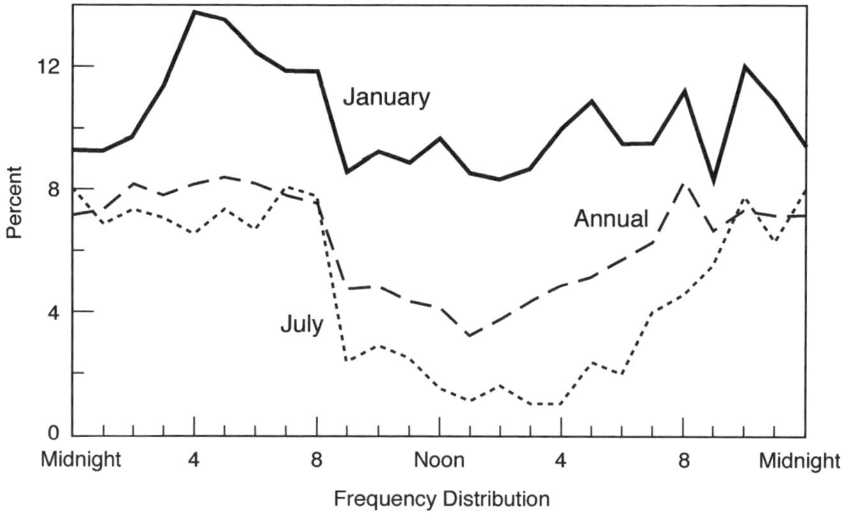


Figure 3.17. Distribution of mean hourly rainfall frequency (0.25 mm = 0.01 in. or more) for the years 1905–1923 in Honolulu, Hawai'i. (Adapted from Loveridge 1924)

of the trade winds by the massive mountains, thus creating an environment that allows local diurnal thermal heating in the afternoon and sea breezes to be effective in promoting rain.

The diurnal rain patterns and trends are detailed for the entire island of Hawai'i in terms of rain frequency (i.e., the probability to rain during a specified hour in a day [Chen and Nash 1994]). Their results (Hawaiian Rainband Project) suggest the location of the initiation and the spatial progression of the rains. For example, the phenomenon of the late afternoon maximum was shown to occur not only in Kailua-Kona but also on most of the island except along the windward coast.

The pattern of diurnal rains on the windward slope near Hilo has attracted different theories regarding the nature of its occurrence. A recent numerical modeling (Rasmussen and Smolarkiewicz 1993) and the Hawaiian Rainband Project data indicate that the highest rain frequency occurs in the inland area rather than near the shoreline and after 7 or 8 P.M. the night before rather than during the early morning hours. The interpretation is that nocturnal rain over the windward lowland starts in situ rather than develops offshore and then drifts inland, as Leopold (1949) and others believed. The diurnal rain is a composite result of three interacting phenomena: orographic lifting, thermal forcing, and the aerodynamics of blocking.

## Intense Rains

The practical interest in intense rains relates to the possibility of flooding. Hawai'i experiences several types of storms that may result in intense rains and flooding: cold-front storms, Kona storms, upper tropospheric disturbances, and hurricanes.

### Storm Types

Cold-front storms commonly occur in the winter. Typically they have a frontal rainband pattern and travel with brisk winds from midlatitudes across Kaua'i, then O'ahu, and even the southeasternmost islands (Figure 3.19*a*). In an average winter, fifteen frontal systems may pass Kaua'i. Of these, thirteen may reach Honolulu; ten, Maui; and nine, East Hawai'i (Schroeder 1993a). The rains they bring are usually light to moderate but can be heavy under unusual atmospheric disturbances. Cool air follows the frontal passage.

Kona storms are also common winter episodes. The name is acquired from the winds that blow from the *Kona* or leeward direction. Originally, the term was in-

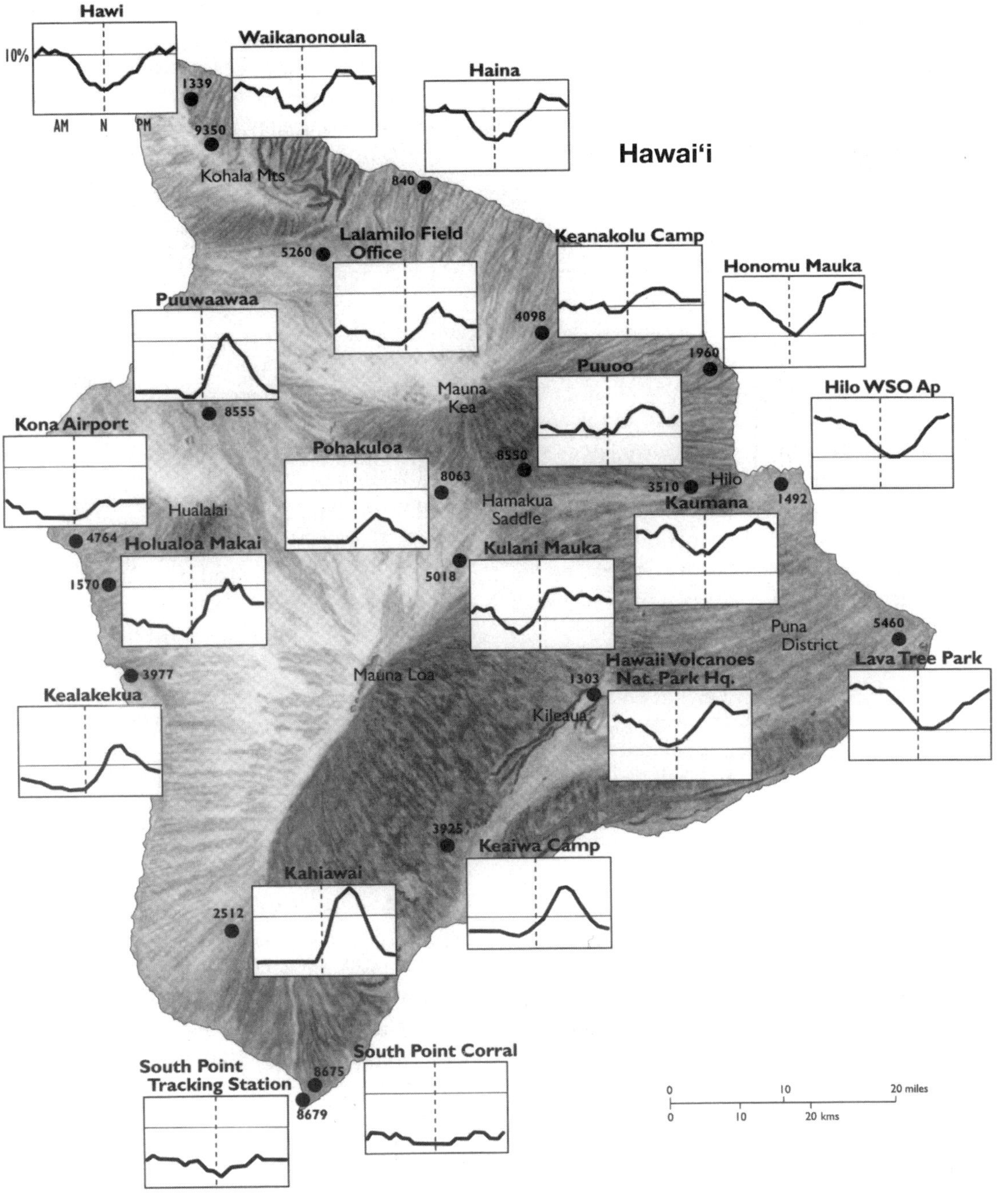


Figure 3.18. Frequency of mean hourly rainfall equaling 0.25 mm (0.01 in.) or more for the years 1962–1973, island of Hawai'i. (Based on Schroeder et al. 1977a and 1977b; reprinted from Sanderson 1993 with permission from University of Hawai'i Press)

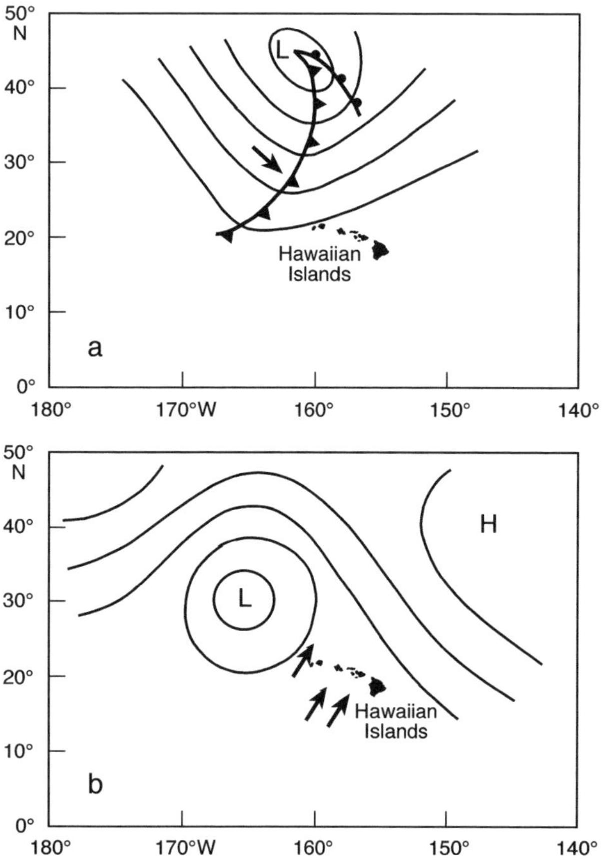


Figure 3.19. Synoptic patterns that produce *a*, frontal storm and *b*, Kona storm. L = low and H = high. (Adapted from Chu et al. 1993)

tended for low-pressure systems developed in the upper troposphere that gradually descend to lower altitudes (Figure 3.19*b*). But the term is applied by the public to any widespread rainstorm with winds coming from a direction other than trade winds (Blumenstock and Price 1967). A Kona storm can bring widespread rain lasting from several hours to several days—more widespread and prolonged than cold-front rains. Deep convection can additionally induce torrential rainfall such as the severe episode in November 1955 (Blumenstock and Price 1967).

Surface trade winds may underlie markedly different upper tropospheric systems. These systems are troughs, an elongated area of relatively low pressure in the upper troposphere that develops near or passes over Hawai'i throughout the year. Often unnoticed, they may at times penetrate downward as a Kona low. At other times, the upward motion associated with the trough may lift or even eliminate the trade-wind inversion and result in heavy rains (Schroeder 1993a). The episode of a severe O'ahu thunderstorm in April 1974 is an example of the latter (Schroeder 1977). A similar example is a May 1965 episode (Blumenstock and Price 1967). Both events caused havoc on O'ahu and Kaua'i. In the 1965 episode, the thunderstorms were developed beneath an upper-level trough and were augmented by the ascent of the coexisting northeasterly trade winds over the Ko'olau Range. The result was intense rains on O'ahu's windward coast.

Hurricanes are rare in the central Pacific. By definition, they are tropical cyclones—intense storm and huge cloud masses that spin as a vortex at high speed (maximum sustained surface wind exceeding 33 m/s [74 mph]) while advancing at a slower speed (4 to 7 m/s [10 to 15 mph]). A "typical" system has a diameter of 650 km (400 mi) with immense convective activities as indicated by great vertical development of cumulonimbus towers over 12 km (7.4 mi) high. At an early stage of development with slower wind speed (less than 17 m/s [38 mph]), they are called tropical storms.

The genesis of hurricanes is not yet fully understood. They form over warm oceans beyond 5 degrees of latitude from the equator. Their formation requires many conditions that are both necessary and sufficient. A warm ocean having a sea-surface temperature exceeding 27°C (80°F) is a necessary condition. Other conditions include an initial atmospheric disturbance, energy sources such as latent heat from condensation, and an upper troposphere anticyclone cell. There have been speculations that hurricanes may be associated with El Niño events, because the Hawai'i hurricanes 'Iwa and 'Iniki occurred in El Niño years. The birthplace of Pacific tropical disturbances is in the eastern and western, but rarely the central, Pacific Ocean. Those born in the eastern Pacific may travel westward, but they usually encounter unfavorable atmospheric and oceanic conditions that cause most of them to decay. A few manage to go past 140° W and enter the central Pacific.

Records indicate that in the period 1905–1992, hurricanes occurred six times in the Hawai'i region. The more recent ones were Dot (August 1959), 'Iwa (November 1982), and 'Iniki (September 1992). 'Iniki caused the worst recorded property damage in Hawai'i. It resulted in $3 billion in damages caused by rains, wind and tornadoes, and coastal surges with high surf. Hurricane Daniel (July 2000) approached from the east, an unusual direction for hurricanes, then veered off in a northwest direction, and harmlessly came no closer than 210 km (125 mi) to the Hawaiian Islands. The season for tropical storms and hurricanes spans from July through November.

## Two Examples

Among the many storms analyzed and documented are the episodes of April 1974 and December 1987. Neither was historically the most intense in the Hawaiian Islands, but their flood damages were severe in urban O'ahu. These analyses helped to identify better instruments and strategies for forecast and warning systems that were finally realized in the 1990s — weather radars, satellites, and telemetered rain gages. The 1974 event suggested an urgent need for radar, and the 1987 event helped prove the efficacy of radars even without state-of-the-art technology.

Rainstorms and the resulting floods in Hawai'i are connected to three essential natural factors: (1) meteorological disturbances on a synoptic scale, (2) ample supply of moisture, and (3) anchoring, which is a discontinuity in surface roughness such as a mountain range or a coastline (Ramage 1971).

The thunderstorm of the April 19, 1974, event lasted only 5 hours and inflicted relatively moderate property damage ($3.9 million), but five lives were lost. The surface trade winds were complicated by an upper tropospheric trough at 250 mbar (approximately 10 km [32,800 ft]), which produced deep convection, cumulonimbus seen by satellite, and thunderstorm. Two centers of high rainfall over the leeward slopes of the Ko'olau Mountains caused severe flooding. One of the two centers was located in urban Honolulu (Schroeder 1976, 1977).

The New Year's Eve flash flood (December 31, 1987–January 1, 1988) inflicted $34 million in property damages in the suburban areas of eastern O'ahu. The meteorological capabilities then available could not forecast the deluge, nor could the flood-control structures control the flood. At Maunawili (State gage no. 787.1), the rainstorm produced a 6-hour rainfall of 398 mm (15.7 in.). Such an event has a recurrence interval of 200 years. Recurrence interval is the average of the intervals, in years, between those rare events when a specified magnitude (rain depth in a given duration) is equaled or exceeded. The downpour registered a 3-hour rainfall of 264 mm (10.4 in.) and a 24-hour rainfall of 581 mm (22.9 in.), which exceeded the 100-year values based on Type I Gumbel frequency distribution (Giambelluca et al. 1984). The isohyets resembled the topographic contours (see Figure 3.20). The meteorological condition was a combination of three events. The first was an extended 12-day-long Kona storm with relatively uniformly distributed rain over O'ahu that brought about only minor flooding. Then a cold front, which had been weakened into a slow-moving shear line, brought low-level north-northeast winds and orographic cumuli clouds. The clouds were lifted over the Ko'olau Mountains and caused considerable orographic rain over windward O'ahu on New Year's Eve. A shear line is a center of strong low-level convergence of winds from different directions. It is a substantial cloud and rain producer. The final aggravating event was a midtropospheric trough at 500 mbar altitude (approximately 6 km [18,000 ft]) lying to the west of O'ahu and moving with a strong jet stream. The strong south-southwesterly winds brought by the trough sheared off the orographic cumuli or curled their top back over to windward O'ahu. The combination of these events is rather typical of usual winter rains in the Islands, but the rainfalls were exceptional (National Research Council 1991b).

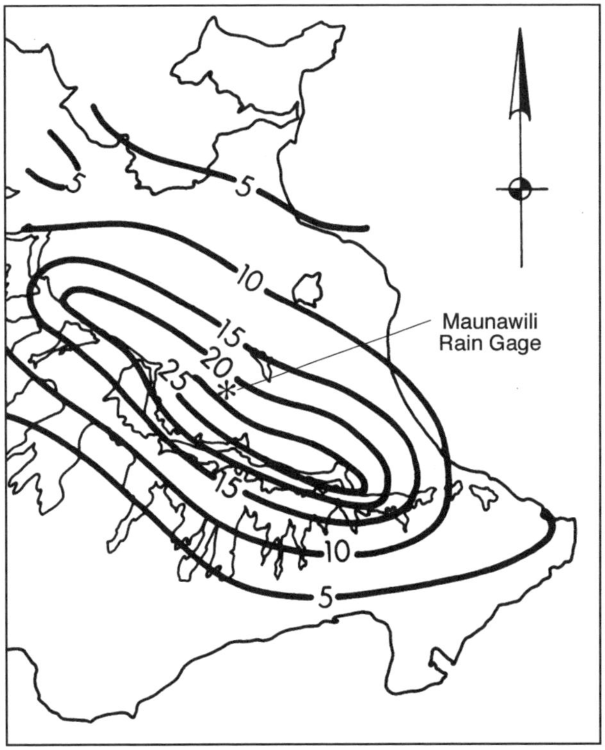


Figure 3.20. Twenty-four-hour isohyets (8:00 A.M., December 31, 1987, to 8:00 A.M., January 1, 1988) of the New Year's Eve rainstorm, eastern Oʻahu, Hawaiʻi. 10 in. = 254 mm; isohyet in inches. (Adapted from National Research Council 1991b)

## Use in Applied Hydrology

Intense rainstorms are pertinent in design storms and flood forecasting. Design storms are synthesized from analysis of historical records of intense rainstorms. The typical end products of the analysis are intensity-duration frequency and depth-area-duration; they are needed in transforming rainfall to surface runoff for which hydraulic structures such as culverts, channels, and basins of a storm drainage system are designed. Design storms are also used as input in computer modeling of rainfall-runoff for the assessment of hydrologic effects of land-use development. Flood forecast requires current (real-time) data recorded during the course of an intense storm from satellite and Next Generation Weather Radar (NEXRAD) systems or a network of telemetered rain gages. The data reveal the storm path and the intensity, duration, and areal distribution of the rainfall and are used as input to forecast or simulate river discharge.

Intensity-duration frequency is employed in the Rational formula for computing the peak discharge of a design flood. The method is used in a traditional way in Hawaiʻi (Lau 1982; City and County of Honolulu 1988).

Frequency is a study of probability dealing with the chance that an event may occur, such as the chance of getting a face value of 5 in rolling a die. In hydrology the question posed is the exceedance probability. It means the chances of getting a face value equal to or larger than 5 in a single toss, that is, getting either a 5 or a 6. With reference to rainfall, it is the chance that the maximum rain depth during a storm would be equal to or exceed, say, 50 mm (1.97 in.) in a 1-hour duration. Data reduction is done by scanning the entire record consisting of many rainstorms. The reciprocal of exceedance probability is the return interval in years.

The methodology for computing intensity-duration frequency from a station record is well documented (Chow et al. 1988). For a large area, intensity-duration frequencies are presented as isohyetal maps that show rain intensity, *I*, for a specified duration, *D*, and return interval, *T*. Such maps are available for Hawaiʻi. An early version was TP43 for all major islands (U.S. Weather Bureau 1962), and a recent one is R73 for the island of Oʻahu only, as shown in Figure 3.21 (Giambelluca et al. 1984). An effort to revise TP43 was reported to be in progress in 2005.

R73 (1984) contains rain frequencies for Oʻahu based on 156 rain gages at 139 different stations, including eight stations with a 60-year or more record, and 58 recording gages. Only stations with a 10-year or more record were used. The database for R73 is longer and more spatially extensive than that for TP43. Spatial extrapolation was done to augment the primary station network. Annual exceedence series, which is a series of values, the numbers of which are equal to the number of years of the record (one from each year) and the magnitude of which exceeds a predefined base magnitude, were extrapolated from the

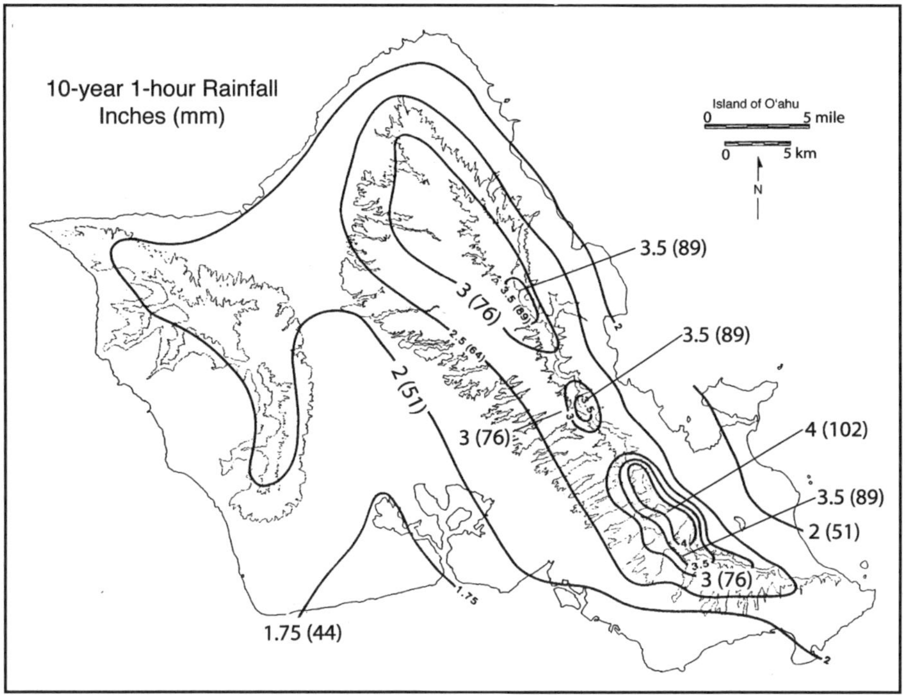


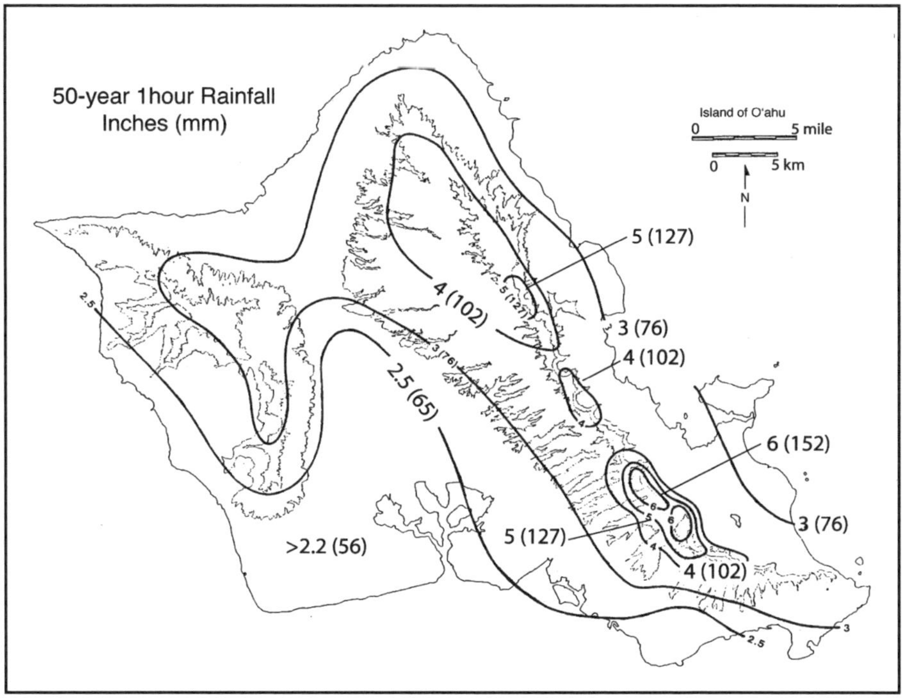


Figure 3.21. Rainfall intensity-duration frequency presented as isohyetal maps, island of O'ahu. (Adapted from Giambelluca et al. 1984)

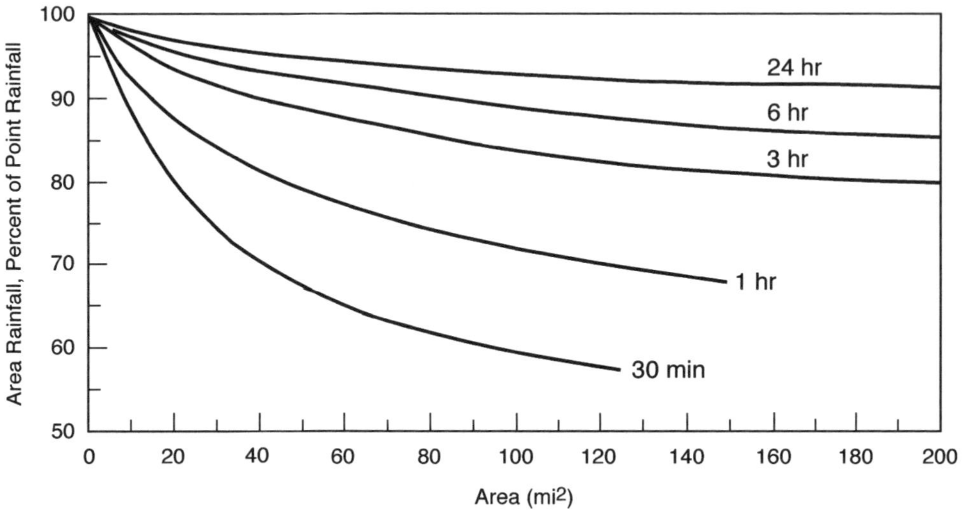


Figure 3.22. Depth-area curves for rain frequency data. 1 $mi^2 = 2.59$ $km^2$. (Adapted from U.S. Weather Bureau 1962)

records for seven durations (30 minutes, 1 hour, 2 hours, 3 hours, 6 hours, 12 hours, and 24 hours). Annual series were converted to partial duration series. The Gumbel extreme value (Type I) distribution was selected after testing selected data for two other distributions, Pearson Type III and log normal. Rainfall depths were computed with Chow's (1954) frequency factor method for three durations and four return periods. Rainfall depth for durations of less than 1 hour was computed with factors developed with other O'ahu data. Depth-area analysis, however, was based on the diagram by the U.S. Weather Bureau (1962).

A hyetograph indicates the time distribution of rain intensity. The U.S. Soil Conservation Service (1986) developed synthetic storm hyetographs, including one (Type I) intended for Hawai'i for storms of 24-hour duration. These graphs were derived from the national exercises on rain frequencies by the U.S. Weather Bureau (Chow et al. 1988).

Theoretical frequency analysis of areal rainfall is apparently not as well developed as that of point rainfall (Chow et al. 1988), but the rainfall depth *D*, distributed over an area *A*, for various durations *D* of an intense rainstorm can be characterized by a set of depth-area-duration curves, which can be empirically deduced from the data of dense networks (Linsley et al. 1982). The widely accepted depth-area-duration curves are based on twenty dense rainfall networks in the U.S. mainland and are used for the Hawaiian Islands (U.S. Weather Bureau 1962). The ordinates of the curves are expressed as a percentage of point rainfall values in the area (see Figure 3.22). These two relations, depth-area and depth-area-duration, are the characteristics of a design storm in computing probable maximum precipitation and probable maximum flood, for which major flood-control facilities such as dams, reservoirs, spillways, and floodways are designed. Probable maximum precipitation for Hawai'i was prepared as a part of a nationwide effort and published in a report (U.S. Weather Bureau 1963).

## Low Rainfall

The most common concerns about low rainfall are how much below normal it is and how long the deficiency will last. Severity is defined as the product of magnitude (intensity) and duration. An event is recognized as a drought when a certain set of severity criteria is exceeded. Three common types of drought are meteorologic (low rainfall), hydrologic (low streamflow), and agricultural (low soil moisture). They are related but not identical.

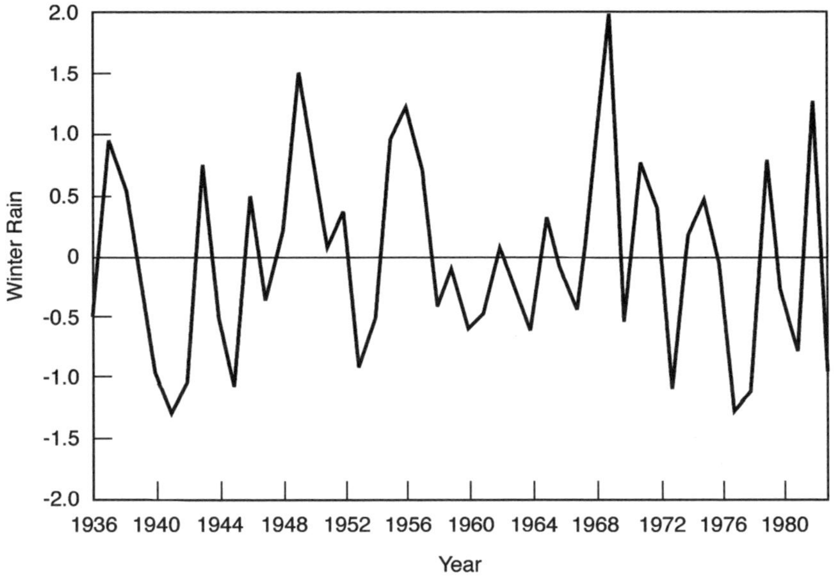


Figure 3.23. Observed time series of Hawaiʻi winter rainfall (1936 – 1983). Seasonal total of 3 months — winter of 1983 denoted as December 1982 and January and February 1983; values presented as departure from the mean then divided by standard deviation; average of twenty-seven stations on Kauaʻi, Oʻahu, and Hawaiʻi. (Adapted from Chu 1989)

## Hawaiʻi's Concerns

In Hawaiʻi, even a short dry spell of 2 to 3 weeks may be long enough to be recognized as a drought of some type. An early problem is often the need to haul water to many homes in rural areas that depend on rainwater cisterns for drinking and domestic water supply. Severe dry spells may raise anxieties about reduced flow in streams and ditches and about low water levels in the small storage reservoirs on which farming, ranching, and rural communities depend. Fire hazards in normally dry areas are heightened during droughts.

## Natural Processes

Hawaiʻi rainfall in winter months is notable for its large interannual variations, for example, as indicated for an average of twenty-seven stations on Kauaʻi, Oʻahu, and Hawaiʻi in Figure 3.23. Meteorologists have long been attempting to identify the causal meteorological processes (Yeh et al. 1951a; Schroeder 1993a; Chu 1995). Current efforts have been much aided by satellites and other observational instruments. By current understanding, these variations may be attributed to two large-scale atmospheric circulation changes: the annual shift of the Pacific Subtropical Anticyclone and the incursion of ENSO. The shift of the Pacific Subtropical Anticyclone in the southeast direction in the winter is expected to occur every year, but what is uncertain is the variable extent and timing from one year to the next. The shift generally weakens the trade winds and the rainfall they bring. Dry winters in Hawaiʻi very likely follow ENSO periods. But unlike the shift of the Pacific Subtropical Anticyclone, ENSO is not an annual event. These two related changes, the Pacific Subtropical Anticyclone and ENSO, together account for a substantial degree but not all of the interannual variations (Chu and He 1994). Additional atmospheric circulation processes are currently being examined, including the subtropical jet stream and strong local Hadley-type circulation (Chu 1995).

Hawaiʻi's appreciation of teleconnection may have begun with Meisner (1976). He reported that winter rainfall in Hawaiʻi was negatively correlated with sea-surface temperature at Canton Island, one of the Line Islands in the eastern equatorial Pacific. Lyons (1982) confirmed with empirical orthogonal function analysis that Hawaiʻi experiences anomalously low trade-wind rainfall in most

Table 3.1. El Niño Events in the Twentieth Century

| |
|---|
| 1902–1903 |
| 1905–1906 |
| 1911–1912 |
| 1914–1915 |
| 1918–1919 |
| 1923–1924 |
| 1925–1926 |
| 1930–1931 |
| 1932–1933 |
| 1939–1940 |
| 1941–1942 |
| 1951–1952 |
| 1953–1954 |
| 1957–1958 |
| 1965–1966 |
| 1969–1970 |
| 1972–1973 |
| 1976–1977 |
| 1982–1983 |
| 1986–1987 |
| 1991–1992 |
| 1997–1998 |

SOURCES: Schroeder 1993b, Chu 1995, Piechota and Dracup 1996.

NOTE: Listed identically by both Chu 1995 and Piechota and Dracup 1996 except for the 1902–1903 event, which is listed only in Schroeder 1993b. El Niño occurred in the first of the 2 years listed by Schroeder.

but not all El Niño episodes. Horel and Wallace (1981) suggested that the dry winters were related to the Southern Oscillation. These fragmentary recognitions of ENSO were consolidated after the 1982–1983 episode for which the linkage of El Niño with Southern Oscillations was firmly established with convincing documentation (Rasmusson and Wallace 1983). El Niño events in the twentieth century are listed in Table 3.1.

## Traits of Low Rainfall

In monitoring rainfall trends and issuing advisories, the state uses multiple-month moving averages of departure from the median (Hawai'i Commission on Water Resource Management 1992). Moving averages remove short-term variations that may mask trends. Nine-month moving averages of departures were used to examine long-term rainfall trends for two stations (Kahana and 'Ewa Mill) on O'ahu (Haraguchi and Matsunaga 1985).

Drought indices for hindsight analysis such as the Palmer Drought Severity Index (PDSI) and Bhalme and Moody Drought Index (BMDI) are available in the literature. Both indicate a relative wet or dry condition with an identical set of indices (–2.00 to –2.99 for moderate drought, –3.00 to –3.99 for severe drought, –4.00 or less for extreme drought). BMDI is based on rainfall only whereas PDSI is based on water balance through computation of sixty-eight terms that describe soil moisture, evapotranspiration, precipitation, and runoff.

According to a Hawai'i BMDI study (Giambelluca et al. 1991), twenty-seven statewide droughts were identified for the record period of 107 years. No statewide droughts lasted longer than 6 months, but individual island droughts apparently lasted longer. Historical accounts of drought events, which were compiled by Price and Matsunaga, compare favorably with the BMDI analysis, although there are some inconsistencies.

Frequency analysis has been applied to Hawai'i monthly rainfall records to compute the minimum consecutive-month rainfall for various durations (3, 6, 9, and 12 months) and various return intervals (2, 3, 5, 10, 20, 30, 50, 100, and 200 years) (Giambelluca et al. 1991). A sample is shown in Figure 3.24. The generalized isohyetal patterns resemble those for annual rainfall, reflecting similar mesoscale influences of mountains and trade winds (Figure 3.25).

## Rainwater Quality

Rainwater in Hawai'i is virtually pure, but it contains trace amounts of sea salts and dissolved atmospheric gases and is moderately acidic. It also extracts from the atmospheric a few isotopes (radiocarbon [$^{14}C$], tritium [$^{3}H$], deuterium [$^{2}H$], and oxygen-18 [$^{18}O$]) that have special hydrologic relevance. Typical values of major ions and some other parameters are provided in Table 3.2.

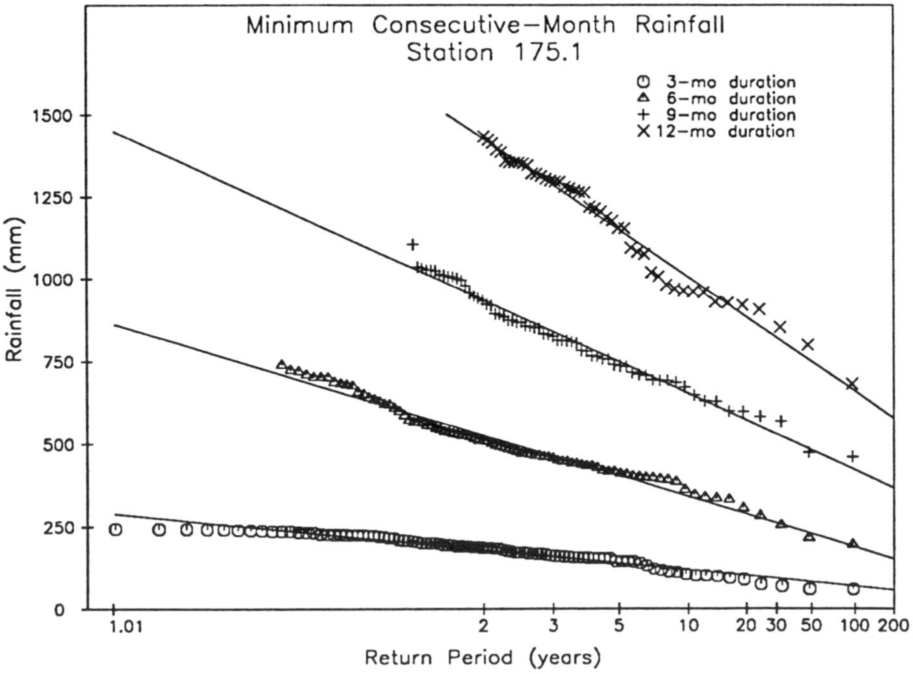


Figure 3.24. Minimum consecutive-month rainfall frequency for various durations and return periods: Kohala Mission (Station 175.1), island of Hawai'i. 1,000 mm = 39.37 in. (Reprinted by permission from Giambelluca et al. 1991)

Rainwater quality in Hawai'i is influenced by sea spray and volcanic emissions. Sodium ($Na^{+1}$) and chloride ($Cl^{-1}$), the two most abundant constituents in seawater, dominate the dissolved constituents in rainfall. Their concentrations decrease with the distance inland and higher elevation (Visher and Mink 1964). Additional data are provided for the island of Hawai'i (Seto et al. 1969; Harding and Miller 1982) and for O'ahu (Dugan and Ekern 1984).

Nitrogen in nitrates ($NO_3^{-1}$) as well as in other forms is found in Hawai'i's rainwater, as evidenced in a 1983–1984 sampling on O'ahu (see Table 3.3). Typically, only $NO_3^{-1}$, the stable form, is analyzed and reported. The O'ahu results indicate that organic N constitutes a major portion (about 40 percent) of total N in rainwater. Similar occurrences of organic N have also been reported for California (Lake Tahoe), Florida, Tennessee, and South Carolina.

Other important parameters are pH and sulfate ($SO_4^{-2}$), which are associated with acid rain. Sulfates in rainwater originate from the fume ($SO_2$) that is abundant in Kīlauea volcanic eruptions. Concentrations as high as 36 mg/l in Kīlauea on the island of Hawai'i (Harding and Miller 1982) and 20 mg/l on O'ahu (Dugan and Ekern 1984) have been reported. Coupled with the high level of $SO_4^{-2}$ in the Kīlauea area is the low pH—as low as 3.0, with a median of 3.6. Far from the Kīlauea Volcano, O'ahu rainwater may be strongly acidic when southerly winds bring the fume of volcanic eruption (vog) from Kīlauea. Acidity ranging from 4.2 to about 5.0 was recorded during a period of southerly winds on February 1–2, 1984, when Kīlauea was erupting.

In 1987–1989, over 2,000 water samples from many of the 7,000 rainwater cisterns on the island of Hawai'i were examined for lead (Hawai'i Department of Health 1989). About 24 percent of the samples exceeded the maximum contaminant level of 20 μg/l proposed by the U.S. Environmental Protection Agency, and 11 percent exceeded the existing 1989 maximum contaminant level of 50 μg/l. Statistical correlation between pH and lead concentration of this data set was only moderate ($r = -0.40$), indicating

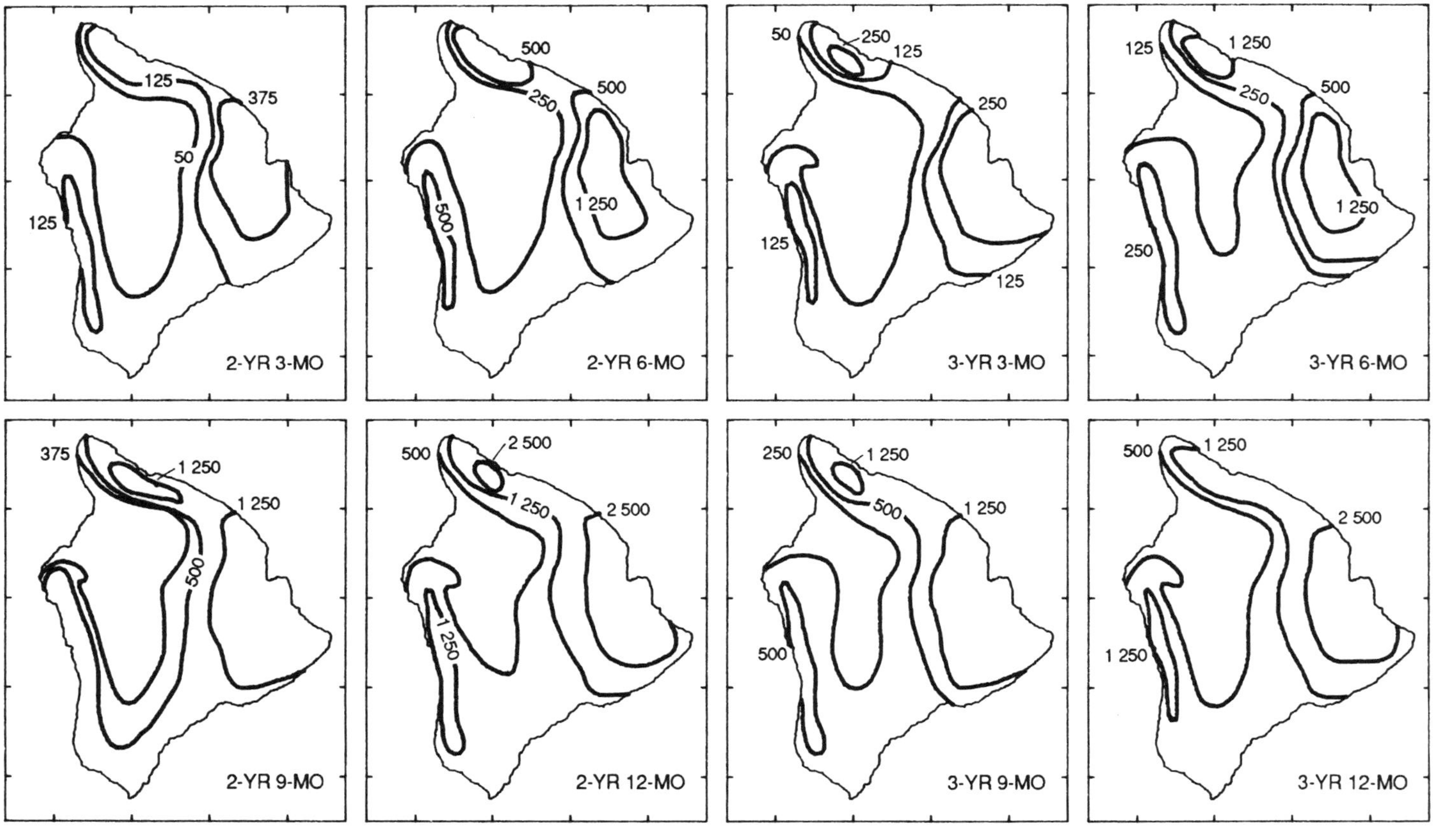


Figure 3.25. Computed minimum rainfall for consecutive months and return period, island of Hawai'i. Return period (first number), consecutive months (second number), rainfall in mm. 1,000 mm = 39.37 in. (Reprinted by permission from Giambelluca et al. 1991)

Table 3.2. Average Chemical Composition of Rainwater and Seawater

| | Rainwater | | |
|---|---|---|---|
| Constituent | Hilo Coast[a] | Mauna Kea Slopes[b] | Seawater[c] |
| $Ca^{+2}$ | 1.11 | 0.39 | 400 |
| $Mg^{+2}$ | 1.54 | 0.45 | 1,272 |
| $Na^{+1}$ | 6.4 | 1.62 | 10,561 |
| $K^{+1}$ | 0.66 | 0.11 | 380 |
| $SO_4^{-2}$ | 2.66 | 0.89 | 2,649 |
| $Cl^{-1}$ | 11.0 | 3.11 | 18,980 |
| $NO_3^{-1}$ | 0.61 | 0.37 | 0.05–3.1 |
| pH | 5.4 | 5.1 | — |
| $F^{-1}$ | — | — | 1.3 |

SOURCE: Visher and Mink 1964.

NOTE: All values in mg/l as the parameter itself except pH.

[a]Hilo elevation 9.2 m (20 ft), distance from coast 1 km (0.6 mi).

[b]Mauna Kea elevation 404–802 m (1,325–2,630 ft), distance from coast 6.3–12.2 km (3.9–7.6 mi).

[c]Source: Sverdrup et al. 1946.

that acidity of rainwater was not the only determining factor for the elevated lead concentrations. The department has since required compliance with standards and has prohibited the use of building materials containing lead for cistern construction. Lead is highly toxic and is also considered a probable carcinogen (Sawyer et al. 1994).

## Fog

Fog (cloud water) occurs in high mountains in Hawai'i, generally above 750 m (2,460 ft), and varies seasonally. Water dripping from and through the spaces in the canopy of trees onto the ground is known as canopy throughfall. Throughfall consists of fog water and rainwater. The amount collected in a network of rain gages set beneath a Norfolk Island pine tree at Lānaʻihale (elevation 838 m [2,750 ft]), island of Lānaʻi, was 2.6 times greater than that collected in similar gages set in the open over a 3-year period (Ekern 1964).

Fog can be measured with a standardized louvered-screen fog gage invented by McKnight and Juvik in 1975 (see Appendix 3.1). Fog and rain in the open were measured at eighteen stations—fifteen in Mauna Loa and three in Hualālai—in 1974–1976 (Juvik and Ekern 1978). A fog belt is well defined on windward Mauna Loa between elevations of 1,500 and 3,000 m (4,920 and 9,840 ft). Variations in fog frequency and amount are attributed to the interaction between trade winds and local land and sea breezes. The fog-to-rain ratio is highest (0.68) at about 2,500 m (8,200 ft), and the annual fog catchment is 706 mm (28 in.). In leeward Mauna Loa, fog increases with elevation up to 2,000 m (6,562 ft), with fog amounts equivalent to one-fourth the rainfall or about 250 mm (9.8 in.) on an annual basis. For Hualālai, the fog-to-rain ratio is highest (1.57) at the summit (elevation 2,496 m [8,189 ft]), and the fog amount was 832 mm (33 in.) over a 17-month period. Throughfall from tree canopy was not a part of the study, but it was observed and modeled for a leeward Mauna Kea site (see following section on models).

# Hawai'i Data and Models

## Data

Rainfall data obtained with conventional rain gages located at specific sites are abundant and well utilized in Hawai'i. Satellites and weather radars have been put into service since the 1990s. Radars can be used to detect, and even estimate, rainfall over a large area. Satellites can be most useful to obtain data in regions where surface rain gages are inadequate or nonexistent, such as the oceans.

Government sources of information on climate and weather in Hawai'i are listed in Appendix 3.2 (Price 1993). Scientific and educational institutions publish their findings irregularly in journals and reports. More recently,

Table 3.3. Median Rainwater Quality, O'ahu, Hawai'i (December 25, 1983 – April 27, 1984)

| | Station[a] | | | | | | | |
|---|---|---|---|---|---|---|---|---|
| Constituent | Huelani | | Holmes Hall | | Kapahulu | | Hawai'i Kai | |
| pH | 4.9 | (69) | 5.0 | (23) | 5.2 | (5) | 5.1 | (15) |
| Chloride (mg/l) | 8.7 | (71) | 9.4 | (24) | 6.3 | (5) | 9.0 | (14) |
| Kjeldahl N (mg/l) | 0.097 | (14) | 0.15 | (10) | 0.33 | (3) | 0.34 | (3) |
| Organic N (mg/l) | 0.07 | (8) | 0.07 | (6) | 0.24 | (3) | 0.16 | (3) |
| Ammonia N (mg/l) | 0.06 | (10) | 0.12 | (6) | 0.07 | (2) | 0.20 | (3) |
| $NO_2^{-1} + NO_3^{-1}$–N (mg/l) | 0.028 | (9) | 0.031 | (6) | 0.022 | (3) | 0.298 | (2) |
| Total nitrogen (mg/l) | 0.183 | (9) | 0.171 | (6) | 0.281 | (2) | 0.728 | (2) |
| Site description | | | | | | | | |
| Elevation (ft) | 350 | | 75 | | 20 | | 70 | |
| Downwind distance (ft) | | | | | | | | |
| Ko'olau crest | 11,500 | | 18,300 | | 23,000 | | 6,600 | |
| Windward coast | 39,370 | | 49,200 | | 49,200 | | 13,100 | |
| Median annual rainfall (in.) | 104 | | 36 | | 24 | | 40 | |

SOURCE: Dugan and Ekern 1984.

NOTE: Total number of samples analyzed in parentheses. 1 ft = 0.305 m, 1 in. = 25.4 mm.

[a]Site description in lower part of table.

the database has become commercially available, such as on CD-ROMs, through the National Climate Data Center and private vendors such as Earth Info Inc.

Point data are the basic and essential building blocks of a database in time and place. But many applications for either water science or water management require more than point data; they need data for large land areas measured in square kilometers (square miles). Besides rain depth, rain intensity and frequency are also important parameters. Applications in water management include irrigation to supplement rain, prediction of flood-producing rainstorms, storm water as a nonpoint source of pollution, and infiltration to recharge groundwater.

Three analyses of Hawai'i rainfall data serve to indicate the methodologies necessary to check, adjust, and extrapolate the data. The first concerns Hawai'i annual and monthly (mean and median) rainfalls (Giambelluca et al. 1986) and was issued by the Hawai'i Division of Water and Land Development as Report No. 76 (R76). This study used 1,202 stations (330 on Hawai'i, 131 on Maui, 203 on Moloka'i, 49 on Lāna'i, 309 on O'ahu, and 180 on Kaua'i), of which 83 served as base stations with long records (53 to 68 years). The station locations are reported in a separate volume, as Report R42 (Hawai'i Division of Water and Land Development 1973a). The second concerns annual median rainfalls only but includes dispersion and skewness parameters (Hawai'i Division of Water and Land Development 1982) and was issued as Circular No. 88 (C88). The third concerns O'ahu rainfall intensities in terms of duration and frequency (Giambelluca et al. 1984) performed for the U.S. Army Corps of Engineers and issued by Hawai'i Division of Water and Land Development as Report No. 73. This study used 156 rain gages, including 58 recording gages located at 139 different sites. All three studies were commissioned work done by the University of Hawai'i.

Standard rain gages and climate instruments such as those described in Brakensiek et al. (1979) are used in Hawai'i for routine observations. Special and research instruments including a rain-intensity gage, fog gage, radar, and satellite-borne sensors are discussed in Appendix 3.1. The first two gages were invented in Hawai'i.

## Models

### Winter Low-Rainfall Forecast Models

A simple regression analysis was made to correlate Hawai'i's winter rainfall with the preceding summer's (June, July, and August) Southern Oscillation Index (Chu 1989). The correlation coefficient was 0.54 for the 1936–1983 record of twenty-seven stations on Kaua'i, O'ahu, and Hawai'i for the 3-month total (December, January, and February). This effort was followed by the use of canonical correlation analysis and an additional atmospheric pressure parameter, which is determined to be the first four eigenmodes of the summer sea-level atmospheric pressure over the northern Pacific Ocean. The correlation coefficient was improved to 0.60 (Chu and He 1994).

### Mesoscale Circulation Models: Trade-Wind Rainfall

#### *Conceptual and observational models*

Hawai'i's winter rainfall is markedly correlated with the locational and kinematic parameters of the jet stream, whereas the summer rainfall is not (Yeh et al. 1951b). Yeh et al. acknowledged that it is more difficult to explain than to correlate.

Leopold (1949) presented four conceptual models on the interaction between trade wind and sea breeze based on observations of cloud-line formation and limited measurements of surface pressure gradients. Because the west coast of the island of Hawai'i is protected from the trade winds by Mauna Kea, sea breeze is able to blow the convective cloud inland and upslope in the afternoon, bringing about Kona-type rains. On Lāna'i and Moloka'i where the mountain range is low, clouds that form over land in the lee of the mountain are blown off in the resultant direction of the converging winds. Other varieties of interactions are the Maui-type circulation cell and Mauna Kea–type land-wind front (see Figure 3.26).

Preceded by several observational studies, the Hawaiian Rainband Project provides detailed delineation of surface wind-stream lines, indicating whether they are directed over the saddle between mountains or around the high mountains, whether they are converging or diverging, and whether they are hugging the ground or remaining aloft (see Figure 3.10). Besides being visually indicative, they are quantitative (wind velocity, air pressure, air temperature, and rainfall) and finely temporal in nature (15-minute readings). The data have been used in numerical modeling (Rasmussen and Smolarkiewicz 1993).

#### *Parametric models*

Physically based models are often expressed in parametric (but not regression) equations. Two examples are those by Stidd and Leopold (1951) and by Mink (1960). The hypothesis advanced by Stidd and Leopold is that the monthly rainfall on high islands is a superposition of orographic rainfall and ocean rainfall and that the orographic effects are readily indicated on isohyetal maps. Their equation is as follows:

$$y = a(x - 30) + b \tag{3.1}$$

where $y$ is the mean monthly station rainfall; $x$ is the mean annual station rainfall; 30 in. (762 mm) is the annual ocean rainfall; $a$ is the rainfall gradient factor to account for orography and is determined by the quotient of monthly rainfall gradient and annual rainfall gradient, as measured from the monthly and annual isohyet maps, respectively; and $b$ is the monthly ocean rainfall. Values of $a$ and $b$ were determined for Hawai'i, O'ahu, Maui, and Kaua'i.

Rainfall is hypothesized to increase as a function of the elevation or distance to the mountain crest from a reference site (Mink 1960). Mink determined that the logarithm function of rainfall is proportional to the dis-

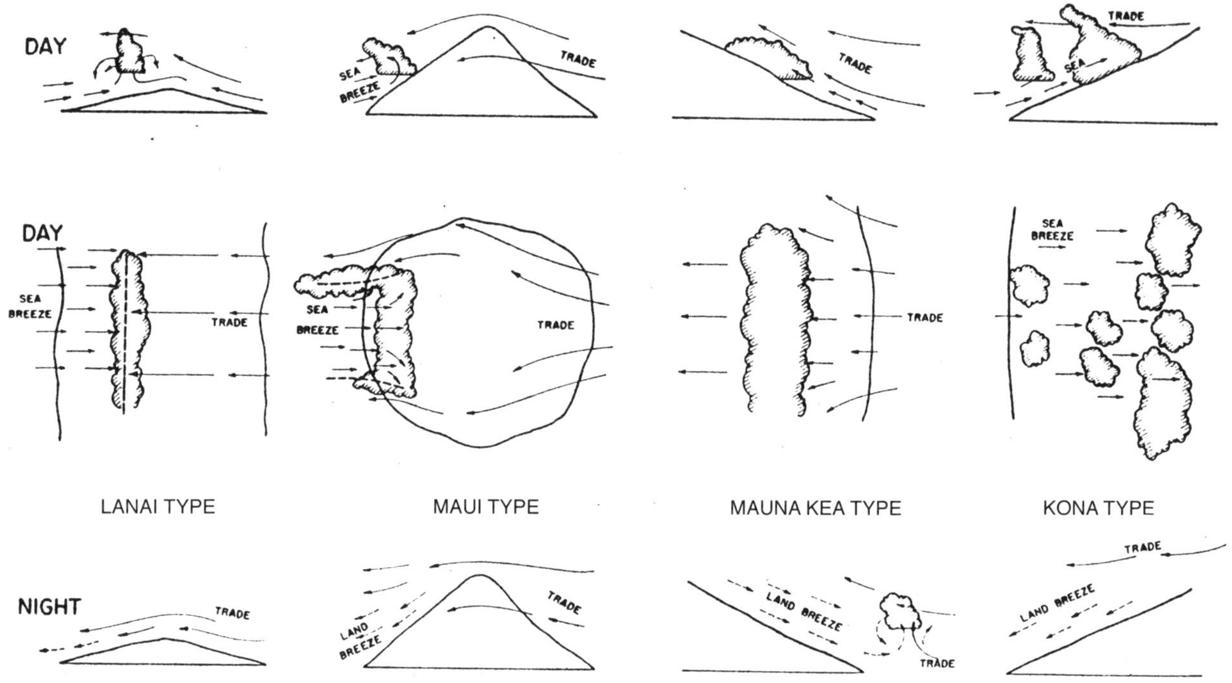


Figure 3.26. Schematic representation of four types of interaction of sea breeze and trade wind. (Reprinted by permission from Leopold 1949)

tance to the crest. The parametric values in the equation were developed for a rainfall transect along the ridge of Kīpapa drainage basin, island of Oʻahu. The basin has a long, narrow, deep gulch between sharp parallel ridges, a shape that is typical of many basins in the leeward Koʻolau Range, Oʻahu. The specifics of Kīpapa are 11.1 $km^2$ (4.3 $mi^2$) in area, 7,500 m (4.7 mi) in length, 5 percent gradient of the ridge, 350 m (1,148 ft) in elevation at the lowest gage and 850 m (2,788 ft) at the highest gage. The log function fits nearly perfectly for trade-wind rainfall but also includes rainfall of all origin because non-trade-wind rainfall is likely to be uniformly distributed, suggesting utility of the function to account for the influence of orography regardless of types of rain.

### *Numerical models*

The first mesoscale numerical model of trade-wind rainfall on Oʻahu was probably that by Lavoie (1974). The model accounts for the effects of terrain, land roughness, land heating, and evaporation on a single well-mixed air layer for which the upper bound is the inversion and the lower bound is the surface-friction shear. Conservations of momentum, heat, and water (specific humidity) were formulated as partial differential equations that were solved with a finite-difference model. The results match the isohyets of observed average daily rainfalls for June and July 1955. However, the model was unable to simulate diurnal rainfall and weak sea breezes.

The formation of cloud bands offshore of the Hilo Bay area during the night, which was first reported by Leopold (1949), was analyzed and confirmed with numerical experiments (Rasmussen and Smolarkiewicz 1993). The numerical (finite-difference) model is that by Clark and Farley (1984) with an interactive nesting scheme. The physics problem is three-dimensional stratified airflow around a mountain, as formulated by Drazin (1961). The

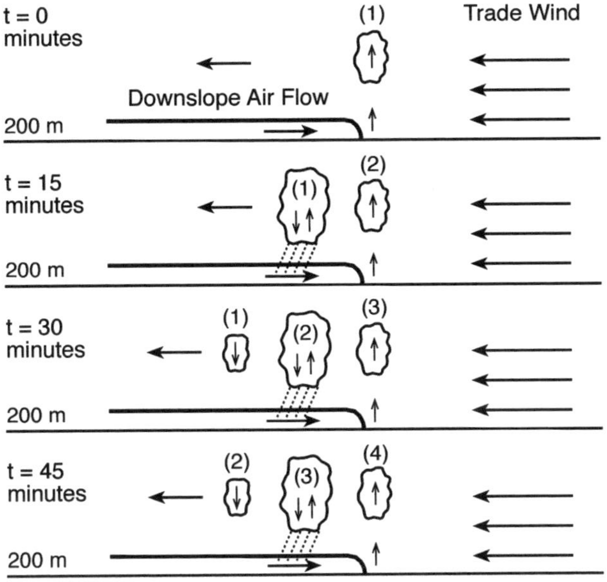


Figure 3.27. Schematic of the evolution of clouds forming over a stationary convergence zone in a typical trade-wind regime. 1 m = 3.28 ft. Dotted lines indicate rain. Numbers in parentheses identify clouds occurring in succession. (Adapted from Rasmussen and Smolarkiewicz 1993)

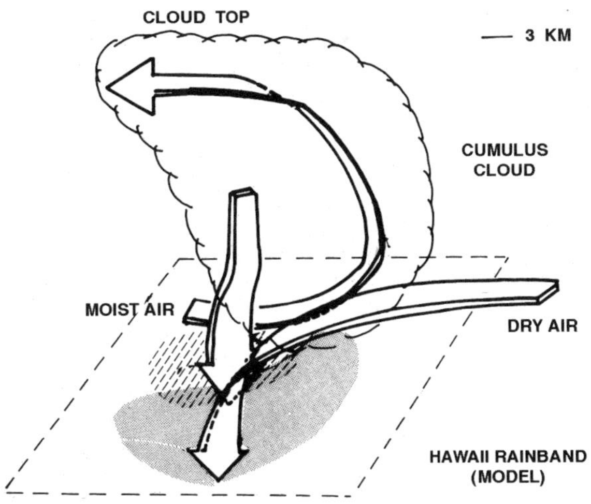


Figure 3.28. A Hawaiian Rainband model that produces very long-lasting rainfall. 1 km = 3,281 ft. (Adapted from Takahashi 1988)

numerical experiment results indicate a stationary low-level line of convergence of nocturnal downslope air flow and the opposing morning trade winds at the upwind edge of the shoreline with a stagnation point located just offshore. The convergence results in a quasi-steady bank of clouds (Figure 3.27). The simulated phenomena vary with the trade-wind velocity as embodied in the Froude number. The simulated results are a reasonable reproduction of the measurements of wind velocity, air temperature, and water vapor mixing ratio from aircraft sounding (Smolarkiewicz et al. 1988; Rasmussen et al. 1989). The numerical study benefited from laboratory water-tank experiments previously conducted by Hunt and Snyder (1980).

The microphysics of Hawai'i shallow convective cumulus clouds (specifically aerodynamics and thermodynamics of clouds and raindrop size) was formulated as partial differential equations that were solved with a three-dimensional numerical model (Takahashi 1988). Recirculation of raindrops within the clouds enhances the raindrop growth rate and the ease of rain. The long duration of rain (up to 50 minutes) is explained as a combination of the resistance to intruding dry air and the sustained flow of low-level moist air (Figure 3.28).

## Canopy Throughfall Model

Fog interception by tree canopy produces canopy throughfall that wets the ground. A throughfall model is formulated by considering the mass conservation of water in the tree canopy on an event basis (Juvik and Nullet 1995). Essentially, fog interception by canopy plus open-site rainfall minus evaporation from canopy minus stem flow (water reaching the ground by running down the branches and trunk) minus canopy throughfall is equal to water storage in the canopy. In solving the equation for throughfall, canopy fog interception is approximated by fog-gage interception, evaporation is computed with the Penman-Monteith equation (see Chapter 4), stem flow is considered negligible, and canopy-water storage is evaluated with an empirical exponential function of the open-site rainfall. Throughfall was measured with trough gages deployed under a tall, mature *māmane* tree at Puʻu Lāʻau (elevation 2,600 m [8530 ft]), leeward Mauna Kea. The model was validated for most of the observed events. The total amount of throughfall during the 1-year period was 194 mm (7.6 in.) or 75 percent of the open-site rainfall. At this site the amount is small for ensuing hydrologic processes, but it is of ecological importance.

# Appendix 3.1. Special Precipitation Instruments

Hawai'i has one of the densest networks of rain gages in the world (Gilman 1964). The number of rainfall observation stations in Hawai'i has increased since 1964 by almost four times to a total of 1,121 (Hawai'i Division of Water and Land Development 1982). The current network still appears as a coarse mesh in high-rainfall mountains for many practical purposes.

David Raymond and Kerry Wilson invented a rain-intensity gage in the course of studying high-intensity showers at the University of Hawai'i at Hilo, island of Hawai'i (Raymond and Wilson 1973; Fullerton and Raymond 1973; Fullerton and Wilson 1974). The basic component of the gage is a small inclined trough with a triangular cross section in which the collected rainwater flows down. Two brass plates are affixed to the trough as electrodes and wetted by the varying water depth. The electrical resistance of water sensed by the electrodes is inversely proportional to the cross-sectional area of the flowing water and directly proportional to the electrical resistance of the water. The gage is capable of directly measuring the instantaneous rain intensity and compensating for the rainwater salinity. The most minute time variations of a shower are recorded (see Figure 3.29). Shown for comparison in Figure 3.29 is a parallel trace that is based on the record of the tipping-bucket recording gage and appears only as a truncated version of the Raymond-Wilson trace.

A fog gage is a louvered-screen cylinder that intercepts fog droplets (smaller than 100 μm [0.0039 in.]) as the fog passes. The amount of rain is accounted for by deploying a rain gage nearby and outside the canopy. The first fog gage was designed, constructed, and tested in Hawai'i (McKnight and Juvik 1975), and many transects of gages were deployed on the island of Hawai'i (Juvik and Ekern 1978).

Radar emits a pulse of electromagnetic energy as a beam that travels at the speed of light in a direction determined by a movable antenna. It also receives the radiated wave as reflected by cloud or precipitation particles. The echo or return power is related to the range (distance) and the size distribution of the particles. An ordinary radar can detect the intensity of change of precipitation echoes and their size, location, and relative motion. Doppler radars can measure the absolute and instantaneous velocity of the raindrops. For quantitative use, radars require calibration with rain gages and with other radars observing the same area (Smith et al. 1996). The Z-R (reflectivity-rainfall rate) relationship by Rosenfelt et al. (1993) is in operational use by the Honolulu Weather Forecast Office for radar-derived rain estimates for O'ahu and provides best match with the rain-gage data for a heavy rain on O'ahu (Lyman et al. 2005).

A 3-cm (1.2-in.) marine radar belonging to the University of Hawai'i, Department of Meteorology, was used and appreciated in the analysis of the New Year's Eve rainstorm and flood (December 31, 1987–January 1, 1988) on O'ahu (National Research Council 1991b). In the early 1990s, four Weather Surveillance Radars—1988 Doppler (WSR-88D)—were deployed in Hawai'i as a part of the national NEXRAD system for real-time use in rainstorms and flash floods.

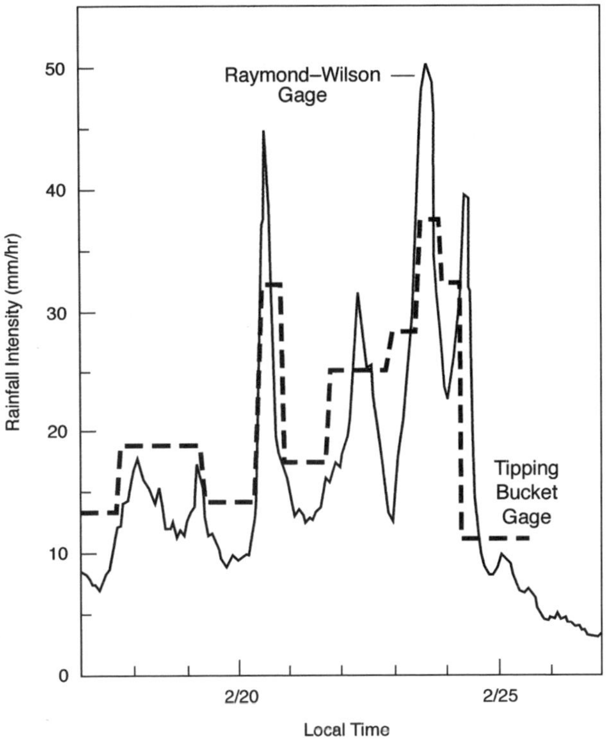


Figure 3.29. Comparison of rainfall intensities recorded by a tipping-bucket gage and a Raymond-Wilson gage, August 24, 1973. 10 mm = 0.39 in. (Adapted from Fullerton and Wilson 1974)

Remote sensing from satellites measures the electromagnetic spectrum to characterize the landscape and certain hydrologic-related parameters. Visible and infrared images provide data about the clouds: type, coverage, and temperature of their upper surface. The data are used to infer cloud-top height and rainfall. A technique was developed to correlate visible convective cloud cover, which is very bright and reflective, as seen by the 1973 National Environmental Satellite Service, to surface rainfall measurements in Pacific Ocean atolls. The correlation coefficient of the regression line is +0.75 and is significant at the 1 percent level (Kilonsky and Ramage 1976). Direct measure-

Figure 3.30. Hydrometeorological analysis of the October 29, 2000, storm, Hāna, Maui: GOES-10 satellite, 1 km visible imagery. An upper-level trough to the north of the Hawaiian Islands destabilized the atmosphere, generating the organized convection to the northeast of the islands. Note the thin bow cloud north of the island of Hawai'i extending to Hāna and manifesting the convergence associated with the deceleration of the trade-wind flow (see Figure 3.27). (Reprinted by permission from Lyman et al. 2005)

ments of rainfall from satellites are beyond the realm of current scientific feasibility (Engman 1993).

Satellites have been present in Hawai'i for storm rain and flood forecast since the early 1990s. A Geostationary Operational Environmental Satellite, GOES-9, was deployed at 135° W and the equator in 1995. It was followed by the deployment of GOES-10. The visual images of clouds seen in the daily TV weather report are transmitted from satellites and radars. A heavy rainstorm that produced 700 mm (27 in.) of rain in 7 hours in the Hāna region of Maui on October 29, 2000, was tracked by satellite and radars, producing images shown in Figure 3.30, Plate 3.1, and Plate 3.2 (Lyman et al. 2005). The rain amount exceeded the value for 100-year recurrence interval for the area (U.S. Weather Bureau 1962).

Hurricane 'Iniki, which devastated Kaua'i in 1992, was tracked in real time by satellite and reported by the U.S. National Weather Service, Central Pacific Hurricane Center, Honolulu. The National Weather Service notifications about hurricanes that may cause damage by high winds, heavy rains, flooding, and surf include watch (expected within 36 hours) and warning (expected within 24 hours). For flash floods that are caused by heavy rains the notifications include watch (flooding possible) and warning (flooding imminent or occurring).

# Appendix 3.2. Sources of Data on Climate and Weather in Hawaiʻi

## Current Data

All materials are published monthly by the National Environmental Satellite and Data Service through the National Climatic Data Center. They are available at any library that is a federal depository.

### Local Climatological Data

Published monthly and annually for Hilo, Honolulu, Kahului, and Līhuʻe.

Monthly issues contain daily values of temperature, rainfall, wind, sunshine and other weather elements, and hourly rainfall amounts.

Annual issues contain monthly data for the current year and tables of normals, means, and extremes for the period of record.

### Climatological Data: Hawaiʻi and Pacific

Published monthly and annually.

Monthly issues contain daily rainfall amounts and maximum/minimum temperatures for locations throughout the state.

Annual issues contain monthly and annual rainfall totals and departures from normal and mean monthly maximum and minimum temperatures.

### Hourly Precipitation Data: Hawaiʻi and Pacific

Published monthly and annually.

Monthly issues contain daily, hourly, and maximum short-duration rainfall amounts for the recording rain-gage network.

Annual issues contain monthly and annual rainfall totals and annual maximum rainfall data by selected time categories.

All of the preceding listed publications include station location maps.

### Storm Data

Published monthly.

Contains a brief description by states of the month's storms and unusual weather, together with deaths, injuries, and estimated property damage.

## Historical Data

The first three items are publications of the U.S. Weather Bureau.

### Bulletin W

Contains monthly rainfall totals, number of rainy days, temperatures, and related data from the beginning of record.

### Climatic Summary of Hawaiʻi, Supplement for 1919–1952

### Climatic Summary of the United States, Supplement for 1951–1960, Hawaiʻi and Pacific

Updates the previous tabulations and adds the average frequency of days per month with rainfall amounts equal to or greater than 0.10 and 0.50 inches.

### Solar Radiation in Hawaiʻi, 1932–1975

Published by the Hawaiian Sugar Planters' Association, 1978 (see How 1978).

Monthly and annual solar radiation values and other statistics since the beginning of record. Includes station location tables and maps.

### Pan Evaporation: State of Hawaiʻi, 1894–1983

Published by Department of Land and Natural Resources, Division of Water and Land Development, State of Hawaiʻi, 1985 (see Ekern and Chang 1985).

Monthly and annual pan evaporation values and other statistics since the beginning of record. Includes station location tables and maps.

### Climatologic Stations in Hawaiʻi

Published by Hawaiʻi Department of Land and Natural Resources, Division of Water and Land Development, State of Hawaiʻi Report R42, 1973a.

A catalog, arranged by station name and state key number, of all known climatologic stations in Hawaiʻi. Includes station location maps.

CHAPTER FOUR

# Evaporation

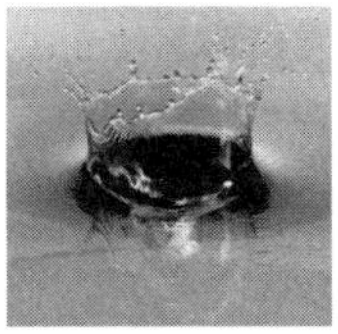

Evaporation is an invisible, slow, but virtually continuous process by which liquid water is transformed to water vapor from surfaces containing water, including open water, soil, and vegetation. Mean annual evaporation from an evaporation pan on O'ahu varies from 508 mm (20 in.) in the cloud-shrouded rain forest to 2,286 mm (90 in.) along the sun-parched leeward shores. These values are equivalent to 1.39 to 6.25 mm/d (0.05 to 0.25 in./d). For the island of O'ahu as a whole, approximately 2.7 million $m^3$ (720 million gal) of the mean daily rainfall evaporates into the air. This is about twice as much as all freshwater use on O'ahu in 1990.

## Nature of Evaporation

Evaporation requires absorption of heat by water. The heat requirement is known as latent heat of vaporization, $\lambda$, which is the heat energy absorbed in joules per kilogram of water to do the work necessary to break up the hydrogen bonds that hold the water molecules together in liquid form. $\lambda$ is virtually a constant but decreases slightly with increasing water temperature, $T$ (°C), as follows:

$$\lambda = 2.501 - 0.002361\ T \qquad \text{MJ/kg} \tag{4.1}$$

where J is the standard International System of Units measurement for unit of energy (1 joule = 0.2389 calorie, MJ = $10^6$ J).

The conventional measure of evaporation rate, $E$, is the loss of water depth over time or loss of water volume per unit evaporative surface area over time. The corresponding mass evaporation of water, $E'$, is equal to water density $\rho$ times $E$. The corresponding flow of latent heat is equal to $\lambda E'$ or $\lambda \rho E$. Conversion from latent heat flow to evaporated water depth is virtually a constant: 100 W/$m^2$ = 3.54 mm/d (100 cal/d/$cm^2$ = 1.72 mm/d [0.07 in./d]). Computations are presented in Appendix 4.1.

Evaporation manifests escape of water vapor molecules from water. These molecules can be reabsorbed by water as they strike the water surface. As a result, evaporation actually reflects the net amount of the exchange or escape minus absorption.

Given a finite source of heat, the rate of evaporation is controlled by the ease with which the water vapor can diffuse away from the evaporative surface. Evaporation takes place continuously because a gradient of vapor pressure can be created easily by the extremely fast molecular diffusion in the air — 500 m/s (1,640 ft/s).

If the vapor molecules are not allowed to diffuse away, the concentration of molecules will increase and so will the vapor pressure and the absorption of molecules until the absorption rate equals the evaporation rate. At this point, the air is called saturated: the air holds the maximum amount of molecules and exerts saturated vapor pressure, $e_s$. The value of $e_s$ is a constant for a given water temperature, $T$, but increases with $T$ (see Appendix 4.1). The average gradient, $\Delta$, of this $e_s - T$ function (curve) is often used in evaporation models.

Upon striking a surface, radiation is either absorbed or reflected. While the water absorbs the heat and evaporates, the reflected heat warms the air by thermal conduction and thermal convection. The heat involved in warming the air is known as sensible heat — so-labeled because humans sense the temperature change. The best-known analysis of the interplay between latent heat content and sensible heat content is that by Penman (1948). Monteith (1965) provided an account of the Penman model followed by a presentation of his own model.

Actual evaporation that involves soil and vegetation is complex when compared with evaporation from an open water surface. The phenomenon is commonly known as evapotranspiration. Transpiration is evaporation from vegetal surfaces. Evaporation from most soil or vegetation differs from open water evaporation by an important condition: that of the unlimited water supply present in the open water body. However, this condition presents itself in perennial wetlands and well-watered crops. In the case of bare soil, climate favorable for evaporation must overcome the resistance of the soil to give up soil water held by surface tension operating in unsaturated soil (see Chapter 5). Vegetation heightens the competition for water because roots can extract water by exerting osmotic pressure, a negative pressure that exceeds about 15 atmospheres.

The physical and physiological processes of evaporation from land and vegetal surface can be treated as diffusional transfer (Shuttleworth 1991), which includes the transfer of water mass by vapor pressure gradient, transfer of heat by temperature gradient, and transfer of momentum (shearing stress) by wind gradient. These transfer processes are enhanced by wind-forcing convection but are countered by resistances to vapor diffusion. The resistances include eddy diffusion in the air, boundary layer in the air, the substrate (soil) surface, canopy surfaces, and bulk stomatal in vegetal leaf. The generalized form of the model is an analogy to Ohm's law of electricity (current × resistance = potential difference).

The rate of flow of latent heat, $\lambda\rho E$ or $\lambda E'$, is proportional to the concentration gradient of the water vapor normal to the evaporative surface or the water vapor deficit $e_2 - e_1$ (see equation 4.2). In the analogy to electricity, $\lambda E'$ is the current, $\int dz/D_v$ is the resistance, and $e_2 - e_1$ is the potential difference (see equation 4.3).

$$\lambda E' = \frac{-\rho C_p}{\gamma} D_v \frac{\partial e}{\partial z} \tag{4.2}$$

$$\lambda E' \left[ \int_{z_1}^{z_2} \frac{dz}{D_v} \right] = \frac{-\rho C_p}{\gamma} (e_2 - e_1) \tag{4.3}$$

where $\gamma$ is the psychrometric constant, which has a value of 0.067 kPa/°C at 25°C and which combines several parameters as follows:

$$\gamma = \frac{C_p p}{\varepsilon \lambda} \tag{4.4}$$

$C_p$ is the specific heat of moist air, taken to be 1.005 kJ/kg/°C; $p$ is the atmospheric pressure (101.3 kPa); $\varepsilon$ is the ratio of the molecular weight of water vapor to that for dry air, being 0.622; and $D_v$ is the molecular diffusion constant.

The conventional approach to modeling actual evaporation is to first determine a standard evaporation rate — potential evaporation, $\lambda E_o$, or reference crop ET, $\lambda E_{RC}$. The latent heat flow requirement, $\lambda E$, for the crop of interest is obtained by means of a crop coefficient, $K_c$, and the equation $\lambda E = K_c (\lambda E_o)$ or $\lambda E = K_c (\lambda E_{RC})$. The main rationale for this approach is practicality based on reasonable data requirement. This approach yields conservative estimates.

## Traits of Hawai'i Evaporation

### Evaporation Climate

In the case of evaporation climate, the three primary controls that govern rain climate are paramount: the marine position of the major Hawaiian Islands, the Hadley cell, and the high mountains. The indicative magnitude of annual net radiation is high and ranges from 100 to 150 W/m² (206 to 309 cal/cm²/d) with approximately 20 percent monthly variation. The highest values are found

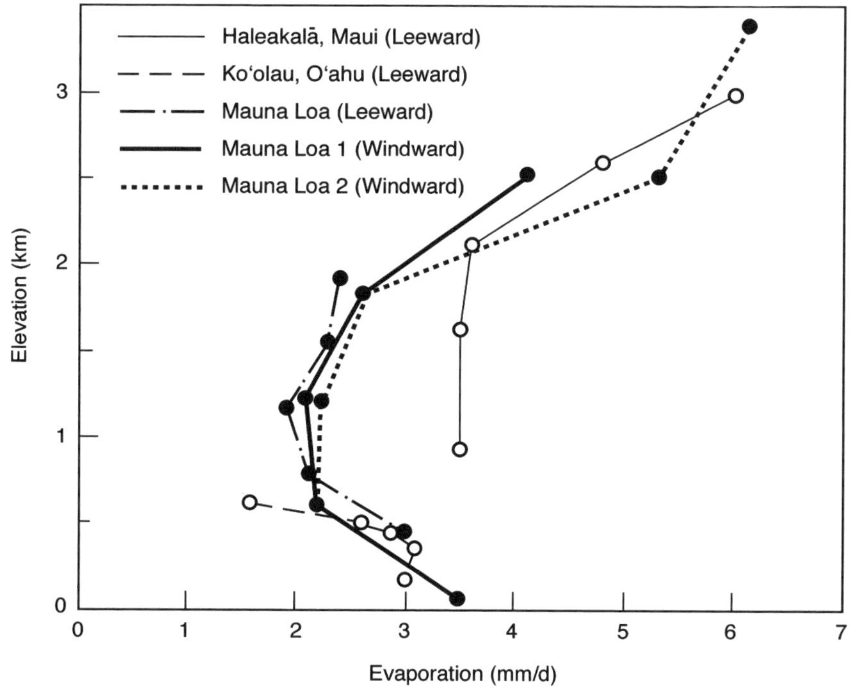


Figure 4.1. Evaporation profiles in Hawai'i. Haleakalā data are annual averages from Giambelluca and Nullet 1992a; Ko'olau data are annual averages from Ekern 1983. Evaporation minima occur near 1,200 m. 1 km = 3,281 ft, 1 mm = 0.04 in. (Redrawn from Bean et al. 1994)

along dry leeward coasts and the lowest values along wet windward slopes (Nullet and Sanderson 1993).

Trade winds and temperature inversion are two principal features of the Pacific Subtropical Anticyclone (see Chapter 3). Their interaction with the high mountains accounts for the spatial variation of the evaporation climate. The trade-wind inversion reveals itself by the nearly flat cloud tops. The cloud base hovers around 1,200 m (3,940 ft), and the inversion base fluctuates between 1,800 and 2,400 m (5,900 and 7,870 ft). The moist trade-wind air flow is wholly or partially blocked by the high mountains and detained beneath the inversion layer. Thus, the windward slopes and mountain summits are shaded by clouds from solar radiation, which is the primary heat source for evaporation. Below the inversion layer, the air is humid, 60 to 80 percent relative humidity, with higher values for the windward areas. In the high mountains that rise above the inversion layer, the air is dry—40 percent, often 10 percent, and even 5 percent relative humidity (Schroeder 1993a). On the leeward lower slopes, the air becomes drier and warmer after condensation occurs as orographic rain on the windward slopes and higher leeward elevations. In the leeward coast region, the clouds dissipate to allow high radiation and evaporation. Although trade winds predominate, they are modified by local winds in the highly varying local topography of the high islands.

The land-air interactions are manifested by climate zones that vary markedly with altitude along the upper slopes of the high mountains. The climate zones on land are categorized as marine, up to 1,200 m (3,940 ft); fog, 1,200 to 1,800 m (3,940 to 5,900 ft); transition, 1,800 to 2,400 m (5,900 to 7,870 ft); and arid, above 2,400 m (7,870 ft) (Giambelluca and Nullet 1991). The corresponding atmospheric layers are subcloud, cloud, temperature inversion, and free atmosphere (Riehl et al. 1951).

In the marine zone, evaporation decreases with elevation; however, evaporation increases with elevation above 2,000 m (6,560 ft). Minimum evaporation occurs at about 1,200 m (3,940 ft), approximately the cloud base, as in-

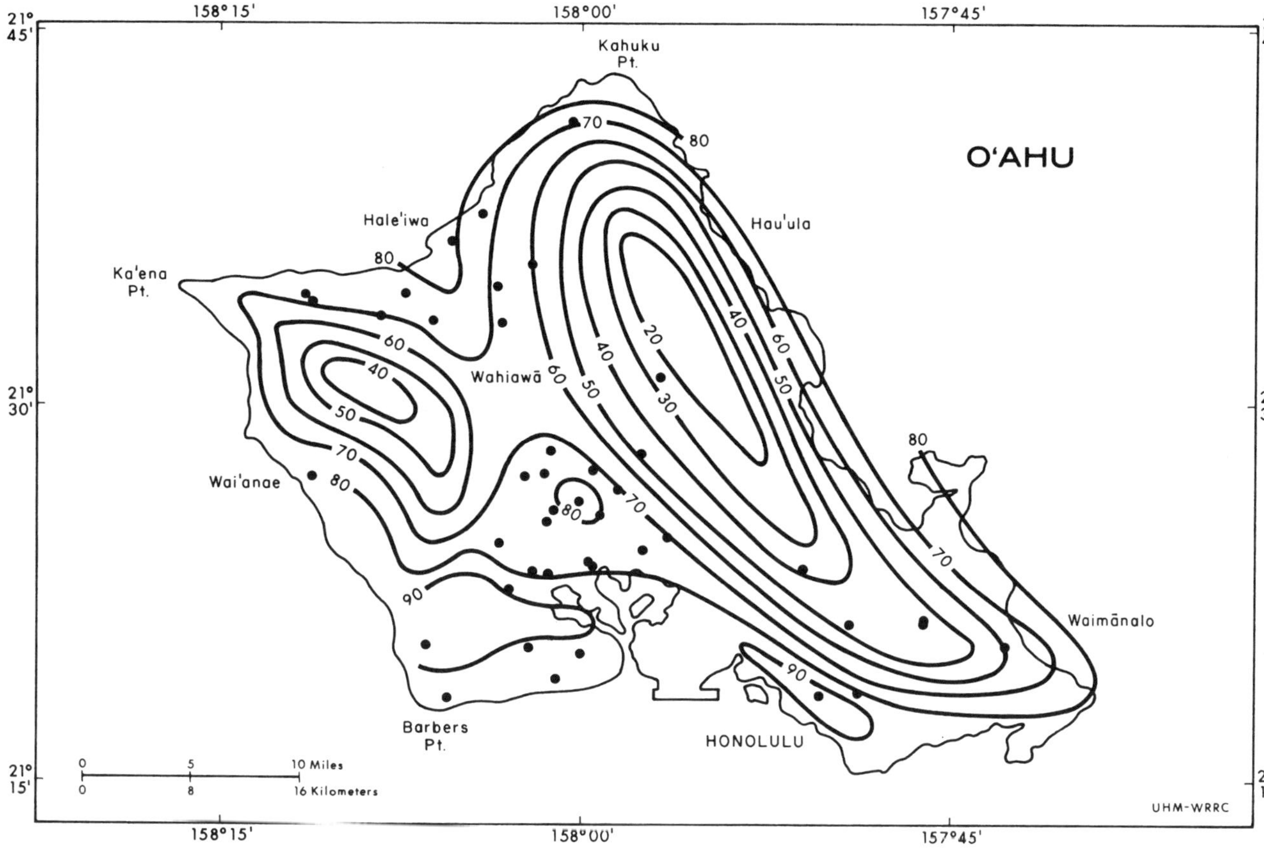


Figure 4.2. Adjusted annual pan evaporation for O'ahu. Indicated numbers are in inches (1 in. = 25.4 mm). (Reprinted by permission from Ekern and Chang 1985)

dicated in Figure 4.1 (Ekern 1983; Nullet and Giambelluca 1990; Giambelluca and Nullet 1992a; Bean et al. 1994; Nullet and Juvik 1994).

Advection of heat is effective near shoreline and dry lee areas. This heat source is in addition to solar radiation and can affect evaporation (Nullet 1987; Ekern 1993).

## Evaporation Patterns and Trends

Pan evaporation is the amount of water that evaporates from the open water contained in a standardized pan, usually measured daily. The most comprehensive pan data source for Hawai'i is given in Ekern and Chang (1985), a source that is readily available and that has been critically evaluated. The data are plotted as evaporation contours (lines of equal evaporation) (Figures 4.2, 4.3, 4.4, and 4.5 for O'ahu, Kaua'i, Maui, and Hawai'i, respectively). An analysis of the patterns of evaporation has been indexed to evaporation from the open ocean surrounding Hawai'i to indicate land influences. The reference value of ocean evaporation, 2,007 mm/yr (79 in./yr), is computed with the Priestley-Taylor model (1972).

In general, evaporation on the windward coasts is much the same as evaporation over the ocean. An exception is where the massive bulk of Mauna Kea on the island of Hawai'i blocks the trade-wind flow and reduces evaporation on the windward coast of the island to 10 to 20 percent below that over the ocean. As the trade

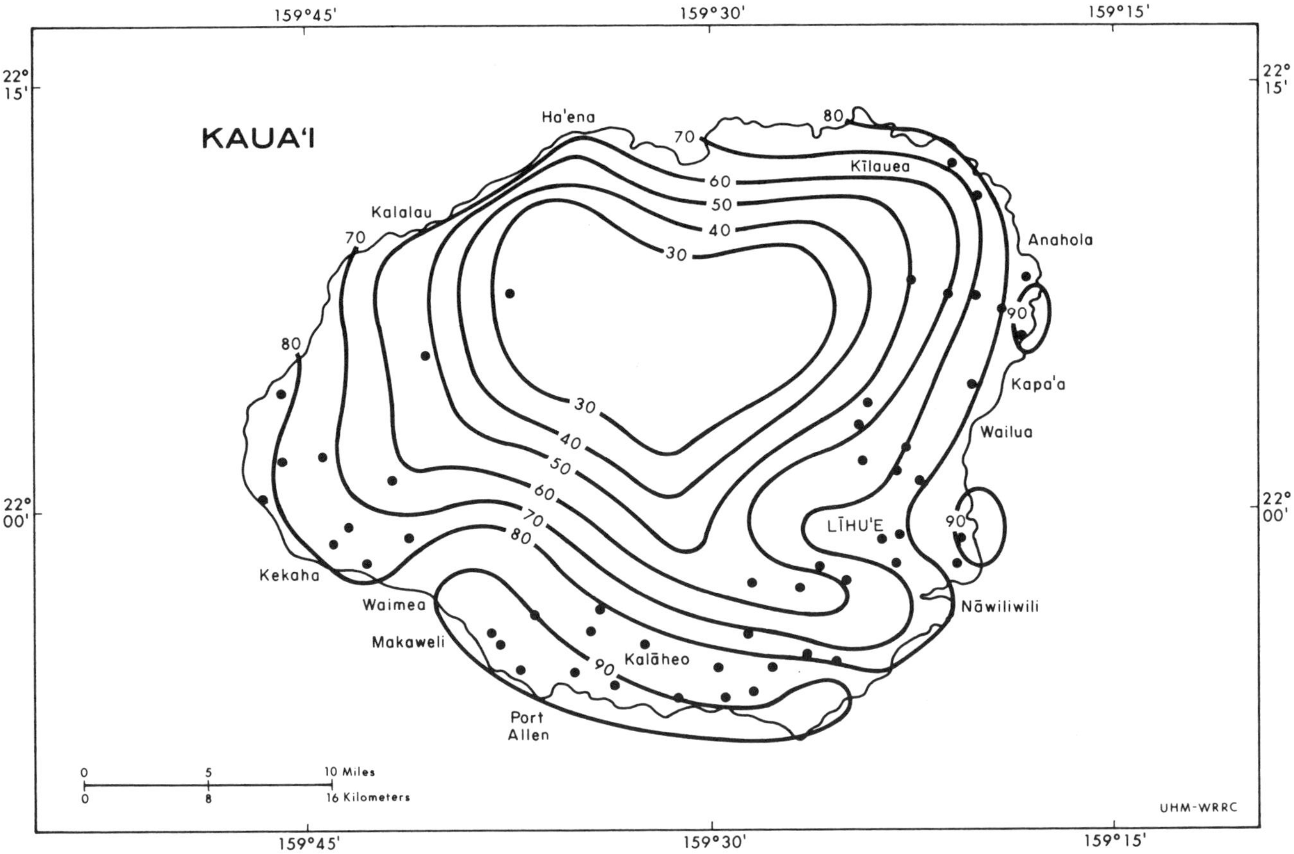


Figure 4.3. Adjusted annual pan evaporation for Kaua'i. Indicated numbers are in inches (1 in. = 25.4 mm). (Reprinted by permission from Ekern and Chang 1985)

winds move onshore, the orographic cloud reduces radiation and evaporation rapidly to less than 30 percent of the ocean rate. Beneath the orographic cloud, which persists more in the summer than in the winter, monthly evaporation rates become nearly constant throughout the year. Evaporation increases 10 to 20 percent where winds increase around the margins of the islands and in saddles between mountains where cloudiness is not pervasive. Summer evaporation rates are greater than 304.8 mm/mo (12 in./mo) on the dry leeward coastal plains where positive advection of heat increases annual pan evaporation rates to 30 to 40 percent above those over the nearby ocean. Only in the leeward Kona area of Hawai'i Island is the sea-breeze cloud sufficient to reduce evaporation. There, evaporation decreases at higher elevations below the trade-wind inversion. Above the inversion, evaporation on high mountain slopes increases to equal or exceed the rates near sea level.

Monthly variations of evaporation at selected stations are illustrated for the major islands (Figures 4.6, 4.7, 4.8, 4.9, 4.10, and 4.11). The variations are due to the evaporation controls. Also, solar radiation is seasonal — high during the summer months through September (How 1978). Although the long-term average annual insolation is high — 242 $W/m^2$ (500 $cal/cm^2/d$) — seasonal fluctuations range from 194 to 339 $W/m^2$ (400 to 700 $cal/cm^2/d$). Net radiation is lower; as a first approximation, it ranges from 121 to 145 $W/m^2$ (250 to 300 $cal/cm^2/d$) (Nullet and

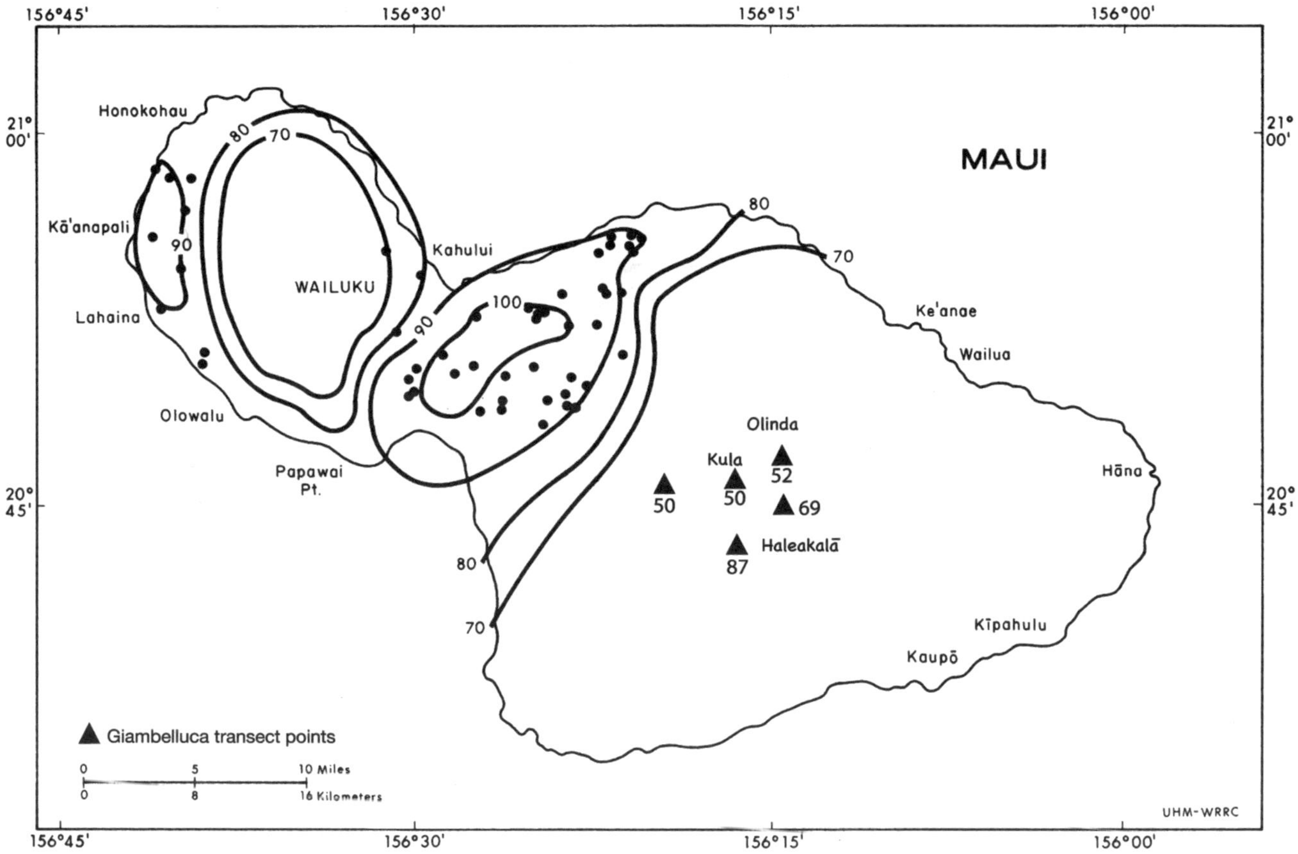


Figure 4.4. Adjusted annual pan evaporation for Maui. Indicated numbers are in inches (1 in. = 25.4 mm). (Adapted from Ekern and Chang 1985, Nullet and Giambelluca 1990, and Giambelluca and Nullet 1992a)

Sanderson 1993). Moreover, trade winds are not uniform throughout the year. They persist in the summer, 93 percent of the time in August, but weaken to 50 percent in January in response to the seasonal shift of the Pacific Subtropical Anticyclone (Blumenstock and Price 1967; Schroeder 1993a).

## Open Reservoir Evaporation

Long-term (annual and monthly) evaporation from large water bodies such as reservoirs and lakes is a subject of practical interest in water resources planning and operations. The general methods for determining open reservoir evaporation are presented in engineering hydrology textbooks such as that by Linsley et al. (1982). The methods are intended for reservoirs already in place and being planned.

For existing reservoirs, the well-established method is the use of pan coefficient, which is defined as the ratio of reservoir evaporation to pan evaporation. An estimate of reservoir evaporation is obtained by multiplying the pan data by coefficients.

Documented Class A pan coefficients for annual evaporation range from 0.64 to 0.81. The most referenced value is 0.70. In general, the coefficient is an attempt to reflect the overall environmental differences between pan and natural water body evaporation, including water volume and water flux. These values are the results of

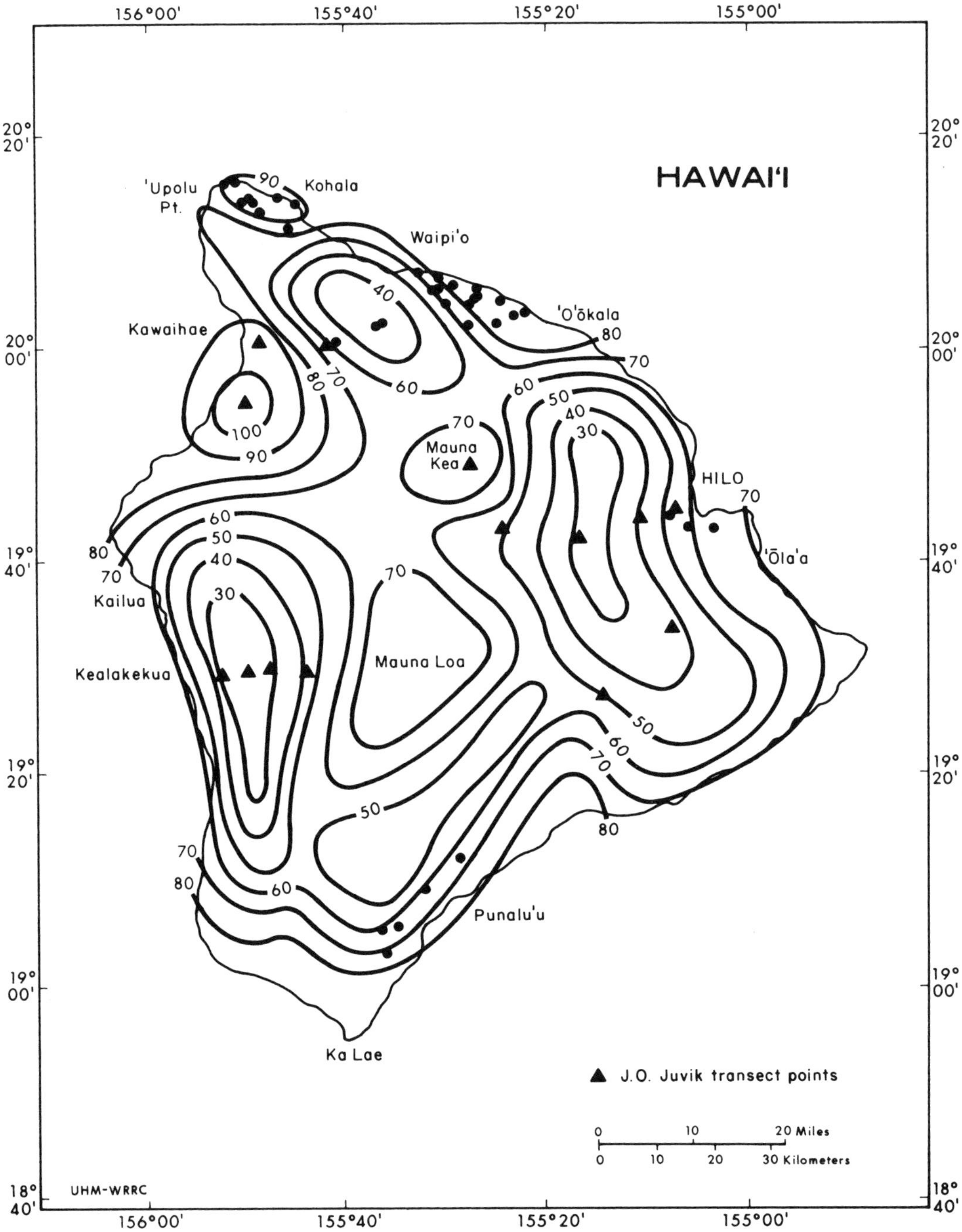


Figure 4.5. Adjusted annual pan evaporation for Hawai'i Island. Indicated numbers are in inches (1 in. = 25.4 mm). (Reprinted by permission from Ekern and Chang 1985)

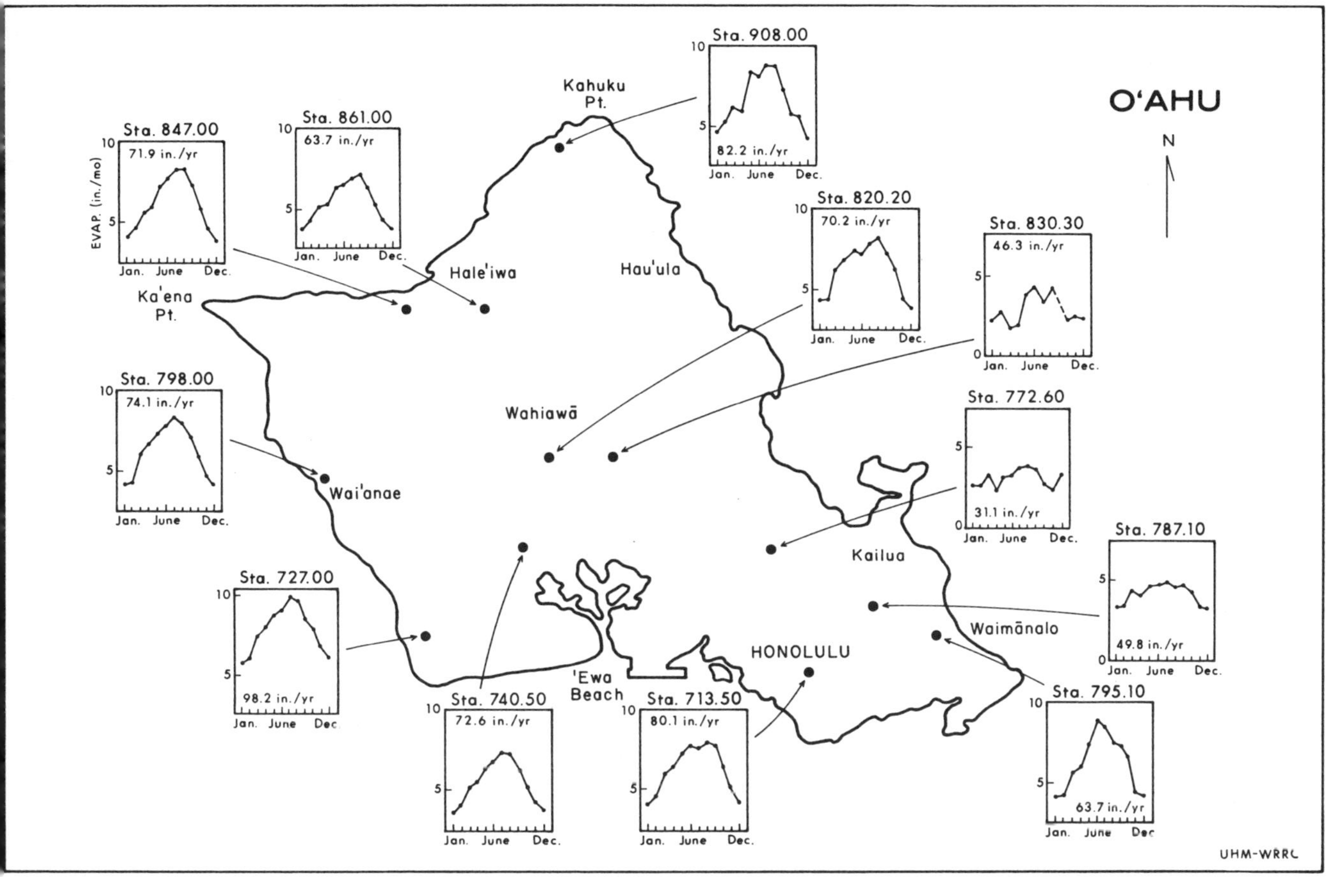


Figure 4.6. Pattern of monthly pan evaporation for selected stations, O'ahu. 1 in. = 25.4. mm. (Reprinted by permission from Ekern and Chang 1985)

investigations performed on small lakes where water and energy budgets could be determined and aerodynamics evaluated. Lake Hefner, Oklahoma, is among the best-known and most-studied sites. Investigations of this kind are generally too costly to be attempted for most water projects.

A judicious choice of the pan coefficient from the literature can provide a good first approximation of reservoir evaporation. In Hawai'i, the pan coefficient is probably about 0.80 (Ekern and Chang 1985).

For the purpose of designing reservoir capacity, Linsley et al. (1982) recommended the use of the Penman model coupled with a pan coefficient value of 0.70 and presented a nomograph (Fig. 4.12) for easy determination of evaporation from data on net radiation, air temperature, relative humidity, and wind speed for the site.

Wahiawā reservoir, a large impounding reservoir on O'ahu ($11.4 \times 10^6$ m$^3$ [$3 \times 10^9$ gal]), was studied in the interest of water-quality management. In the course of the studies, selected values of pan coefficient were used: 0.70 for monthly water balance evaluation (Young et al. 1975), 0.85 for the summer months, and 0.70 for the winter months (Moore et al. 1981).

Many small surface-water reservoirs serve various purposes in Hawai'i: at one time storage reservoirs constructed by sugarcane plantations to store water over-

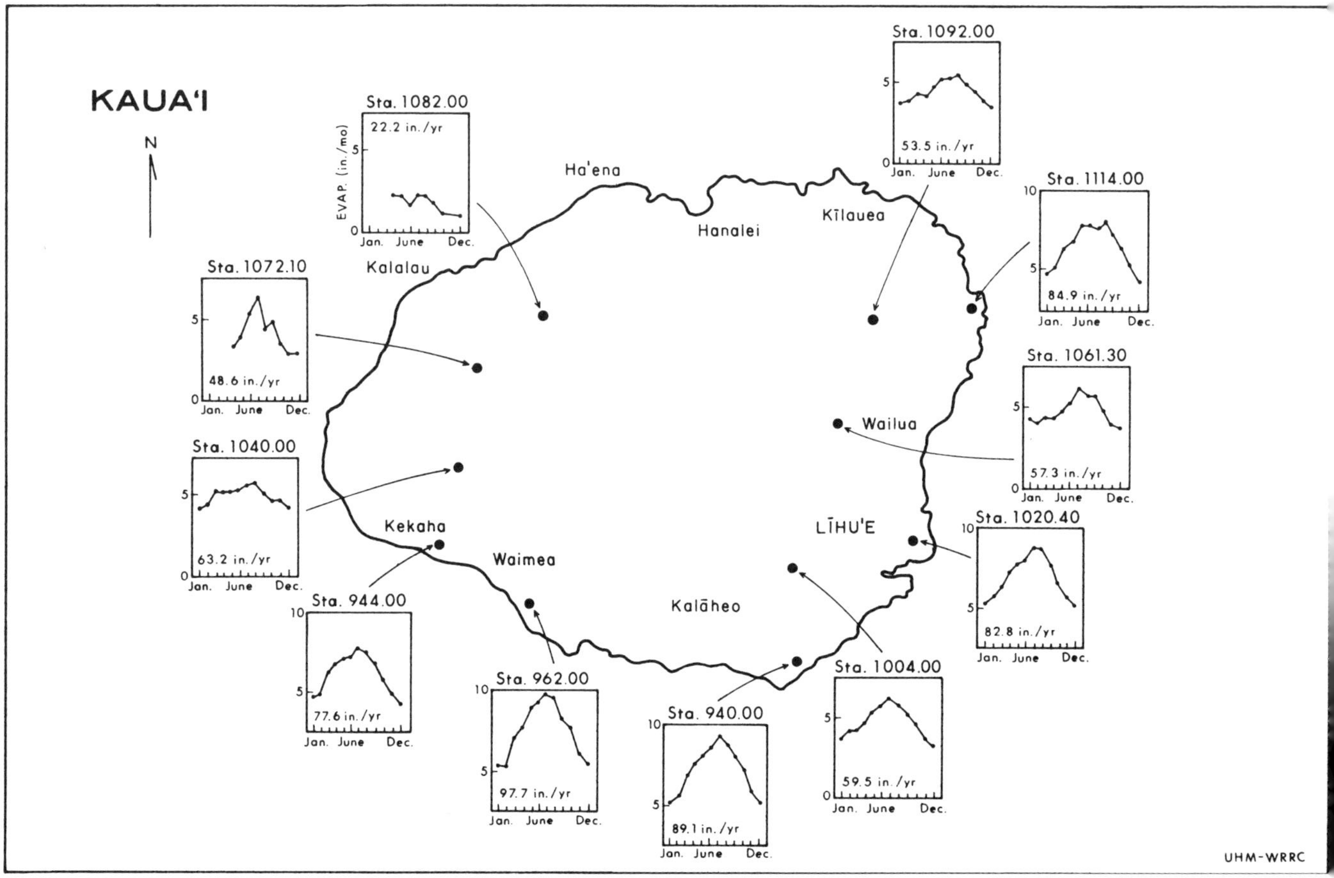


Figure 4.7. Pattern of monthly pan evaporation for selected stations, Kaua'i. 1 in. = 25.4 mm. (Reprinted by permission from Ekern and Chang 1985)

night for irrigation on the following day; a flood-control reservoir on O'ahu; and several impounding reservoirs used to store surface water for drinking water after treatment. Although evaporation must have been an involved element, no known evaluative documents have been published.

## Crop Water Management

### Potential Evapotranspiration

Potential evapotranspiration is the water requirement for a crop grown under nonrestricting water supply conditions. In planning agricultural crops, this ample water condition is usually strived for but not usually possible in operations. Two distinctly different approaches are available for potential evapotranspiration determination. United Nations Food and Agricultural Organization (FAO) guidelines consider the Penman model to be the best; the pan method is rated the second best among four methods that were evaluated (Doorenbos and Pruitt 1977). Use-to-pan ratio, based on a general approach similar to that of FAO, is much simpler to use (Ekern and Chang 1985). A summary of the FAO guidelines is presented in Appendix 4.2.

The use-to-pan ratio, when multiplied by pan evapo-

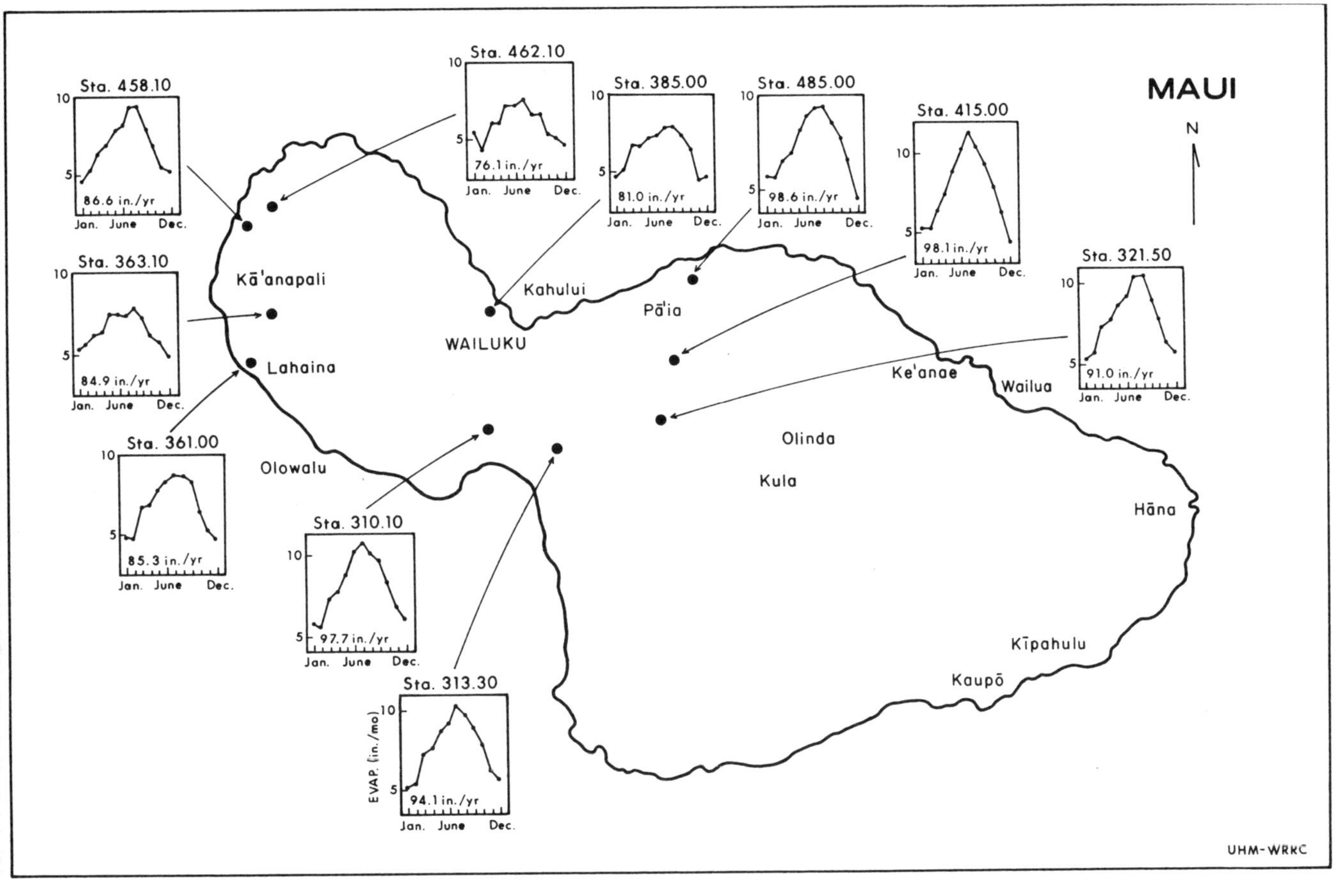


Figure 4.8. Pattern of monthly pan evaporation for selected stations, Maui. 1 in. = 25.4 mm. (Reprinted by permission from Ekern and Chang 1985)

ration, provides an estimate of water use by a crop. The ratio was determined by measurements for several crops in Hawai'i: sugarcane, Bermuda grass, pineapple, and some truck crops as summarized by Ekern and Chang (1985).

For sugarcane, the ratio increases from 0.40 to 1.01 during the first 5 months from germination to the establishment of a full canopy of sugarcane (Jones 1980). After that, the ratio reaches a peak of 1.20 at 10 months and then declines gradually to 0.98 at 17 months. For furrow-irrigated or sprinkler-irrigated mature sugarcane employing a pan placed at ground level, the ratio is 1.00. Drip-irrigated sugarcane uses an average of 0.80 of the annual ground surface pan evaporation or 0.70 of the pan elevated to 1.52 m (5 ft) height. Drip-irrigated sugarcane uses about 15 percent less water than furrow-irrigated or sprinkler-irrigated sugarcane.

Evapotranspiration by Bermuda grass is essentially the same as pan evaporation when soil moisture stress (negative pressure) is small (Ekern 1966a). As the soil moisture stress increases, the grass sod still maintains a high rate of water use until the soil moisture stress exceeds 1 bar ($1 \times 10^5$ Pa), but it is unable to maintain these rates as the soil moisture stress increases toward the 15-bar level.

Pineapple evapotranspiration rate is only one-fifth that of pan evaporation (Ekern 1965b). The greatly reduced

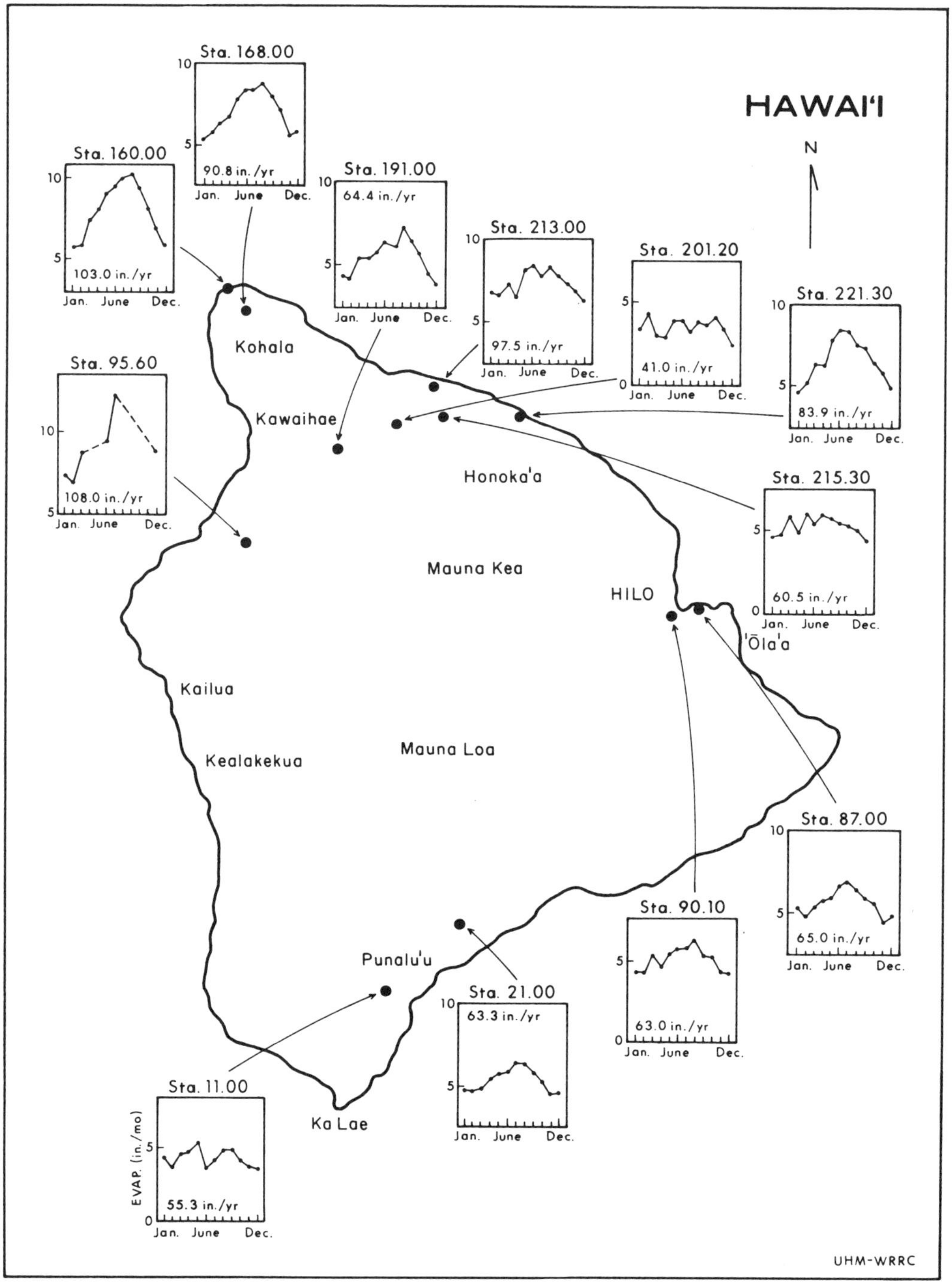


Figure 4.9. Pattern of monthly pan evaporation for selected stations, Hawai'i Island. 1 in. = 25.4 mm. (Reprinted by permission from Ekern and Chang 1985)

transpiration rate is largely due to daytime suppression of water vapor exchange by the pineapple leaf.

Lettuce and Chinese cabbage in the winter season have a pan ratio of 1.00. Under drip irrigation the use-to-pan ratio is about 0.60 for lettuce and 0.70 to 0.80 for Chinese cabbage (Wu 1972). Water requirements for many vegetable crops grown under furrow or surface irrigation in Hawai'i are available in an informal document (Valenzuela 1994). The requirements vary from crop to crop within a range of 3.0 to 8.3 mm/d (0.1 to 0.3 in./d) with an average of 5.1 mm/d (0.2 in./d).

Evaporation from bare Oxisols, a common agricultural soil in Hawai'i, at field moisture content is only one-third the rate from a pan (Ekern 1966b).

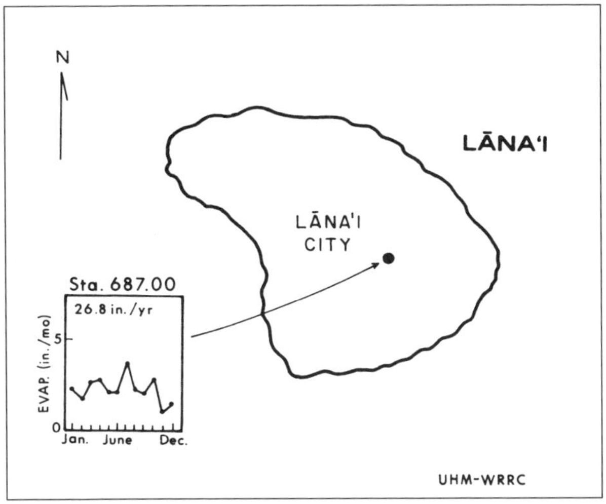


Figure 4.10. Pattern of monthly pan evaporation for selected station, Lāna'i. 1 in. = 25.4 mm. (Reprinted by permission from Ekern and Chang 1985)

## Actual Evapotranspiration

When water supply is limited, such as in a drying soil, the evapotranspiration amount as demanded by atmospheric condition and vegetation cannot be fully met. Actual

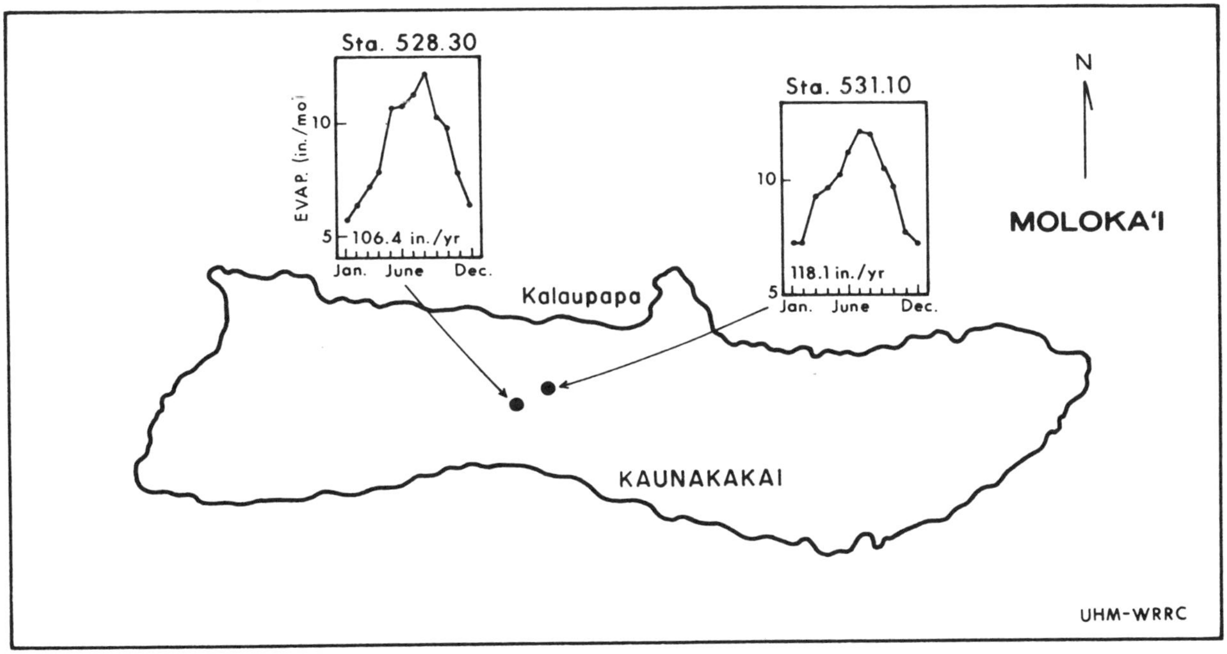


Figure 4.11. Pattern of monthly pan evaporation for selected stations, Moloka'i. 1 in. = 25.4 m. (Reprinted by permission from Ekern and Chang 1985)

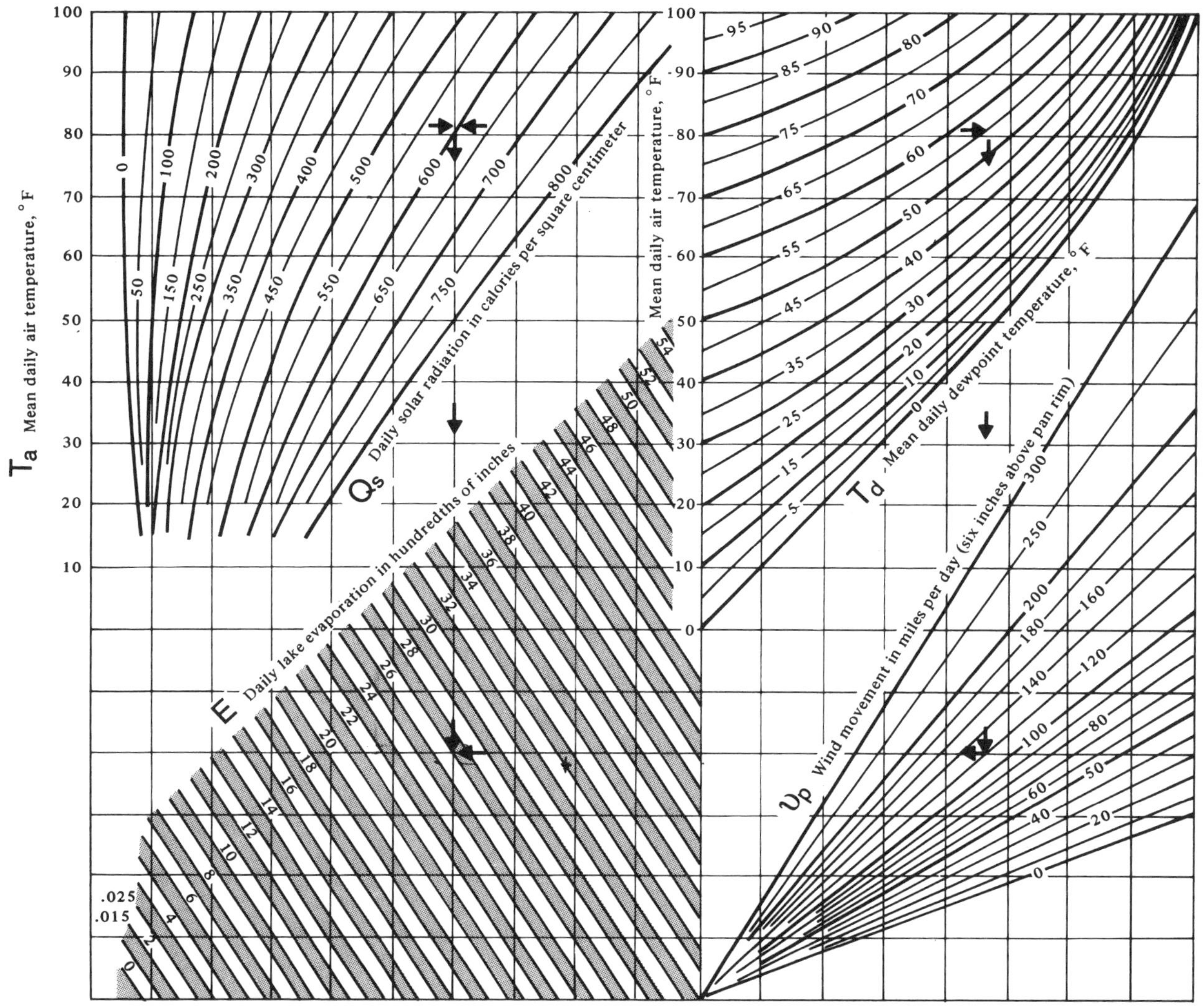


Figure 4.12. Shallow-lake evaporation as function of solar radiation, air temperature, dew point, and wind movement. By Penman's model and a pan coefficient of 0.7. Example of use: Enter this plot with sea-level observation $T_a = 70°F$, $T_d = 45°F$, $Q_s = 530$ cal/cm$^2$/d, and $\upsilon_p = 140$ mi/d to obtain $E = 0.22$ in./d. (Adapted from Linsley et al. 1982)

evapotranspiration prevails when evapotranspiration drops below the potential level. In part, actual evapotranspiration depends on the available water in the soil. Actual evapotranspiration can continue at the potential evapotranspiration rate until the soil moisture drops to a critical point. Then actual evapotranspiration declines as a function of soil moisture content. Measurements of the evapotranspiration of Bermuda grass in a lysimeter provide a data set (Fig. 4.13) to support these processes (Ekern 1966a).

Soil moisture storage and rooting depth are the controlling parameters of the soil-vegetation system for actual evapotranspiration determination. Soil moisture storage is a function of soil texture and is commonly measured as available water content, which is equal to field capacity minus wilting point (see Chapter 5). For ex-

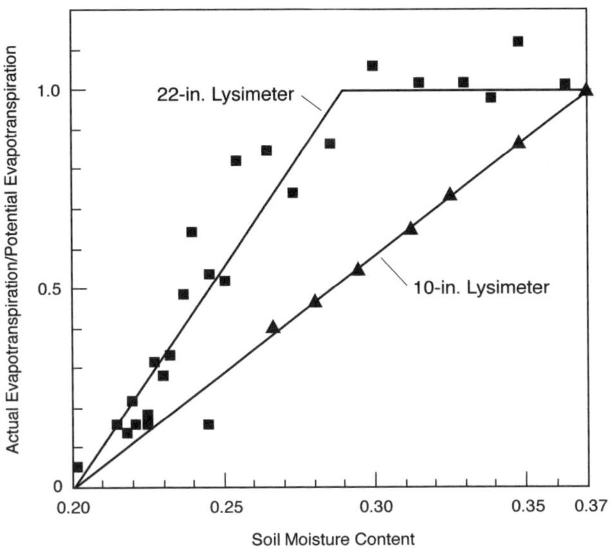


Figure 4.13. Evapotranspiration of Bermuda grass sod under varying soil-moisture depletion. (Based on Ekern 1966a; redrawn from Giambelluca 1983)

ample, Wahiawā silty clay, a typical Hawai'i agricultural soil, has an available water content value of 0.15 (= 0.45 – 0.30). A refinement suggests the use of a zero extraction point in place of the wilting point because of evidences of continued water extraction after the soil moisture content drops below the wilting point. An example of the value of a zero extraction point is 0.06 less than the wilting point (Black 1979).

Rooting depth in Hawai'i varies by crop, with estimates such as 40 to 51 cm (16 to 20 in.) for sugarcane, 31 cm (12 in.) for pineapple, 31 cm (12 in.) for grass(es) in urban parks and golf courses, and 15 to 61 cm (6 to 24 in.) for natural vegetation (Giambelluca 1983).

## Hawai'i Data and Models

### Data

Pans are the most common evaporation instruments used in Hawai'i. In addition to pans, various lysimeters including the accurate weighing lysimeters have been used for measurement of evapotranspiration. An atmometer was developed in Hawai'i especially for use in high-rainfall areas (see Appendix 4.3).

The first pan measurement in Hawai'i was made in 1894 (Hawai'i Division of Water and Land Development 1973b). Measurements made in dry, sunny leeward areas where sugarcane and pineapple were cultivated constitute the bulk of the database. The database is poorly distributed spatially and is inadequate for determining water balances in watersheds, water requirements for diversified crops, and water pollution control. Some data have been added to fill voids for high slopes and summits on Hawai'i and Maui as a result of the many studies conducted by the University of Hawai'i.

All pan evaporation data in Hawai'i were evaluated by Ekern and Chang in 1985. The basic presentation of the results is the monthly evaporation values for 212 stations that were operated for various periods from 1893 to 1983–1984, when 85 stations were in service. Statistics (means, standard deviation, and coefficient of variation) are also presented.

The annual pan evaporation for three selected stations of long record (16 to 24 years) nearly follows the normal frequency distribution. The coefficient of variation depends on the duration of measurement — annual, monthly, or daily. Based on a limited sample, the coefficient of variation approximately doubles from the annual (7 percent) to monthly (15 percent) value and doubles again to the daily (30 percent) value.

Evaporation maps for O'ahu, Kaua'i, Maui, and Hawai'i are based on adjusted annual pan evaporation measurements. Adjustments were made to compensate for several pan variables (composition, paint, elevation above ground, screening) and for long-term (1920–1984) variation of measured sunlight in Hawai'i. Maps were not prepared for Lāna'i with its single station and for Moloka'i with only three stations. The delineation of the evaporation contours was partially guided by the annual and diurnal rainfall patterns that reflect cloudiness, which consequently affects sunlight and evaporation potentials. Monthly evaporation is graphed for selected stations for each major island to demonstrate the pronounced seasonal pattern of evaporation.

## Models

The pragmatic approach to meet Hawai'i data needs by measurement of pan evaporation was complemented by modeling after the 1960s. Investigations confirmed that solar radiation is the primary variable controlling evaporation. Additional information gained includes determination of the altitudinal evaporation minimum along high mountain slopes, establishment of the relevance of advected heat in the land-ocean interface, and suggestion of nonradiative heat sources above the temperature inversion layer. Simulation of evaporative-diffusional processes that require setting up and solving mathematical formulations as a boundary value problem has not been done yet in Hawai'i.

Empirical and physical models for pan evaporation that are essentially statistical correlations of pan evaporation with air temperature and rainfall have been developed in Hawai'i. Tempered and narrowed by the mild maritime climate, temperature does not mirror sufficiently the solar heat received and is a rather poor evaporation predictor in contrast to its normal adequacy in many continental regions (Ekern and Chang 1985). Rainfall exhibits an empirical relationship with pan evaporation primarily because of the reduced insolation by clouds in the high-rainfall mountains. Transects in the Ko'olau Range on O'ahu register high evaporation in the low-rain and sunnier slopes and low evaporation in the high-rain and cloudier mountains. The annual evaporation for 1957 through 1959 was postulated as proportional to the reciprocal of the exponential function of distance toward the crest of the Ko'olau Range (Mink 1962b). Later, wind was added as a parameter to the evaporation-rainfall correlation involving a few selected stations from several islands (Takasaki et al. 1969). The considerable scattering of data points around the regression lines probably reflects the few stations used and their different evaporative environments.

Literature is replete with evaporation models that were developed elsewhere to meet practical needs, including irrigation water requirements and loss in water-storage reservoirs. A summary of the models is provided by Veihmeyer (1964) and Shuttleworth (1993). Shuttleworth (1991) broadly classified the models in terms of data requirements as energy balance (e.g., Penman), radiation (e.g., Priestley-Taylor), temperature (e.g., Blaney-Criddle), humidity (e.g., Dalton), and plant canopies (e.g., Monteith).

The Penman model was derived to predict open water evaporation. Also known as the combination model, it takes into account radiation and aerodynamics (ventilation), which are respectively the first and the second terms in the Penman (1948) equation as follows:

$$E_o = \left(\frac{Q^*}{\lambda \rho_w} \frac{\Delta}{\Delta + \gamma}\right) + E_a\left(\frac{\gamma}{\Delta + \gamma}\right) \tag{4.5}$$

where

$E_o$ = open water evaporation (mm/d)
$Q^*$ = net radiation ($MJ/m^2/d$)
$\lambda$ = latent heat of vaporization (MJ/kg)
$\Delta$ = gradient of the saturated vapor pressure versus temperature curve (mbar/K)
$\rho_w$ = density of water $\approx$ 1,000 $kg/m^3$ = 1 $g/cm^3$
$\gamma$ = psychrometric constant (mbar/K)
$E_a$ = aerodynamic term (mm/d)
$= (0.263 + 0.138\ u)\ (e_s - e_d)$ (4.6)

where

$u$ = wind speed at 2 m height (m/s)
$e_s$ = saturation vapor pressure of air (mbar)
$e_d$ = vapor pressure of the air (mbar)

The Penman model was tested in Hawai'i to be reliable for estimating open water evaporation and potential evapotranspiration (Ekern 1965a; Chang et al. 1967; Chang and Okimoto 1970). The Penman model provides the best estimate of potential evapotranspiration in comparison with other models (Noguchi 1982).

The Priestley-Taylor model is intended for moist natural surfaces. It is a simplification of the Penman model treating the aerodynamic term as a constant fraction (0.26) of the energy term. In the tropics, this simplification appears justified by observations (Chang 1993). The model is (Priestley and Taylor 1972):

$$E_p = 1.26\left(\frac{\Delta Q^*}{\lambda \rho_w}\right)\left(\frac{1}{\Delta + \gamma}\right) \tag{4.7}$$

where

$E_p$ = evaporation from moist natural surface (mm/d)

Sample computations with equations (4.5), (4.6), and (4.7) are demonstrated with Waimānalo data in Appendix 4.1 (see Example 4.2).

Besides the Penman and Priestley-Taylor models, the Van Bavel (1966) and Monteith (1965) models were also tested in Hawai'i. Van Bavel improved on Penman in the evaluation of the aerodynamics term; his model is suited for open water surfaces and wet soil. Monteith considered the vegetal canopy effects on evaporation in some detail. His model is intended for wet vegetal surfaces. All four models account for the influence of solar (net) radiative energy and the effect of aerodynamics (ventilation) on evaporation. All four models can be used for short-interval prediction such as a day or an even shorter interval, but none of these models adequately accounts for advection from other heat sources.

An extensive test of the Priestley-Taylor model was conducted with long-term monthly data from pans at forty-nine lowland sites that are located in open agricultural areas and below 300 m (984 ft) in elevation on four major islands (Nullet 1987). Another test involved all four models with short-term data recorded with atmometers on Haleakalā mountain slopes on Maui during November 1987 and January 1988 at six sites at elevations from 290 to 1,650 m (951 to 5,413 ft) and from July to September 1988 at three sites at elevations from 950 to 2,130 m (3,116 to 6,988 ft) (Nullet and Giambelluca 1990; Giambelluca and Nullet 1992a). Later, an evaporation investigation was performed on both the windward and leeward slopes of Mauna Loa at sites covering a wide range of altitudes. For the highest site, the Mauna Loa Observatory at 3,400 m (11,155 ft) above sea level, the pan data from February to July 1987 were compared with three models: Penman, Penman-Monteith, and Priestley-Taylor (Bean et al. 1994). Also, the Penman-Monteith model was tested against the period averages (September 1991 to August 1993) at three lower-altitude stations on the windward slope of Mauna Loa (Juvik and Nullet 1994). Still another testing of six models — Penman, Penman 1963, Priestley-Taylor, Hargreaves, Jensen-Haise, and Kohler — was performed against 3-year (1990–1992) daily data at a single site in Waimānalo on the windward coast of O'ahu. All methods but Kohler's were concluded to be satisfactory. Only the Kohler method does not take radiation into consideration (Kong 1996).

Advection effects were isolated and quantified on several islands by Nullet (1987). He reasoned that because the Priestley-Taylor model is regarded as either advection-free or minimally affected by advection (Priestley and Taylor 1972; Brutseart and Stricker 1979), the difference between Priestley-Taylor–predicted evaporation and the measured pan evaporation could be attributed to advection. The amount of the difference can be as high as +0.84 mm/d (+0.03 in./d) in November (positive advection indicates measured pan data higher than model prediction) and −0.71 mm/d (0.03 in./d) in June. The average values for the twenty sites most affected were +0.63, +0.34, −0.10, −0.21, −0.52, −0.71, −0.51, −0.60, −0.25, +0.34, +0.84, and +0.74 mm/d (+0.03, +0.01, 0.00, −0.01, −0.02, −0.03, −0.02, −0.02, −0.01, +0.01, +0.03, and +0.03 in./d) for the months of January through December (Nullet and Giambelluca 1990). Ocean influences diminish in inland areas and wet leeward areas. Advection from land heat sources can be appreciable, especially in dry leeward areas during the summer. An example indicates a deviation as high as +2.8 mm/d (+0.11 in./d). On an annual basis, Ekern (1993) determined negative advection from the Waimānalo station and strong positive advection from dry leeward surroundings. No original advection model has been created specifically for Hawai'i.

Predictions using models may be justified in Hawai'i as

first approximations. Their use should be guided first by the ground elevation of the location concerned: lowland (less than 300 m [984 ft]) versus above and within the inversion. For the lowland sites tested, which are well below the cloud base of the temperature inversion, prediction with the Priestley-Taylor model is generally adequate as a first approximation. However, adjustments are necessary for coastal areas because of sensible heat advection from or to the ocean, and for dry leeward areas.

Above and within the inversion, prediction becomes more complicated and less certain because of additional but not-well-known nonradiative heat sources, including those associated with the temperature inversion. In a short-term test on Haleakalā, Maui, all four models underestimated pan evaporation. The greatest deficits were those by the Van Bavel model — 52 percent at an elevation of 1,650 m (5,413 ft) in the winter and 22 percent at an elevation of 2,130 m (6,988 ft) in the summer. The only exception to the underestimation is the Priestley-Taylor model: when used for summer, it actually overpredicted. Among the four models, Priestley-Taylor and Penman render predictions closer to measured values than Van Bavel and Monteith (Nullet and Giambelluca 1990; Giambelluca and Nullet 1992a).

At the Mauna Loa Observatory at elevation 3,400 m (11,155 ft), the Penman and Monteith models provided close estimates (+3 percent and −10 percent, respectively) of the measured pan evaporation rate, and the model estimates had high correlation (regression coefficient of 0.92 and 0.87, respectively) with measured rates. The Priestley-Taylor model, relying primarily on net radiation, was less satisfactory (−20 percent for the estimate and 0.54 for the correlation coefficient) than the other two models (Bean et al. 1994). The performance of these three models disagrees somewhat with the measurements made at Haleakalā. The same study revealed, by correlation computations, that among the climatic parameters (including solar radiation, wind, and air temperature), vapor pressure deficit is highly correlated (0.94) with pan evaporation, thus confirming Chang's findings in 1985 (Nullet and Juvik 1994).

# Appendix 4.1. Basic Computations

## Example 4.1

(a) Given latent heat flow $\lambda E' = 100 \text{ W/m}^2 = 100 \text{ J/s/m}^2$
Compute water evaporation rate $E$ at 25°C
Solution:
Latent heat of evaporization, $\lambda$, at 25°C = 2.442 MJ/kg [see equation (4.1)]
Water density $\rho \sim 1 \text{ g/cm}^3$

$$E = \frac{\text{latent heat flow}}{\lambda \rho_w}$$
$$= \frac{100}{(2.442 \times 10^6)(1)} \frac{\text{J}}{\text{s m}^2} \frac{\text{m}^2}{10^4 \text{ cm}^2} \frac{\text{kg}}{\text{J}}$$
$$\times \frac{10^3 \text{ g}}{\text{kg}} \frac{\text{cm}^3}{\text{g}} \frac{8.64 \times 10^4 \text{ s}}{\text{d}}$$
$$= 0.354 \text{ cm/d} = 3.54 \text{ mm/d} = 0.14 \text{ in./d}$$

(b) Given latent heat flow $\lambda E' = 100 \text{ cal/d/cm}^2$
Compute water evaporation rate $E$ at 25°C
Solution:
1 calorie = 4.187 joules

$$E = \frac{100}{(2.442 \times 10^6)(1)} \frac{\text{cal}}{\text{d cm}^2} \frac{\text{kg}}{\text{J}} \frac{\text{cm}^3}{\text{g}} \frac{4.187 \text{ J}}{\text{cal}} \frac{10^3 \text{ g}}{\text{kg}}$$
$$= 0.172 \text{ cm/d} = 1.72 \text{ mm/d} = 0.07 \text{ in./d}$$

## Example 4.2

Given data from Waimānalo, O'ahu, April 1, 1991
Net radiation $Q^* = 11.57 \text{ MJ/d/m}^2 = 133.9 \text{ J/s/m}^2$
Air temperature $T = 22.9$°C
Saturation vapor pressure $e_s = 2.793$ kPa (from Table 4.1)
Air vapor pressure $e_d$
$e_s - e_d = 8.3$ mbar (1 mbar = 100 Pa)
Wind speed at 2 m height = 3.69 m/s

(a) Compute evaporation rate $E_p$ with Priestley-Taylor model
Solution:

$$E_p = 1.26 \left(\frac{Q^*}{\lambda \rho_w}\right) \left(\frac{\Delta}{\Delta + \gamma}\right)$$

From equation (4.1)

$$\lambda = 2.501 - 0.002361\ T$$
$$= 2.501 - 0.002361(22.9) = 2.447 \text{ MJ/kg}$$

Interpolation from Table 4.1, for $T = 22.9$°C

$$\Delta = 0.169 \text{ kPa/°C}$$
$$\gamma = 0.067 \text{ kPa/°C}$$

Substitution

$$E_p = 1.26 \left(\frac{11.57}{2.447} \frac{0.169}{0.169 + 0.067}\right)$$
$$= 1.26 \times 4.73 \times 0.716$$
$$= 1.26 \times 3.39 = 4.26 \text{ mm/d} = 0.17 \text{ in./d}$$

Note: The numerical value $Q^*/\lambda \rho_w$ gives equivalent water depth if $Q^*$ is expressed as MJ/m²/d and $\lambda$ in MJ/kg, and because $\rho_w = 1{,}000 \text{ kg/m}^3$

$$\frac{Q^*}{\lambda \rho_w} = \frac{\text{MJ/m}^2\text{/d}}{\text{MJ/kg } 1{,}000 \text{ kg/m}^3} \frac{1{,}000 \text{ mm}}{\text{m}} = \text{mm/d}$$

(b) Compute evaporation rate $E_o$ with Penman (1948) model using the same data.
Solution:

$$E_o = \left(\frac{Q^*}{\lambda \rho_w}\right) \left(\frac{\Delta}{\Delta + \gamma}\right) + E_a \left(\frac{\gamma}{\Delta + \gamma}\right)$$
$$E_a = (0.263 + 0.138u)(e_s - e_d)$$
$$= (0.263 + 0.138 \times 3.69)(8.3)$$
$$= 0.772 \times 8.3 = 6.41 \text{ mm/d}$$
$$E_o = \left(\frac{11.57}{2.447} \frac{0.169}{0.169 + 0.067}\right) + \left(6.41 \frac{0.067}{0.169 + 0.067}\right)$$
$$= (4.73 \times 0.716) + (6.41 \times 0.284)$$
$$= 3.39 + 1.82 = 5.21 \text{ mm/d} = 0.21 \text{ in./d}$$

Note: For this data set, the computed aerodynamics (second term) amounts to 54 percent of the radiation (first term), differing from the average value of 26 percent formulated by the Priestley-Taylor model.

Table 4.1. Temperature Dependence of Saturated Vapor Pressure, Its Temperature Gradient, Together with the Psychrometric Constant at Standard Atmospheric Pressure

| Temperature (°C) | (°F) | Saturated Vapor Pressure, $e_s$ (kPa) | Gradient of Saturated Vapor Pressure, Δ (kPa/°C) | Psychrometric Constant, γ (kPa/°C) |
|---|---|---|---|---|
| 0 | | 0.611 | 0.044 | 0.0654 |
| 1 | | 0.657 | 0.047 | 0.0655 |
| 2 | | 0.706 | 0.051 | 0.0656 |
| 3 | | 0.758 | 0.054 | 0.0656 |
| 4 | | 0.814 | 0.057 | 0.0657 |
| 5 | 41 | 0.873 | 0.061 | 0.0658 |
| 6 | | 0.935 | 0.065 | 0.0659 |
| 7 | | 1.002 | 0.069 | 0.0659 |
| 8 | | 1.073 | 0.073 | 0.0660 |
| 9 | | 1.148 | 0.078 | 0.0660 |
| 10 | 50 | 1.228 | 0.082 | 0.0661 |
| 11 | | 1.313 | 0.087 | 0.0661 |
| 12 | | 1.403 | 0.093 | 0.0662 |
| 13 | | 1.498 | 0.098 | 0.0663 |
| 14 | | 1.599 | 0.104 | 0.0663 |
| 15 | 59 | 1.706 | 0.110 | 0.0664 |
| 16 | | 1.819 | 0.116 | 0.0665 |
| 17 | | 1.938 | 0.123 | 0.0665 |
| 18 | | 2.065 | 0.130 | 0.0666 |
| 19 | | 2.198 | 0.137 | 0.0666 |
| 20 | 68 | 2.339 | 0.145 | 0.0667 |
| 21 | | 2.488 | 0.153 | 0.0668 |
| 22 | | 2.645 | 0.161 | 0.0668 |
| 23 | | 2.810 | 0.170 | 0.0669 |
| 24 | | 2.985 | 0.179 | 0.0670 |
| 25 | 77 | 3.169 | 0.189 | 0.0670 |
| 26 | | 3.363 | 0.199 | 0.0671 |
| 27 | | 3.567 | 0.209 | 0.0672 |
| 28 | | 3.781 | 0.220 | 0.0672 |
| 29 | | 4.007 | 0.232 | 0.0673 |
| 30 | 86 | 4.244 | 0.243 | 0.0674 |
| 31 | | 4.494 | 0.256 | 0.0674 |
| 32 | | 4.756 | 0.269 | 0.0675 |
| 33 | | 5.032 | 0.282 | 0.0676 |
| 34 | | 5.321 | 0.296 | 0.0676 |
| 35 | 95 | 5.625 | 0.311 | 0.0677 |
| 36 | | 5.943 | 0.326 | 0.0678 |
| 37 | | 6.277 | 0.342 | 0.0678 |
| 38 | | 6.627 | 0.358 | 0.0679 |
| 39 | | 6.994 | 0.375 | 0.0670 |

SOURCE: Shuttleworth 1993.
NOTE: 1 kPa = 6.895 psi.

# Appendix 4.2. FAO Guidelines For Potential Evapotranspiration

The United Nations Food and Agriculture Organization (FAO) calls the potential evapotranspiration crop water requirement ET-crop (Doorenbos and Pruitt 1977). The FAO guidelines involve a three-stage procedure: (1) climate effects, (2) crop effects, and (3) local condition and agricultural practice. Basically, ET-crop is computed by multiplying reference crop evapotranspiration (ET-O) by a crop coefficient ($K_c$) of the crop concerned. ET-O is defined by FAO as the rate of evapotranspiration from an extensive surface of 8- to 15-cm-tall (3.1- to 5.9-in.-tall) green grass cover of uniform height, actively growing and completely shading the ground, and not short of water. To determine ET-O, FAO presents four methods that are modifications of three common evaporation models and the standard pan measurement. The choice of one from these four is based primarily on available climate data and desired accuracy of the results. The three models are Penman, Blaney-Criddle, and "Radiation." The Blaney-Criddle (1950) model is a predictive formula of monthly evapotranspiration and requires data on temperature, daytime hour, and specificity of the crop. The "Radiation" model is based on work by Makkind (1957), who tested the Penman model by lysimeter. FAO employs mean daily climate data for 30-day or 10-day periods that yield results in ET-O over the same period.

By FAO's evaluation, the modified Penman method offers the best results with a minimal possible error of plus or minus 10 percent in summer and up to 20 percent under low evaporative conditions. The pan method is rated second best with a possible error of 15 percent, depending on the location of the pan. The radiation method, in extreme conditions, involves a possible error of up to 20 percent in summer. The Blaney-Criddle method should only be applied for periods of 1 month or longer in humid, windy, midlatitude winter conditions; an over- and underprediction of up to 25 percent has been noted.

In obtaining ET-O, FAO modifies to various degrees all four methods as they are commonly known. For instance, for the FAO pan method, FAO develops a coefficient to be applied to pan data, and Class A pan data need to be down-adjusted to obtain ET-O. FAO rates the Penman method as its top choice because more measured data are used for this method.

For the second stage, which considers crop characteristics, FAO provides values of crop coefficients for many crops, including sugarcane, for 30-day or 10-day periods and for various conditions including stage of growth, growing season, and prevailing weather conditions.

For the third stage, which considers local conditions and agricultural practices, FAO treats the following factors in great detail for operational purposes: climate variation over time, distance and altitude, size of fields, advection, soil water, salinity, method of irrigation, and cultivation method and practices.

# Appendix 4.3. Evaporation Instruments

## Pans

The standard Class A evaporation pan has been used extensively in Hawai'i. The pan is 118.1 cm (46.5 in.) in diameter and 25.4 cm (10 in.) deep and is made of either galvanized iron or monel metal. It is usually read daily.

Standard use of the pan must conform to the rigorous description of siting, installation, and operation. In Hawai'i, the use of a Class A pan is often modified. For instance, the pan should be mounted on an open wooden platform elevated 15 cm (5.9 in.) above the surrounding ground. But to help offset the effect of exposure on a pan in a small enclosure surrounded by high-rising sugarcane, many pans are elevated to 1.52 m (5 ft) above the ground, resulting in higher measured evaporation values than those for pans placed at standard elevation. For instance, at the former Hawaiian Sugar Planters' Association Kunia Substation on O'ahu, for the years 1964 through 1981, pan evaporation at 1.52 m (5 ft) was 1,885 mm/yr (74.22 in./yr); evaporation from a pan at 0.3 m (1 ft) was expectedly lower at 1,688 mm/yr (66.46 in./yr) (Ekern 1977). Another Hawai'i deviation from the standard use of a Class A pan concerns the requirement of maintaining the water level at 5.1 cm (2 in.) below the rim plus or minus 2.5 cm (1 in.). For rainy areas in Hawai'i, adhering to this requirement presents a practical problem. To counter the problem, a pan evaporimeter with a constant water level was developed in Hawai'i, although it was not adopted for routine use (van't Woudt 1960, 1963).

## Lysimeters

The basic methodology of lysimetry entails growing a crop in a tank. The tank is lowered into a tightly fitted pit, positioned with its surface flush with the natural ground, repacked with the excavated soil to simulate natural conditions, and operated to simulate field cultivation conditions. The most elaborate and accurate type is the weighing lysimeter; the entire content is weighed continuously or intermittently and measurements are made of the amount of irrigation water applied, rainfall, and any percolate collected at the bottom of the lysimeter. Evapotranspiration is the only unknown and can be computed with a water-balance equation written for the lysimeter. The nonweighing type requires ample and frequent watering to keep the variation of soil moisture content to a minimum or negligible. The weight change of the growing plant, if appreciable, needs to be accounted for.

Accurate lysimeters are elaborate research instruments not suited for routine operational use. Water requirements for sugarcane, pineapple, and some turfgrasses have been established with lysimetric measurements on O'ahu and Maui (Ekern 1972; Ekern and Chang 1985).

## Other Instruments

Ekern (1983, 1993) developed an atmometer (porous-surface evaporimeter) that is made with a porous Carborundum stone and is equipped with a rain shield. It was developed for and extensively tested in high-rainfall areas on O'ahu. Pan evaporation adjusted to an equivalent galvanized surface pan is 0.53 times the Ekern evaporimeter rate (Ekern and Chang 1985). Ekern's atmometer was later modified for precise measurement and automatic recording. The new meter involves the use of a view tube of the type used in medical intravenous systems and an infrared LED-photodetector pair to add and count the drops of water feeding the water-intake line connected to the stone. It was calibrated and successfully field-tested on Maui (Giambelluca et al. 1992; Giambelluca and Nullet 1992b) and in the Mauna Loa (Bean et al. 1994) investigations. Other evaporimeters are not routinely used in Hawai'i including the Piche evaporimeter, Livingston atmometer, and Bellani atmometer (Ekern and Chang 1985).

# CHAPTER FIVE

# Wetting the Surfaces

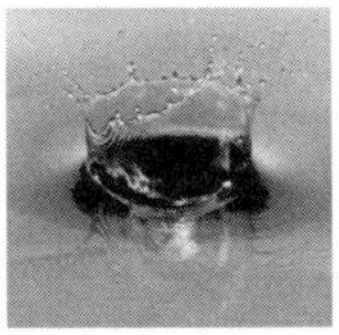

Upon contact with land surfaces, rainwater is partitioned in many ways. Consider the observations made after turning on a preset lawn sprinkler. Initially, much of the simulated rain is intercepted by vegetation, and some disappears by infiltration beneath the surface. As the event continues, depressions may become puddles. Water ponded on the surface spreads over land and flows toward lowland and rills. Not visible is the soil water that is redistributed, some of which percolates or evaporates long after the rain ceases. These phenomena of wetting the surfaces are simulated in many deterministic, physically based digital models, one of the earliest being the Stanford Watershed model, originated in 1966 by Ray K. Linsley at Stanford University (Figure 5.1).

## Unsaturated Zone in the Subsurface

### Soils and Rocks

Soils have been of special interest since the earliest civilizations. Soils were first classified by their productivity in China some 40 centuries ago (Simonson 1962). The soils of the kingdom were divided into nine classes during the Yao dynasty (2357–2261 B.C.). The best grades were the yellow, soft soils of Yung Chow (Shensi and Kansu provinces), whereas the next best were the red, rich clayey soils of Su Chow (Shantung, Kiangsu, and Anhwei provinces). Size of individual land holdings and tax to be paid the state were contingent on soil productivity.

Soil is a collection of granular particles (grains) that are either loose or aggregated as crumbs (Figure 5.2*a–d*). The solids are mostly (95 percent) minerals of rock origin and the rest is humus, which is organic residue of vegetation and animal origin. Soils are characterized by particle size, texture, structure, and profile.

According to the U.S. Department of Agriculture classification, particles smaller than 2 mm are grouped into three texture sizes: sand (2 to 0.05 mm), silt (0.05 mm to 0.002 mm), and clay (<0.002 mm). Those larger than sand sizes are gravel, and the even larger ones are cobbles.

Depending on the percentage of sand, silt, and clay, by weight, a soil is classified by its texture. The soil-texture triangle in Figure 5.3 shows twelve soil textures. Soil texture determines water retention in a soil. Clayey soils have a high water content and also retain most of it against free (gravity) drainage, whereas sandy soils drain most of the water easily. A desirable agricultural soil is fine loam with large amounts of silt and humus. A soil having a high percentage of clay (>10 percent) is further classified by clay mineralogy because of the practical significance of expansion of certain clays when wetted. Montmorillonites are highly expandable clays with very low permeability and high water retention relative to the nonexpandable clay kaolinite. Clay properties are responsible for minute movements that may detach masses of earth materials. Engineers are interested in other parameters, including liquid limit, the water content beyond which a soil begins to slip under stress.

Soil structure relates to how strongly the soil grains aggregate as crumbs or clods, such as those found in a plow layer after tillage. The stability of the soil structure

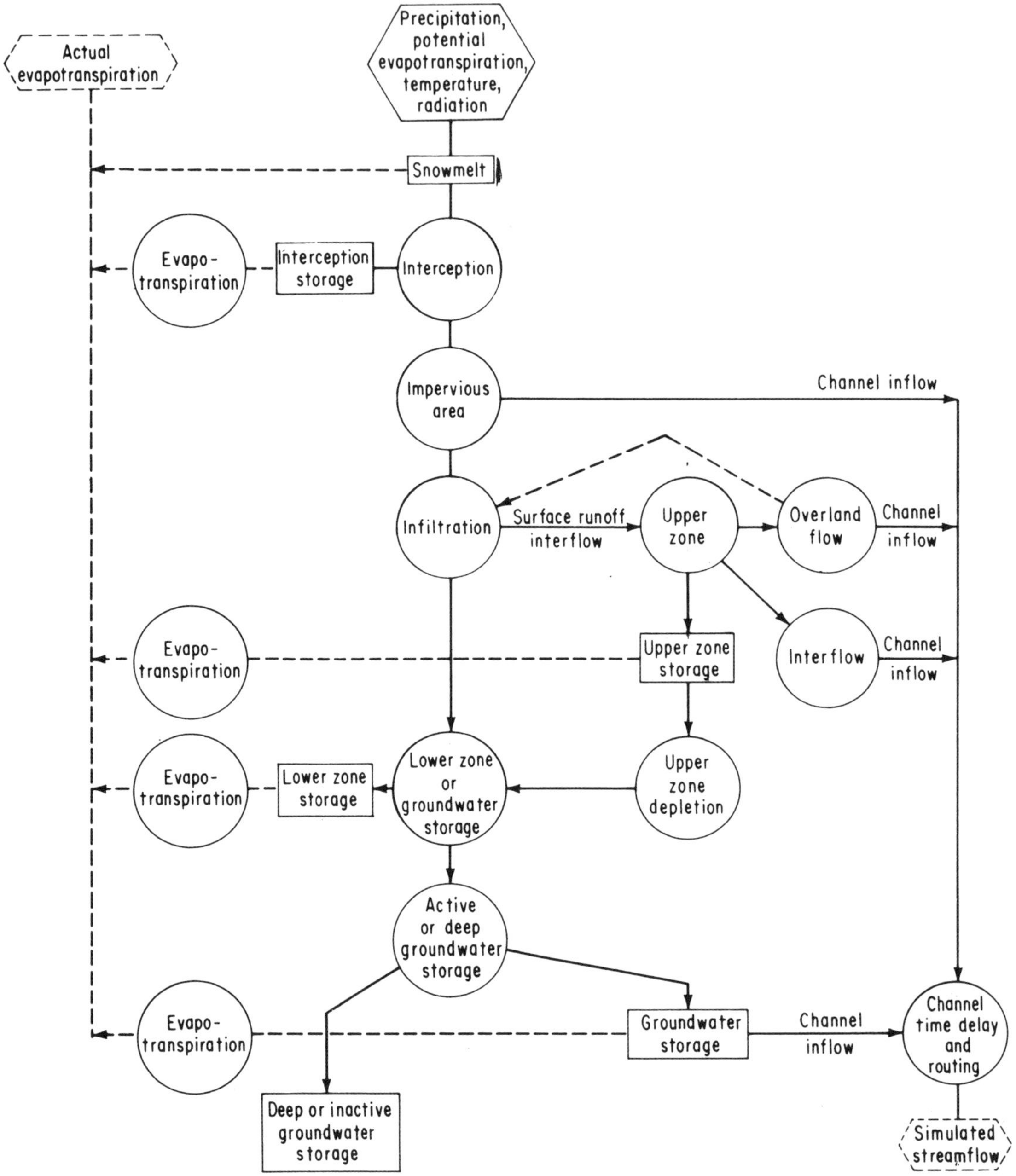


Figure 5.1. Flow diagram of Stanford Watershed model. (Reprinted from Linsley et al. 1982 with permission from McGraw-Hill Companies)

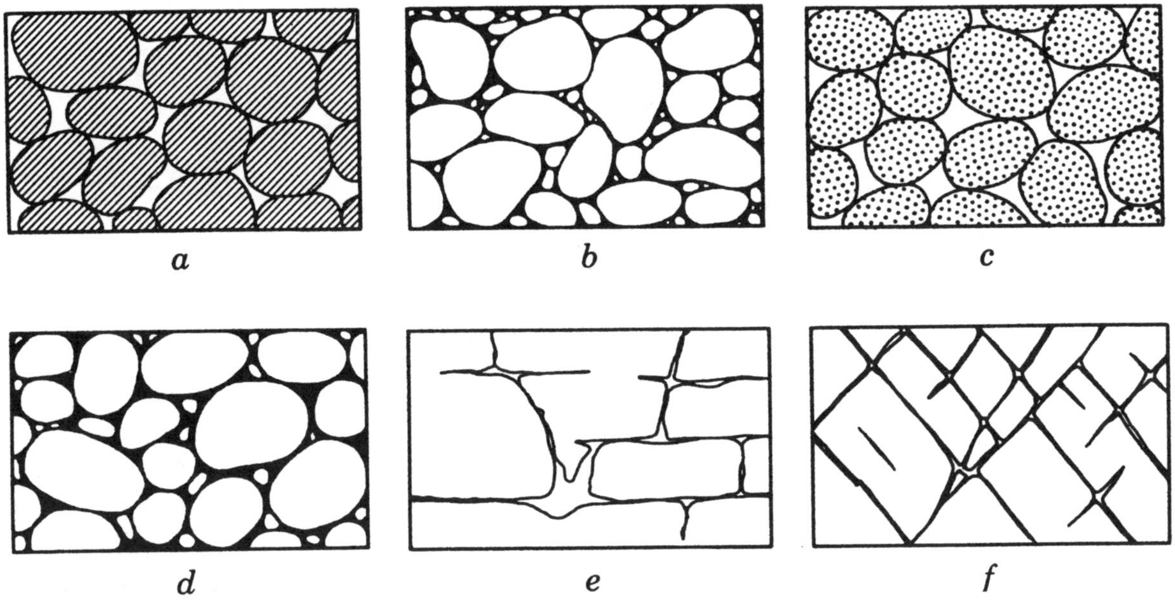


Figure 5.2. Different rock interstices and the relation of rock texture to porosity: *a*, well-sorted sedimentary deposit having high porosity; *b*, poorly sorted sedimentary deposit having low porosity; *c*, well-sorted sedimentary deposit consisting of pebbles that are themselves porous, so that the deposit as a whole has a very high porosity; *d*, well-sorted sedimentary deposit whose porosity has been diminished by the deposition of mineral matter in the interstices; *e*, rock rendered porous by solution; *f*, rock rendered porous by fracturing (after Meinzer 1923). (Reprinted from Todd 1980 with permission from John Wiley and Sons, Inc.)

depends on clay, organic matter, and inorganic cements such as colloids of iron and aluminum oxides and calcium carbonate. The structure can be collapsed by the addition of sodium to deflocculate the clay, by swelling and shrinkage, by beating of raindrops, by runoff scouring, and by excessive tillage and compaction. Soil structure is difficult to measure; however, the size of clods can be measured by dry or wet sieving. Other indirect measures of soil structure include pore-size distribution, mechanical properties, and permeability to air and water. Soils with a strong structure contain large pores for water entry and aeration.

A soil profile reveals horizons (distinctive layers), such as those usually exposed in the top 1.2 m (4 ft) of the vertical section of a pit. The A-horizon is the top layer, which is usually dark in color due to a high humus content and which is subject to maximum leaching. The B-horizon or subsoil accumulates the materials leached from above and contains more clays than the A-horizon. The C-horizon is gradational to rock, with the upper part slightly altered and the lower part transitional to parent rock. Agricultural tillage disturbs the A-horizon and in some instances the B-horizon as well. Disturbances leave a plow pan (layer) that is compressed and nearly poorly permeable at the bottom. In some bare soils there may exist a thin (1.5 to 3.0 cm [0.06 to 0.12 in.]) crust caused by raindrop impact during the first 10 minutes of rainfall. The crust can markedly limit water entry (Ahuja 1983).

Many soils have large noncapillary pores known as macropores. Examples are structural cracks, decayed root channels, and wormholes. They drain water much faster than common pores, often known as matrix pores. They are difficult to quantify, but various descriptive parameters have been used. Soils are commonly treated by

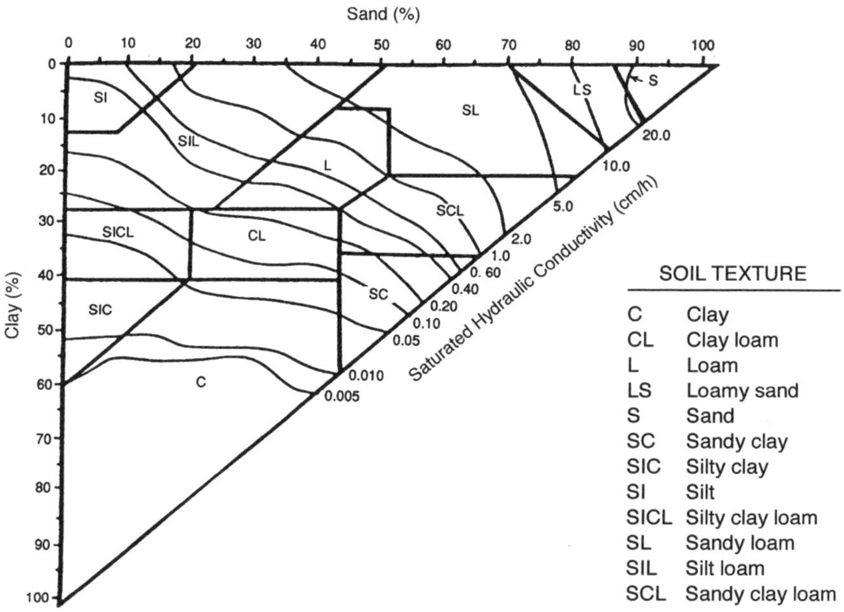


Figure 5.3. Soil-texture triangle according to U.S. Department of Agriculture. The curves indicating values of saturated hydraulic conductivity are empirical. 1 cm/h = 0.79 ft/d = 0.24 m/d. (Adapted from Rawls et al. 1993)

the continuum or equivalent-media approach in dynamics studies (Bear 1972). But other approaches, such as the two-domain model of matrix pores and macropores have been proposed (Chen et al. 1993).

Openings in rocks differ from those in soils. As shown by Meinzer in 1923 (Figure 5.2*e–f*), some rocks, such as consolidated sedimentary deposits, have openings like those in soils, but they also have fracture and solution openings that are not formed between grains. Fractures occur typically in crystalline and volcanic rocks, whereas solution openings occur in limestones. The openings in typical basalt flows, the most common rock in Hawaiʻi, are discussed in Chapter 1 and depicted in Figure 1.4.

Porosity, $n$, is the volumetric ratio of pore volume to bulk (solid plus pore) volume. Bulk density, $\rho_b$, is the ratio of solid weight to bulk volume, whereas particle density, $\rho_p$, is the ratio of solid weight to solid volume. These three parameters are related as follows:

$$n = 1 - \frac{\rho_b}{\rho_p} \tag{5.1}$$

The value of $\rho_p$ is 2.65 g/cm$^3$ for most sands and soils and 2.7 to 2.9 g/cm$^3$ for rock fragments. Porosity depends not only on particle sizes but also on packing and sorting (uniform versus mixed sizes). The respective porosities of a loosely packed soil and a compacted soil may be on the order of 0.67 and 0.50. Macropores are defined by size (30 to 3,000 μm in diameter), by tension (<5 kPa), by volume fraction (0.001 to 0.05), or by infiltration rate (1 to 10 mm/h). Porosity of soils and rocks varies over a wide range, as indicated in Table 5.1 (Davis 1969). For samples of disturbed soils, porosity and water content are determined by standard methods in the laboratory. But, for rocks and undisturbed soils, geophysical methods such as electrical resistivity, neutron logging, and gamma-gamma methods are used on site (de Marsily 1986; Rawls et al. 1993).

The surface area of soil particles is hydrologically significant because many contaminants are surface active and adsorb on the surface of particles. The surface area of particles per unit weight varies greatly with soil texture. Clay surface areas are about three orders of magnitude higher than that of coarse sand.

Table 5.1. Range of Values for Soil and Rock Porosity

| Soil and Rock | Porosity (%) |
|---|---|
| Unconsolidated deposits | |
| Gravel | 25–40 |
| Sand | 25–50 |
| Silt | 35–50 |
| Clay | 40–70 |
| Rocks | |
| Fractured basalt | 5–50 |
| Karst limestone | 5–50 |
| Sandstone | 5–30 |
| Limestone, dolomite | 0–20 |
| Shale | 0–10 |
| Fractured crystalline rock | 0–10 |
| Dense crystalline rock | 0–5 |

SOURCE: Davis 1969.

## Relief and Vegetal Cover

Relief is indicated on a topographic map that shows ground elevation above sea level of the terrain by contours, or lines of equal elevation. Steep terrain is indicated by closely spaced contours. A line drawn perpendicular to the contours traces the slope and the direction of natural overland flow. Spurs and ridges are surface-water divides of a drainage basin. The axis of descending contours is a stream channel (see Figure 5.4).

Vegetal cover and its hydrologic effects vary with land uses. These effects have been identified on local and regional scales for afforestation and deforestation, agricultural intensification, and wetland draining. For instance, afforestation can reduce annual streamflow on a basin scale by increasing interception and transpiration. Quantitative prediction of these effects is uncertain but can be estimated by synthesis of data from watershed experiments or process studies (Calder 1993).

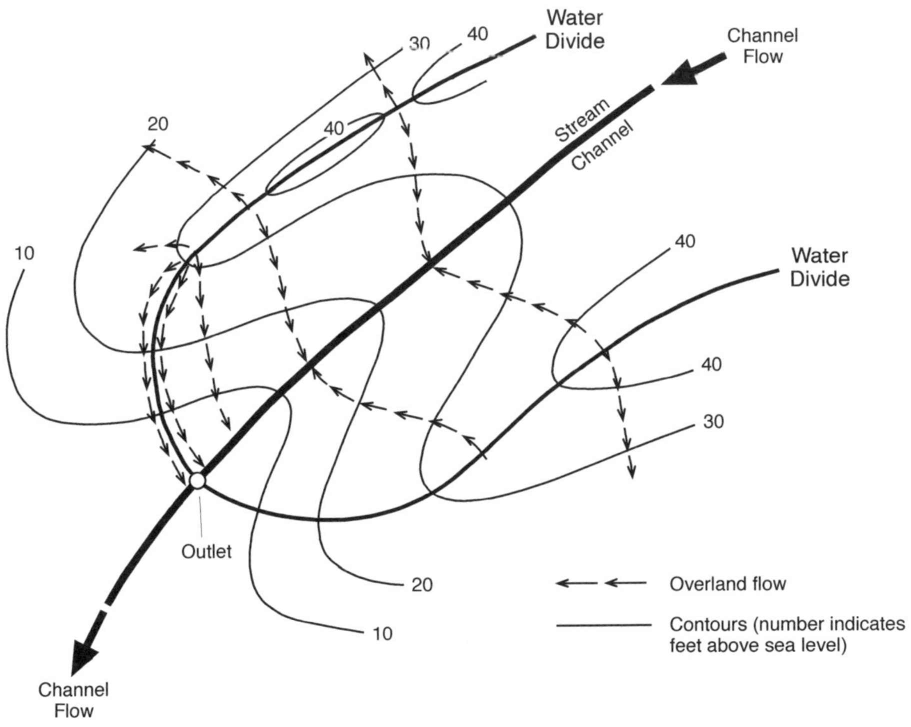


Figure 5.4. Delineation of a drainage basin.

## Water in the Unsaturated Zone

In the subsurface, a vadose zone lies between the land surface and the regional groundwater table. It consists of surface soils, saprolite, and rock to the capillary fringe of the regional water table (see Figure 5.5*a*). It may embody perched water bodies that occur on areally limited and poorly permeable strata such as buried soils. Except for localized perched water, this zone is unsaturated, and the subsurface water pressure is less than the atmospheric pressure. Soil water physics, which explains how a soil holds, releases, and allows water to flow, began with Buckingham in 1907 and was rejuvenated by Richards in 1928 (Sposito 1986).

### Holding and Releasing Water

In the unsaturated state, soil water is held by tension (negative pressure or pressure less than the atmospheric pressure). Water content is expressed volumetrically as a percentage of the bulk volume. In fluid mechanics, tension is inversely proportional to pore size.

For a presaturated soil draining freely under the pull of gravity, after many hours the dripping slows to a stop even though the soil is still moist. The remaining water is held by surface tension, which becomes active at the air-water-solid interfaces (meniscus) as air enters and permeates the soil. Water content remaining in the soil after a prolonged period (2 to 3 days) of draining is called field capacity. The condition is assumed equivalent to subjecting the soil to a suction of 0.33 bar (33 kPa [336 cm]) (Stephens 1996). Vegetal roots can extract water until the water content is reduced further to a level called the wilting point. The wilting point is assumed equivalent to the remaining water content by subjecting the soil to a suction of 15 bars (1,500 kPa [155,000 cm]). The initially freely drained water is called gravity water. The water between the field capacity and the wilting point is called capillary water or available water, and the water remaining at less than the wilting point is hygroscopic water, which can be removed only by drying. These three indices are used in determining irrigation water requirements and establishing recharge of groundwater by percolating water.

The water-retention curve depicts water content as a function of tension. Field capacity and wilting point are two points on this curve. The curve differs with soil texture. Various power functions are used in curve fitting the data. Reference curves and soil water indices are available for all twelve soil textures, thus facilitating first estimates if no on-site data are available (Rawls et al. 1993). The curve for wetting differs from that for draining. Soils hold less water during wetting than during draining for the same value of tension because entrapped air in the empty pores blocks ready entry of the advancing water. This phenomenon is known as hysteresis.

With recent field and experimental measurements, the classical concepts of soil water indices have been questioned (Ahuja and Nielsen 1990). The soil water suction attained after a 2- to 3-day drainage is deemed in the range of 5 to 10 kPa rather than 33 kPa, and movement of water continues internally at an ever-decreasing rate. Plant growth may be reduced before the wilting point is reached. The condition at which plants wilt and cannot recover their turgidity is not wholly dictated by soil water states. The dynamic concept of soil water availability for vegetation requires that the application of irrigation water be flexible and be based on transpiration demand and stage of plant growth in addition to soil water status (see Chapter 4).

### Continuous Water Flow

Subsurface water exists in a continuous state of slow motion induced by the energy level of the water. Energy in the water at each location is indicated by the hydraulic head, $h$, or simply head. Head is the sum of position head, $z$, and pressure head, $\pm p/\rho g$, where $\rho$ is the water density and $g$ is the gravitational acceleration. When $p$ is negative, it indicates tension that prevails in the unsaturated state (see Figure 5.5).

Subsurface water flow is governed by Darcy's Law, which is applicable to both the saturated and unsaturated states:

$$q = KJ \tag{5.2}$$

where specific discharge $q$ is the volumetric discharge of water, $Q$, divided by the bulk cross-sectional area, $A$, per-

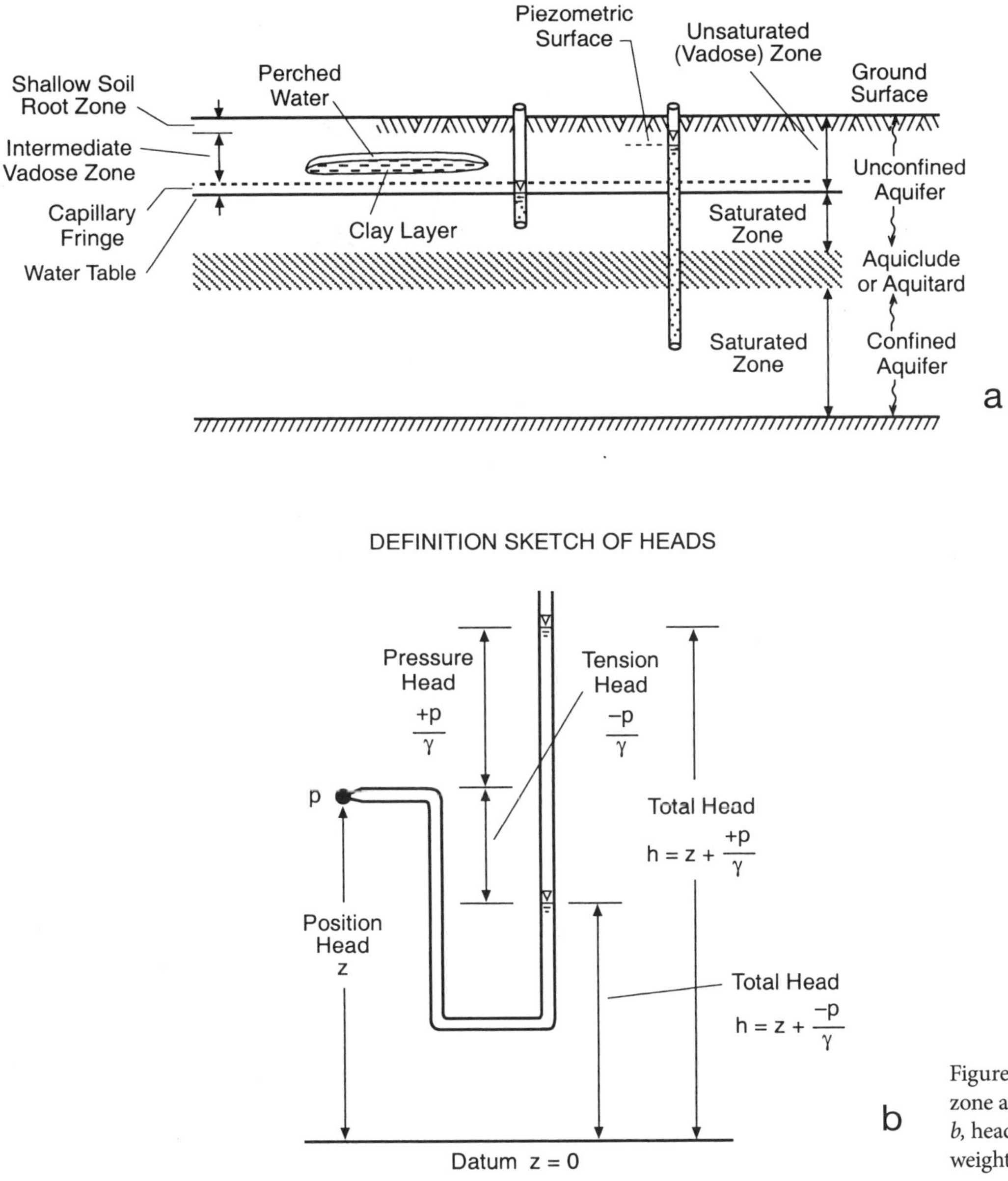


Figure 5.5. *a*, Unsaturated zone and saturated zone; *b*, heads. $\gamma$ = water specific weight = $\rho g$

pendicular to the direction of flow. Hydraulic gradient, *J*, is the difference in head between two points along the direction of flow divided by the linear distance between them. Hydraulic conductivity, *K*, is a measure of the permeability of soil or rock, is a property of the solid and water, and has a dimension of velocity. Specifically, $K = k\,\rho g/\mu$, where *k* is intrinsic hydraulic conductivity, which has a dimension of length and is solely dependent on the solid, and $\mu$ and $\rho$ are kinetic viscosity and density of the water, respectively.

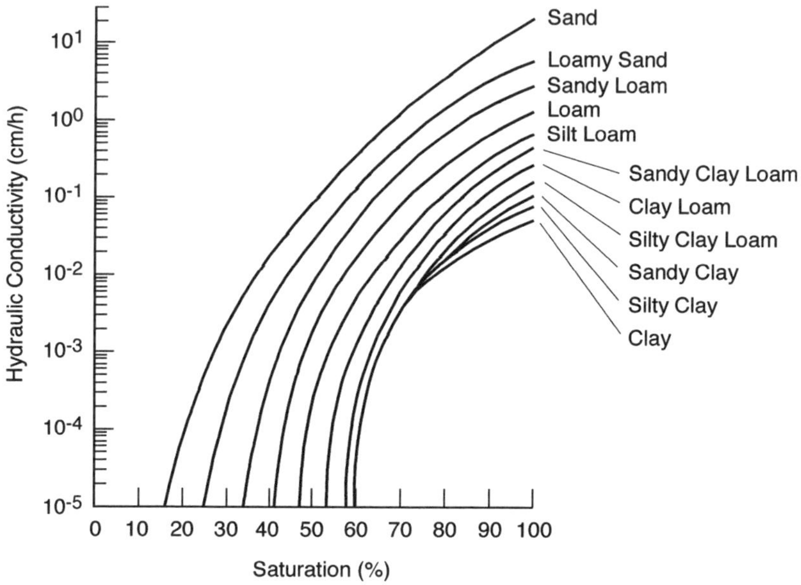


Figure 5.6. Empirical hydraulic conductivity by soil texture. 1 cm/h = 0.79 ft/d = 0.24 m/d. (Adapted from Rawls et al. 1993)

Permeability in the saturated state, $K_{sat}$, varies with soil texture and ranges from $10^{-2}$ to 10 m/d (see Figure 5.6). Rock permeability ranges widely from $10^{-10}$ to $10^{+7}$ m/d (Smith and Wheatcraft 1993). A value of $K_{sat}$ = 1 m/d is generally considered permeable. Permeability decreases sharply as soils and rocks lose saturation.

The velocity, $V$, of subsurface water flow is equal to specific discharge, $q$, divided by porosity, $n$, for the saturated state or by water content, $\theta$, for the unsaturated state. The velocity is the average velocity of all water particles, each with its own microscopic velocity, passing through the flow cross-sectional area.

In the unsaturated zone, the flow is commonly vertical. The gradient can be nearly unity under the steady state, especially in deep soils of uniform texture and also in stratified zones (Yeh 1989). The governing equation for flow is Richards equation (Richards 1931), which incorporates mass conservation, equation (2.7), and Darcy's Law. The equation is difficult to solve analytically for most problems.

## Infiltration and Redistribution of Water

Infiltration is the process by which water enters the ground surface. The infiltration rate depends on soil texture and compaction, initial soil water content, vegetal cover, and rate of water application. Water intake at the beginning can exceed the water application rate until ponding occurs on the ground surface. Infiltration capacity denotes the maximum rate that occurs under ponding conditions. The infiltrometer test is conducted in the field under a ponding condition (see Appendix 5.1). The measured value is called infiltration capacity, which typically decreases gradually with time and approaches a constant known as steady-state infiltration capacity. The test period is typically 1 hour. Soil moisture is redistributed as the infiltrated water advances as a wetting front (Figure 5.7). The front tends to advance more uniformly in coarse-textured than in finer-grained soils.

Two different analyses of the downward-moving fronts are commonly used: Green and Ampt (1911) and Philip (1957). The former assumes a ponding condition and a uniformly advancing wetting front, which is subjected

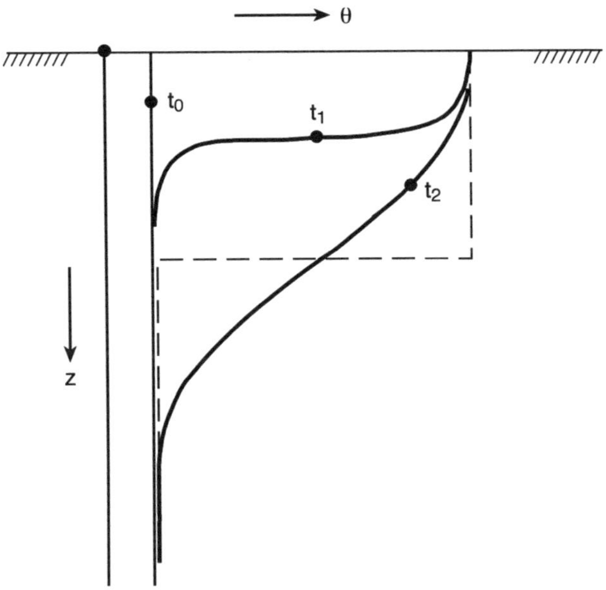


Figure 5.7. Distribution of infiltration water content with time and depth. Dashed line is for Green-Ampt theory. θ, soil water content at different depths *z* and progressive times *t*.

to a soil suction head, $S_f$. The solution is stated as follows (Chow et al. 1988):

$$q(t) = K\left[1 + \frac{(n - \theta_i)S_f}{F(t)}\right] \tag{5.3}$$

$$F(t) = Kt + S_f(n - \theta_i)\ \ln\left[1 + \frac{F(t)}{(n - \theta_i)S_f}\right] \tag{5.4}$$

where $q(t)$ is the infiltration rate, $\theta_i$ is the initial soil water content, and $F(t)$ is the cumulative infiltration volume. Typical values for the involved parameters are available for all soil textures (Rawls et al. 1993). Equation (5.4), being an implicit function of $F(t)$, is cumbersome to use. Explicit expressions have been presented by Fok (1975) and later by Serrano (2001). Besides these two, other infiltration equations also appear in the literature (Green et al. 1982; Rawls et al. 1993).

Philip's work is for the general case, yielding an analytical solution to Richards equation as follows:

$$q(t) = \frac{1}{2}St^{-1/2} + K \tag{5.5}$$

$$F(t) = St^{1/2} + Kt \tag{5.6}$$

where $S$ is sorptivity, which has a dimension of $L/T^{1/2}$. Sorptivity governs early-time infiltration and varies with soil structure, pore-size distribution, and antecedent water content. It can be measured in the field (Chong et al. 1981) or computed with known values of $S_f$, $K$, and $n$ (Rawls et al. 1993).

The Talsma-Parlange equation, (5.7), is similar to the Philip equation but is more accurate for extended times (Green et al. 1982).

$$F(t) = St^{1/2} + \frac{1}{3}K_s t + \frac{1}{9}\frac{K_s^2}{S}t^{3/2} \tag{5.7}$$

Estimating infiltration for a large area is complicated by the heterogeneity of soil and vegetal cover. For an area in which the soil and vegetation are reasonably homogeneous, a common approach is the use of the U.S. Soil Conservation Service runoff curve number method (Chow et al. 1988). Infiltration and initial hydrologic abstractions, including surface depression storage and vegetal interception for a rainfall event, are accounted for by the use of a parameter called runoff curve number. This parameter is used in soil groupings by hydrologic traits as A, B, C, and D, with A denoting low runoff and D high runoff (see Table 5.2). The method is a mix of analytical and empirical approaches (see Appendix 5.2). Originated in 1957 for small agricultural and rural watersheds, this method has been adapted for small urban watersheds and extended for the determination of flood hydrograph and flood peak discharge (see Chapter 7). The method is considered acceptable for the stated uses (Ponce and Hawkins 1996), but its basic assumption is questioned for lack of theoretical justification (Pilgrim and Cordery 1993).

Table 5.2. Runoff Curve Numbers for Selected Agricultural, Suburban, and Urban Land Uses

| Land-Use Description | | Hydrologic Soil Group A | B | C | D |
|---|---|---|---|---|---|
| Cultivated land[a] | | | | | |
| Without conservation treatment | | 72 | 81 | 88 | 91 |
| With conservation treatment | | 62 | 71 | 78 | 81 |
| Pasture or range land | | | | | |
| Poor condition | | 68 | 79 | 86 | 89 |
| Good condition | | 39 | 61 | 74 | 80 |
| Meadow | | | | | |
| Good condition | | 30 | 58 | 71 | 78 |
| Wood or forest land | | | | | |
| Thin stand, poor cover, no mulch | | 45 | 66 | 77 | 83 |
| Good cover[b] | | 25 | 55 | 70 | 77 |
| Open spaces, lawns, parks, golf courses, cemeteries, etc. | | | | | |
| Good condition: grass cover on 75% or more of the area | | 39 | 61 | 74 | 80 |
| Fair condition: grass cover on 50 to 75% of the area | | 49 | 69 | 79 | 84 |
| Commercial and business areas (85% impervious) | | 89 | 92 | 94 | 95 |
| Industrial districts (72% impervious) | | 81 | 88 | 91 | 93 |
| Residential[c] | | | | | |
| Average lot size | Average % impervious[d] | | | | |
| 1/8 acre or less | 65 | 77 | 85 | 90 | 92 |
| 1/4 acre | 38 | 61 | 75 | 83 | 87 |
| 1/3 acre | 30 | 57 | 72 | 81 | 86 |
| 1/2 acre | 25 | 54 | 70 | 80 | 85 |
| 1 acre | 20 | 51 | 68 | 79 | 84 |
| Paved parking lots, roofs, driveways, etc.[e] | | 98 | 98 | 98 | 98 |
| Streets and roads | | | | | |
| Paved with curbs and storm sewers | | 98 | 98 | 98 | 98 |
| Gravel | | 76 | 85 | 89 | 91 |
| Dirt | | 72 | 82 | 87 | 89 |

SOURCE: Chow et al. 1988.

[a]For a more detailed description of agricultural land-use curve numbers, refer to U.S. Soil Conservation Service 1969.

[b]Good cover is protected from grazing and has litter- and brush-covered soil.

[c]Curve numbers are computed assuming that the runoff from the house and driveway is directed toward the street with a minimum of roof water directed to lawns where additional infiltration could occur. 1 acre = 0.4 ha.

[d]The remaining pervious areas (lawn) are considered to be in good pasture condition for these curve numbers.

[e]In some warmer climates of the country a curve number of 95 may be used.

## Transport

Vegetated soil systems behave as natural living filters separating water from many of the solutes. They especially excel in the humid tropics. Transport phenomena in the unsaturated state are complicated. The pore structure, the flow parameters, and the flow are all functions of the soil water content. Transport of nonaqueous-phase liquids and highly volatile organics are multiphase: water, nonaqueous liquid, and vapor. The basic physicochemical properties of the chemicals need to be known for particular soils. Because some chemicals are suspected to impact human health at trace concentrations, rigor is required in analysis.

The simplest transport model for the unsaturated zone consists of a conservative solute involving advection and dispersion with the following mass balance equation:

$$\frac{\partial}{\partial z}\left(\theta D_z \frac{\partial C}{\partial z} - qC\right) = \frac{\partial (C\theta)}{\partial t} \tag{5.8}$$

where $C$ is the concentration, $D$ is the coefficient of dispersion, which has a dimension of $L^2/T$, and $q$ is the specific discharge according to Darcy's Law,

$$q = K(\theta)J_z = -K(\theta)\frac{\partial}{\partial z}\left(\frac{-p}{\rho g} + z\right) \tag{5.9}$$

In equation (5.8) the first term manifests dispersion and the second term accounts for advection with $q = V\theta$.

Advection is induced by velocity. The solute molecules move with microscopic velocities in the indescribably complex porous network. They split up and rejoin at junctions, move fast in narrow passages and midstream locations, and slow down in wide channels and near the solid surface. The variations of microscopic velocity result in nonuniform arrival at a downgradient destination.

If the release is a single dose, the arrival pattern, known as the breakthrough curve, appears in a bell shape. If the release is continuous, the breakthrough curve will occur as a sigmoidal, or S, curve. If there is no dispersion, all molecules will travel with the average velocity, $V = q/\theta$, and arrive at the same time as a step in the breakthrough curve (see Figure 5.8). Although longitudinal transport takes place in the general direction of flow, transverse or lateral transport occurs simultaneously but with a lesser spread.

For all practical purposes, coefficient of dispersion $D$ is equal to $\alpha V$, where $\alpha$ is dispersivity, which is controlled solely by the porous medium itself for either the saturated or the unsaturated state (Bear 1972; Biggar and Nielsen 1976). In a homogeneous material such as uniform-sized sand, the value of $\alpha$ is on the order of the size of the sand grain. However, for on-site heterogeneous media, a scale factor takes effect. Dispersivity far exceeds laboratory-based values (0.5 cm or less). For field soils, it is about 10 cm or larger (Nielsen et al. 1986). This is generally attributed to macrodispersion, a process effective in field soils. For rock aquifers, the value of $\alpha$ ranges from a few centimeters to tens of meters and increases with distance traveled (Gelhar 1986).

Many attenuative processes occur to reduce the contaminant concentration. In natural filtration some sorptive and volatile chemicals are reduced effectively in a relatively thin (1 m [3 ft]) mantle of vegetated soils. Beneath it, reduction decreases sharply because some of the processes (for example, sunlight and plant uptake) virtually cease to exist.

Adsorption is attachment of the solute to solid surfaces. It is especially effective in subsurface transport because of long residence times, large contact surfaces, and presence of sorptive organic matter. Partitioning a solute between dissolved and sorbed phases is described by isotherms. The Freundlich isotherm, $C_s = K_d C$, is commonly used, where $C_s$ is the solute concentration sorbed on the solid in milligrams per kilogram of soil and $K_d$ is the sorption coefficient in liters per kilogram. $K_d$ is soil and chemical specific and is the product of the organic carbon partitioning coefficient and the fraction of organic carbon in the soil. The former parameter depends on chemical features only and not on the soil. In some cases sorption is virtually irreversible.

Volatilization changes a solute from liquid or solid to the gaseous phase. Many chemicals are not especially

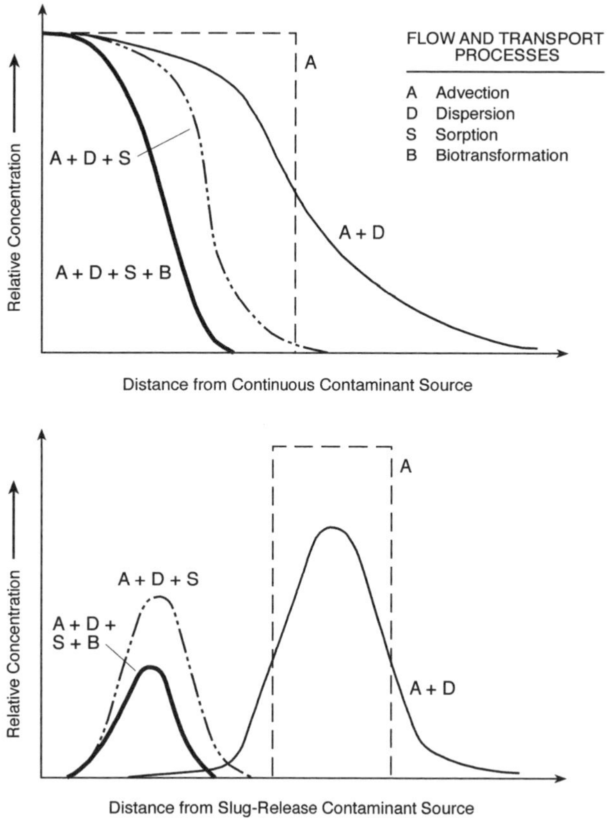


Figure 5.8. Influence of natural processes on levels of contaminant downgradient from continuous (*top*) and slug-release (*bottom*) sources. (Adapted from Barcelona et al. 1987)

volatile and do not pose an environmental concern. Atrazine and diuron, common agricultural chemicals, are poorly volatile, whereas 1,2-dibromo-3-chloropropane (DBCP), ethylene dibromide (EDB), and 1,2,3-trichloropropane (TCP), the trace organics discovered in some potable groundwater sources, are highly volatile. If only water and air are involved, volatilization is measured by Henry's constant, which is defined as vapor pressure in atmospheres divided by water solubility in mol/m$^3$ of the chemical, yielding, for example, $2.59 \times 10^{-13}$ atm-m$^3$/mol for atrazine and $9.10 \times 10^{-3}$ atm-m$^3$/mol for trichloroethylene (TCE). However, in a phase system containing soil, water, air, and chemicals, volatilization can also originate from the solid and the chemical itself. Further, volatilization interacts with sorption because stronger adsorption reduces volatility.

Both adsorption and volatilization retard the transport of solutes in the water. The total effect is evaluated by the retardation factor (Rao et al. 1985):

$$RF = 1 + \frac{\rho_b K_d}{\theta_{FC}} + \frac{n_a K_h}{\theta_{FC}} \tag{5.10}$$

where $RF$ is the retardation factor, $K_d$ is the sorption coefficient, $\theta_{FC}$ is the field capacity, $n_a$ is the soil air-filled porosity, $n$ is the porosity, $n_a = n - \theta_{FC}$, and $K_h$ is Henry's constant. For non- or low-volatile chemicals, the last term in equation (5.10) drops out. To account for adsorption and volatilization, $RF$ is multiplied to the right side of equation (5.8). $RF = 1$ signifies no retardation.

Transformation breaks down a chemical by various processes such as biodegradation by microbes, natural decay, hydrolysis (breaking down the bonds in the presence of water), and photolysis by light. Assessment of transformed chemicals in terms of their toxicity, persistence, and mobility is important.

Biodegradation of organic chemicals depends on many factors, including microorganisms, nutrients for microbial metabolism, water content and temperature of the soil, soil pH, and water solubility of the chemical. Surface soils, especially those in the humid tropics, provide a favorable environment for the first three factors and thus for active biodegradation. Biodegradation is commonly assumed to be a first-order rate reaction: the rate of loss of a chemical is proportional to the chemical concentration present, thus yielding the first-order rate constant, $k$, or half-life, $t_{1/2}$. They are convertible by the relation:

$$t_{1/2} = \frac{\ln 2}{k} = \frac{0.693}{k} \tag{5.11}$$

Chemical half-lives range from a few days (TCE, 4 days) to several thousand days (dichlorodiphenyltrichloroethane [DDT], 3,840 days). For some organic chemicals, a hyperbolic rate (Monod kinetics) may fit the transformation better than the first-order rate. Radionuclide decay perfectly follows the first-order rate reaction. Tritium has a short half-life (12.7 years), whereas that of radiocarbon is long (5,730 years).

To account for transformation, $-kC$ is added to the left side of equation (5.8). Hydrolysis is an exceedingly slow and ineffective process for breaking down an organic chemical. It is usually not explicitly considered in transport in surface soils.

## Natural Ground Surfaces in Hawai'i

### Soils and Vegetation

Surficial rocks that are differentiated by volcanic series and sediment types have been mapped as a part of the fundamental works on the geology and groundwater resources of the Hawaiian Islands: O'ahu (Stearns and Vaksvik 1935), Lāna'i (Stearns 1940), Maui (Stearns and Macdonald 1942), Hawai'i (Stearns and Macdonald 1946), Moloka'i (Stearns and Macdonald 1947), Ni'ihau (Stearns 1947), and Kaua'i (Macdonald et al. 1960). Geological maps also indicate geological features such as dikes and faults and hydrologic features such as streams and springs. The hydrologic characters of the rock types are discussed in Chapter 1. Simplified geological maps are presented in Figure 1.6.

#### Soils

Soils differ from one another for several reasons: length of time of weathering, type of parent materials, slope and drainage conditions, addition of organic debris, and local rainfall and temperature. Remarkably diverse soils are present in the relatively small landmass in Hawai'i. Most of the soil orders, which are the highest category in soil taxonomy based on soil-forming processes, are found in Hawai'i.

The soils of Hawai'i have been classified and mapped twice: the first time by Cline in 1955 and the second time by Foote et al. in 1972 and Sato et al. in 1973 (Uehara 1973; McCall 1973). Reclassification is in progress (2005), especially for the island of Hawai'i and for East Maui (Plate 5.1).

Limited hydrologic information is provided for successive soil taxonomy categories: order, suborder, great group, subgroup, family, and series. The soil series provides the most detailed description about the soil. A sample map by soil series is given in Figure 5.9. Information on soil texture, depth to rock, soil horizons, seasonal high water table, shrink-swell potentials, pH value, and other parameters is reported in a soil survey.

Hydrologic criteria are not the determining factor in soil taxonomy. Estimates of permeability and available water are provided for each soil series, but they are not the principal criteria used in delineating soils into different series (Green et al. 1982). However, soil series are assigned a runoff curve number and classified into hydrologic soil groups for the purpose of evaluating surface-water drainage. The current Hawai'i soil classification by hydrologic soil groups is listed in Table 5.3 (U.S. Natural Resources Conservation Service 1993).

Oxisols and Entisols are of special hydrologic importance in Hawai'i. Oxisols occur in old, gently sloping landscapes. They are strongly weathered, have thick horizons up to 2 m (79 in.), and are composed primarily of kaolinite and oxides of iron and aluminum. They have stable structural aggregates, high permeability, and high resistance to erosion but low natural fertility. They require proper management to serve as productive soils. Oxisols are the principal agricultural soils in Hawai'i.

Much of the mountain slopes in the high-rainfall areas on the geologically older islands (Kaua'i, O'ahu, Moloka'i, and Maui) were classified formerly as miscellaneous land types that include rock and stony lands. Many of these areas are prime groundwater recharge sites. Their permeability and related hydrologic parameters are insufficiently quantified, however. They are currently classified as Entisols (Gavenda et al. 1998), which are very young soils with only A- and C-horizons. Mountain slopes are too steep to develop typical soil horizons.

Andisols are volcanic ash soils that have high phosphorus uptake. Andisols do not appear in the early Hawai'i soil maps. In the latest map, Andisols, Inceptisols, and

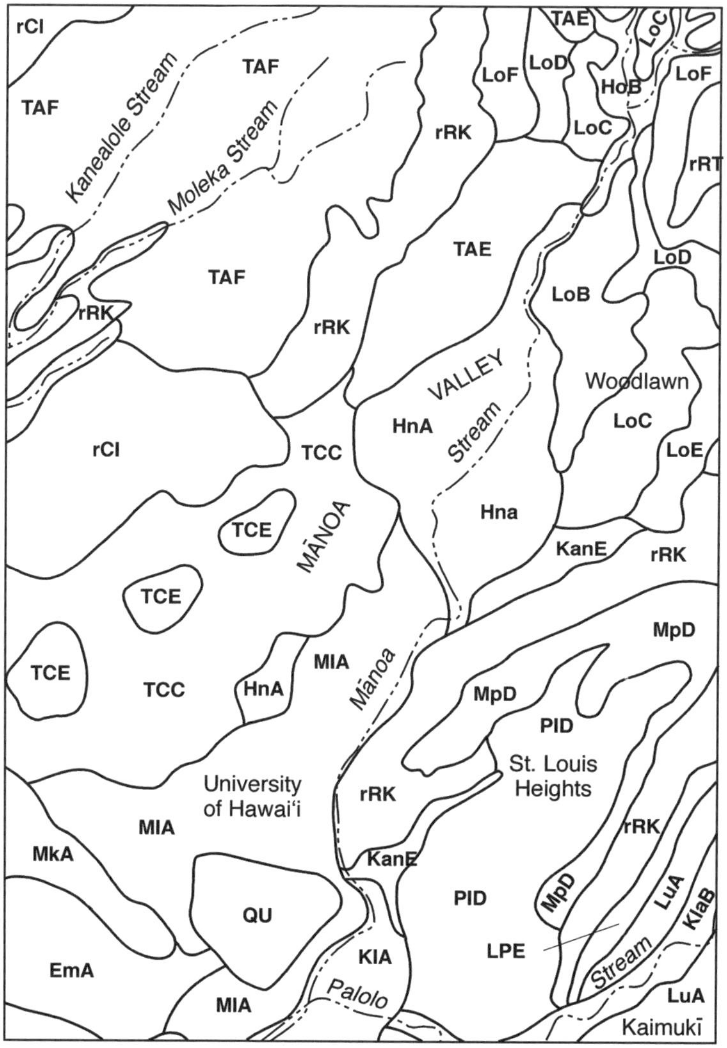


| | |
|---|---|
| **EmA** | Ewa silty clay loam, 0 to 2 percent slope |
| **HnA** | Hanalei silty clay, 0 to 2 percent slope |
| **HoB** | Hanalei stony silty clay, 2 to 6 percent slope |
| **KanE** | Kaena very stony clay, 10 to 35 percent slope |
| **KlA** | Kawaihapai clay loam, 0 to 2 percent slope |
| **KlaB** | Kawaihapai stony clay loam, 2 to 6 percent slope |
| **LoB** | Lolekaa silty clay, 3 to 8 percent slope |
| **LoC** | Lolekaa silty clay, 8 to 15 percent slope |
| **LoD** | Lolekaa silty clay, 15 to 25 percent slope |
| **LoE** | Lolekaa silty clay, 25 to 40 percent slope |
| **LoF** | Lolekaa silty clay, 40 to 70 percent slope |
| **LPE** | Lualualei extremely stony clay, 3 to 35 percent slope |
| **LuA** | Lualualei clay, 0 to 2 percent slope |
| **MkA** | Makiki clay loam, 0 to 2 percent slope |
| **MlA** | Makiki stony clay loam, 0 to 3 percent slope |
| **MpD** | Manana silty clay, 15 to 25 percent slope |
| **rCl** | Cinderland |
| **rRK** | Rock land |
| **PlD** | Pooku silty clay loam, 8 to 25 percent slope |
| **QU** | Quarry |
| **TAE** | Tantalus silt loam, 15 to 40 percent slope |
| **TAF** | Tantalus silt loam, 40 to 70 percent slope |
| **rRT** | Rough mountainous land |
| **TCC** | Tantalus silty clay loam, 8 to 15 percent slope |
| **TCE** | Tantalus silty clay loam, 15 to 40 percent slope |

Figure 5.9. Soils map for Mānoa and Pālolo areas, island of Oʻahu, Hawaiʻi. (Adapted from Foote et al. 1972)

Table 5.3. Hydrologic Classification of Hawai'i Soil Series

| Soil Group | Hydrologic Classification |
|---|---|
| ʻĀinakea | B |
| Akāka | A |
| ʻAlae | A |
| ʻAlaeloa | B |
| Alakaʻi | D |
| Alapaʻi | A |
| Amalu | D |
| Apakuie | A |
| Badlands | C |
| Beaches | A |
| Blown-out land | B |
| Cinder land | A |
| Colluvial land | B |
| Coral outcrop | D |
| Dune land | A |
| Dystrandepts | B |
| ʻEwa | B |
| Fill land | C |
| Fill land, mixed | C |
| Gullied land | C |
| Haʻikū | B |
| Hālawa | B |
| Haleʻiwa | B |
| Hāliʻi | B |
| Hāliʻimaile | B |
| Hāmākuapoko | B |
| Hāna | A |
| Hanalei | C |
| Hanamāʻulu | B |
| Hanipoe (HCD, HDD) | B |
| Hanipoe (HFD) | C |
| Hāwī | B |
| Heake | D |
| Helemano | B |
| Hīhīmanu | B |
| Hīlea | D |
| Hilo | A |
| Holomua | B |
| Hōnaunau | C |
| Honokaʻa | A |
| Honolua | B |

Table 5.3. *(cont.)*

| Soil Group | Hydrologic Classification |
|---|---|
| Honomanū | A |
| Honouliuli | D |
| Honuaulu | A-1 |
| Hoʻolehua | B |
| Huikau | A |
| Hulua | D |
| Hydrandept (rHP) | A |
| Hydrandept (rHT) | B |
| ʻĪao | B |
| ʻIo | B |
| Ioleau | C |
| Jaucas | A-1 |
| Kaʻaluʻalu | A |
| Kaʻena | D |
| Kaʻena variant | D |
| Kahaluʻu | D |
| Kahana | B |
| Kahanui | C |
| Kahua | D |
| Kailua | A |
| Kaimū | A-1 |
| Kainaliu | A |
| Kaipoioi | B |
| Kaiwiki | A |
| Kalaʻe | B |
| Kalapa | B |
| Kalaupapa | D |
| Kalihi | D |
| Kaloko | D |
| Kaloko variant | D |
| Kamakoa | A |
| Kamāʻoa | B |
| Kamaʻole | B |
| Kāneʻohe | B |
| Kānepuʻu | B |
| Kapaʻa | A |
| Kapāpala (KLC, KLD) | B |
| Kapāpala (KMD) | C |
| Kapuhikani | D |
| Kaupō | A |
| Kawaihae | C |

Table 5.3. *(cont.)*

| Soil Group | Hydrologic Classification |
|---|---|
| Kawaihāpai | B |
| Kea'au | D |
| Keāhua | B |
| Kealakekua | C |
| Keālia | D |
| Keaukaha | D |
| Keawakapu | B |
| Ke'ei | D |
| Ke'eke'e | B |
| Kehena | C |
| Kekaha | B |
| Kekake | D |
| Kemo'o | B |
| Kīkoni | B |
| Kīlauea | B |
| Kīloa | A-1 |
| Kilohana | A |
| Kō'ele | B |
| Kohala | B |
| Kōke'e | B |
| Koko | B |
| Kokokahi | D |
| Kolekole | C |
| Kōloa | C |
| Kolokolo | B |
| Kona | D |
| Ko'olau | C |
| Kūka'iau | A |
| Kūka'iau (KWD) | C |
| Kūka'iau | A |
| Kula | B |
| Kunia | B |
| Kumuweia | B |
| Lahaina | B |
| Lāla'au | A-1 |
| Laumai'a | B |
| Lava flows, *'a'ā* | A |
| Lava flows, *pāhoehoe* | D |
| Lāwa'i | B |
| Leilehua | B |
| Līhu'e | B |

Table 5.3. *(cont.)*

| Soil Group | Hydrologic Classification |
|---|---|
| Lithic tropofolists | D |
| Lolekaa | B |
| Lualualei | D |
| Mahana | B |
| Māhukona | B |
| Maile | A |
| Maka'alae | B |
| Makalapa | D |
| Makapili | A |
| Makawao | B |
| Makaweli | B |
| Mākena | B |
| Makiki | B |
| Māla | B |
| Malama | A-1 |
| Mamala | D |
| Manahaa | C |
| Mānana | C |
| Manu | C |
| Marsh | D |
| Māwae | A-1 |
| Mixed alluvial land | B |
| Moa'ula | A |
| Mokulē'ia variant | B |
| Mokulē'ia | B |
| Moloka'i variant | B |
| Moloka'i | B |
| Nā'ālehu | B |
| Nā'ālehu (NhD) | C |
| Nā'iwa | B |
| Niu | B |
| Niuli'i | C |
| Niuli'i variant | C |
| Nohili | D |
| Nonopahu | D |
| Oanapuka | B |
| 'O'ōkala | A |
| 'Opihikao | D |
| Paaiki | B |
| Paaloa | B |
| Pā'auhau | A |

Table 5.3. *(cont.)*

| Soil Group | Hydrologic Classification |
|---|---|
| Pāʻia | C |
| Pākalā | B |
| Pākini | B |
| Palapalai | B |
| Pāmoa | C |
| Panaʻewa | D |
| Pane | B |
| Pāpaʻa | D |
| Pāpaʻi | A-1 |
| Paumalū | B |
| Paʻuwela | B |
| Pearl Harbor | D |
| Piʻihonua | A |
| Pōhākupu | A |
| Poʻokū | A |
| Puaulu | A |
| Puhi | A |
| Puhimau | D |
| Pūlehu | B |
| Puna | A-1 |
| Punaluʻu | D |
| Pūnohu | A |
| Puʻukala | D |
| Puʻuone | C |
| Puʻu ʻŌʻō | A |
| Puʻu ʻŌpae | B |
| Puʻu Pā | B |
| River wash | A |
| Rock land | D |
| Rock outcrop | D |
| Rough broken and stony land | C |
| Rough broken land | C |
| Rough mountainous land | D |
| Rubble land | A |
| Sandy alluvial land | A |
| Stony blown-out land | B |
| Stony colluvial land | B |
| Stony land (rST) | B |
| Stony land (MXC) | C |
| Stony steepland | B |
| Tantalus | A |
| Tropaquods | D |
| Tropaquepts | D |
| Tropofolist | A |
| Tropohumults | C |
| ʻUlupalakua | B |
| Uma | A |
| ʻUmikoa | B |
| ʻUwala | B |
| Very stony land | C |
| Very stony land (Maui) | C |
| Very stony land, eroded | C |
| Wahiawā | B |
| Wahikuli | B |
| Waiʻaha | D |
| Waiakoa | C |
| Waiʻaleʻale | D |
| Waialua | B |
| Waiawa | D |
| Waihuna | D |
| Waikaloa | B |
| Waikāne | B |
| Waikapū | B |
| Waikomo | D |
| Wailuku | B |
| Waimea | B |
| Waineʻe | B |
| Waipahu | C |

SOURCE: U.S. Natural Resources Conservation Service 1993.

several combination orders, namely Andisols-Inceptisols, Histosols-lava, and Histosols-lava-Andisols, are prevalent in the relatively high-rainfall areas on the island of Hawai'i and in East Maui. Histosols are thin, highly organic soils that are formed from decomposed forest litter on young lava flows. These soils are well drained and occur in moderate rain forests. Inceptisols from volcanic ash are young soils with a thin mantle and weakly developed horizons on sloping surfaces. They serve commonly as pastureland.

Ultisols, an association of Oxisols, occur on steep slopes in humid regions and thus are highly subject to erosion. They support timber and pasture. Mollisols are dark-colored, well-structured, and well-drained young soils; they are rich in nutrients and support excellent grassland in subhumid regions. Vertisols are dark-colored clayey soils that swell when wet and crack when dry. They occur mostly in the drier areas on talus and the floors of valleys on O'ahu. Aridisols are desert soils found only in the arid lee areas of the island of Hawai'i. Spodosols and Podzols are infertile forest soils restricted to the wettest mountain on Kaua'i in Alaka'i Swamp. Alfisols occur in subhumid regions, where they support forests.

Soil data given in the Hawai'i Soil Survey have been used for predicting the mobility of contaminants to leach below the root zone in the watershed overlying the Pearl Harbor aquifer, O'ahu (Loague et al. 1989). The results indicate that soil information from the lowest taxonomic category should be employed in the computation of the retardation factor, despite the fact that considerable uncertainty exists in the predicted retardation factor that is used to screen and rank chemicals.

### Relief and Vegetal Cover

The relief of the Hawaiian Islands is recorded in the U.S. Geological Survey base quadrangles (1:24,000). A set of maps indicating major topographic features and streams of the major Hawaiian Islands is provided in Figure 5.10.

Vegetation in Hawai'i is highly varied and dependent on rainfall and elevation. Vegetation is delineated by zones (Ripperton and Hosaka 1942). The zones include *kiawe* and lowland shrubs, guava forest, *'ōhi'a lehua* forest, *koa* forest, and alpine stony desert (see Plate 5.2) (Lamoureux 1973). A different zonal delineation of the terrestrial ecosystems is also available (Pratt and Gon 1998). Some hydrologic impacts of the vegetal cover in Hawai'i have been identified and quantified: interception of fog by Norfolk pines and other trees (Chapter 3) enhances infiltration, surface runoff is reduced in pineapple and sugarcane fields, and contaminants are diminished by reaction with vegetation.

## Hydrologic Characteristics of Surface Soils

### Porosity and Water-Retentive Properties

The values of porosity and water-retentive properties are high for virtually all the great soil groups of Hawai'i soils (Yamamoto 1963). Total porosity in Hawai'i soils ranges from 68 to 74 percent and macroporosity from 10 to 18 percent. Field capacity is within a narrow range of 56 to 58 percent, wilting point from 28 to 38 percent, and available water from 19 to 28 percent. These values when classified by soil texture exceed those for all reference soils of the continental United States (Rawls et al. 1993). The differences may be attributed to the strongly aggregated soil structure and the typically nonswelling clay minerals of the Hawai'i soils. These reference soils are those in a compilation, by the U.S. Department of Agriculture, of published soil water characteristic data for some 1,200 mostly agricultural soils from thirty-four states from coast to coast but not including Hawai'i (Rawls et al. 1983).

The effects of land use in Hawai'i are clearly reflected by those properties. In Yamamoto's work, forest soils exhibit the highest value (28 percent) of available water retention, and the lowest value (19 percent) is associated with pastureland. Similar results are reported for field pits on O'ahu (Wood 1971) and on Hawai'i, Maui, and Kaua'i (Wood 1977). Forest soils have the highest porosity (66 to 90 percent), and soils in agricultural land used for sugarcane, pineapple, and pasture have lower porosity (ranging from a few percent to 8 percent).

The values of the Hawai'i soils (undisturbed cores up to 30.5 cm [12 in.] deep) were determined for twenty-five soils series found in one or several of the four major land-use areas—forest, pasture, agriculture, and idle grassland—on the islands of O'ahu, Kaua'i, and Hawai'i. Over one-half to two-thirds of these soils have clay texture. Silty clay, silty clay loam, and clay loam make up the remaining one-third to one-fifth. These clays are predominantly kaolinitic. Yamamoto assigned 0.06 bar to define field capacity. Green and Guernsey (1981) used both 0.06 bar and 0.15 bar. As a general rule, the value of 0.1 bar may be used for Hawai'i soils in comparison with 0.33 bar for reference soils. The data on available water provided by the U.S. Soil Survey are based on a combination of laboratory tests, soil properties, and field experience.

The saprolites on O'ahu have similar magnitudes as determined from core samples. At Schofield Barracks the porosity range is 45 to 66 percent at depths down to 22 m (73 ft) (Harding Lawson Associates 1996). At Mililani the range is 46 to 72 percent at depths down to 30 m (98 ft) (Miller et al. 1988).

## Permeability

The values of saturated hydraulic conductivity, $K_s$, in Hawai'i soils are typically a few meters per day. However, they are about three orders of magnitude smaller than that for unweathered basalts, the parent rock. By field measurements, $K_s$ for the Lahaina series, Moloka'i series, and Wahiawā series of the Oxisol order decreases with depth: 0.9 to 2.7 m/d (3 to 9 ft/d) in the top 30 cm (12 in.) and 0.6 to 1.2 m/d (2 to 4 ft/d) at depths of 30 to 60 cm (12 to 24 in.) (Green et al. 1982). For the Tantalus silty clay loam series of the Inceptisol order, $K_s$ is about 0.3 m/d (1 ft/d) as determined in a large field plot at Lyon Arboretum on O'ahu (Ahuja and El-Swaify 1979). A value of 1.7 m/d (5.5 ft/d) was measured at the same location with multiple tensiometers and infiltrometers (Ahuja et al. 1976b).

The permeability of Hawai'i soils is sharply reduced at low soil water content, as expected (Ahuja and El-Swaify 1979; Chong et al. 1981). The values for the Moloka'i silty clay series are typical of this behavior (Figure 5.11). The break in the slope at 22-cm negative pressure reflects the effect of macroporosity.

A total range, 0.04 to 12.19 m/d (0.12 to 40 ft/d), is used for reporting soil permeability in the soil survey. This range is based on soil structure, porosity, and limited tests conducted on undisturbed cores. Suitability of soils for septic-tank filter fields in terms of permeability and ground slopes is also discussed in the soil survey.

Reference values of hydraulic conductivity can be computed on the basis of soil texture and percentage saturation (Rawls et al. 1993); however, such computations have doubtful validity for Hawaiian soils. For example, the Wahiawā series, whose texture is 46 percent clay, 27 percent sand, and 27 percent silt (Yamamoto 1963), would have a computed $K_s$ value of 0.01 m/d (0.05 ft/d), which is about two orders of magnitude too small in comparison with on-site measurements. Abundance of clay-sized particles alone can lead to incorrect deductions. Hawaiian clay-texture soils with strong aggregate structure, such as the Wahiawā series, are in fact as permeable as sand.

Saprolite formations are known to be poorly permeable, but their hydraulic behavior can exceed the feeble state, as suggested by the results of laboratory tests on small core samples. A case in point is the infiltration tests that were conducted under ponding conditions in July 1995 in a large basin (20.9 $m^2$ [225 $ft^2$]) excavated in a saprolite formation near a former landfill at Schofield Barracks (Harding Lawson Associates 1996). The wetting front was observed to pass through layers of saprolite and reach a depth of 15.8 m (52 ft) in 24 hours. This apparently fast passage took place despite the low $K_s$ values for the small core samples, as determined in the laboratory, which ranged from $2.1 \times 10^{-2}$ to $1.0 \times 10^{-5}$ m/d ($6.8 \times 10^{-2}$ to $3.4 \times 10^{-5}$ ft/d). The $K_s$ value for the field saprolite layers was subsequently determined to be approximately 8.6 m/d (28 ft/d) for these infiltration events. The rapidity of flow in the tests is attributed to a surface zone of vertical fractures in the saprolite formation, based on simulation modeling. Fractures are required to allow passage of a

A

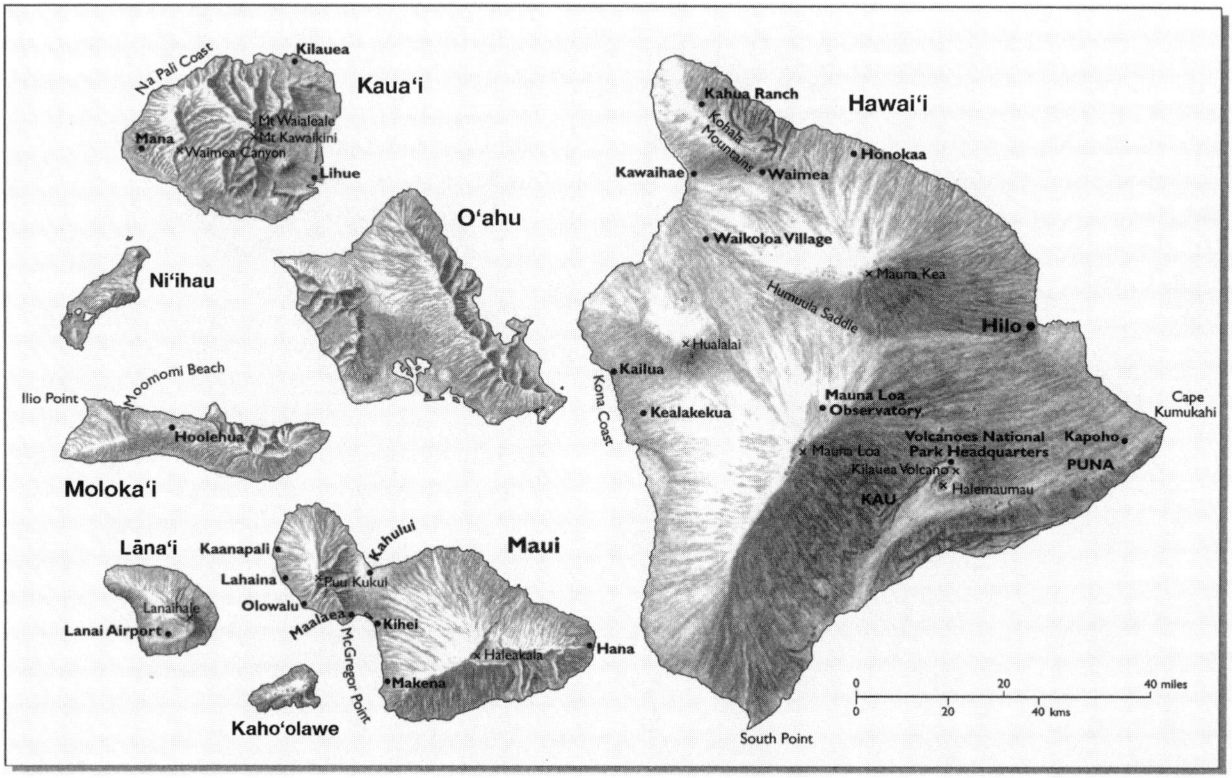


Figure 5.10. Major topographic features, streams, and geographic features of the Hawaiian Islands: *A (above),* Major topograhic features of the Hawaiian Islands (reprinted from Sanderson 1993 with permission from University of Hawai'i Press); *B (facing page),* Streams and other geographic features, island of Kaua'i (adapted from Fontaine et al. 1997); *C (page 112),* Streams and other geographic features, island of O'ahu (reprinted by permission from Oki 1998); *D (page 113, top),* Streams and other geographic features, island of Moloka'i (reprinted by permission from Oki 2000); *E (page 113, bottom),* Streams and other geographic features, island of Maui (adapted from Fontaine et al. 1997); *F (page 114),* Streams and other geographic features, island of Hawai'i (reprinted by permission from Oki 1999).

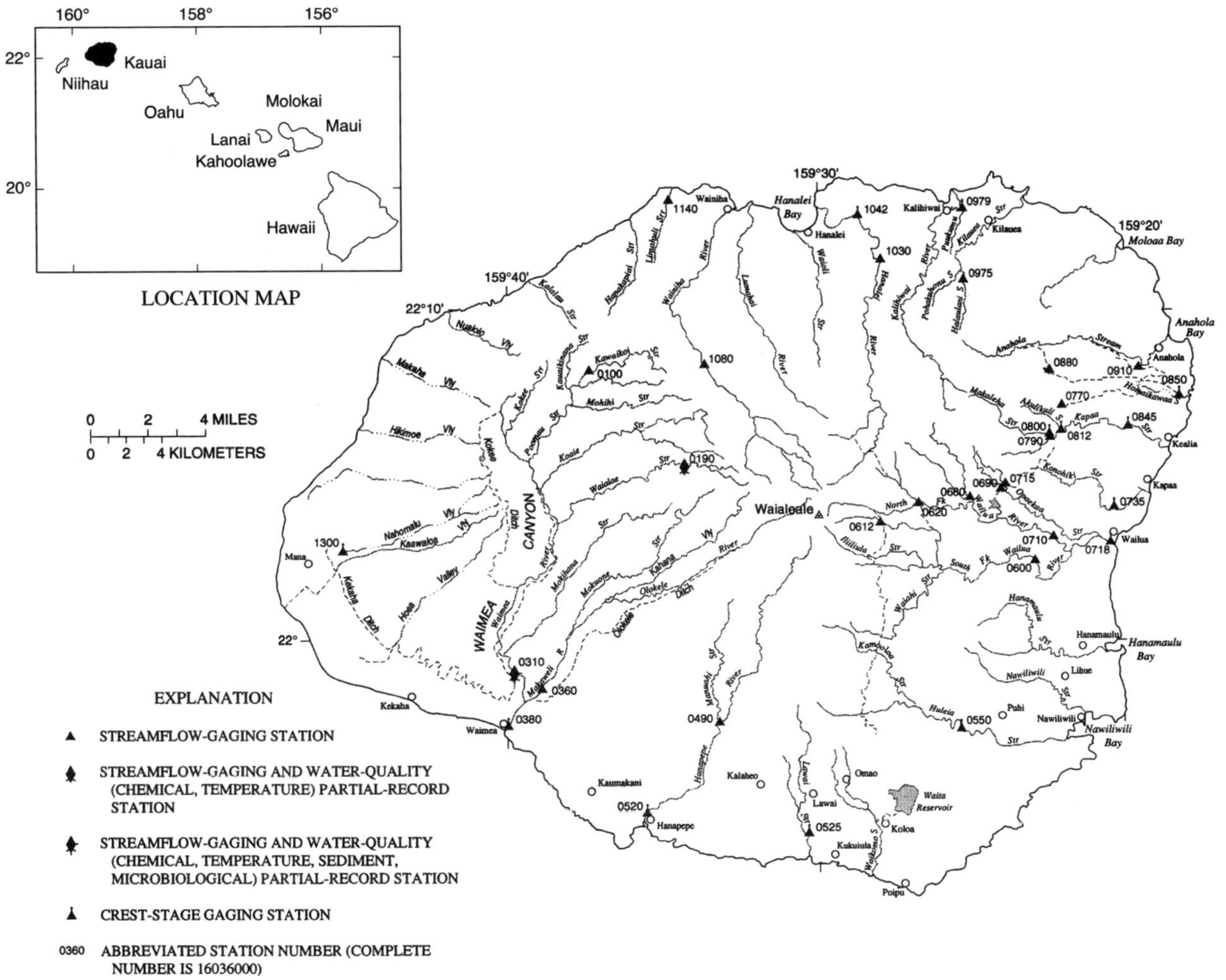
160°
158°
156°
22°
20°
Kauai
Niihau
Oahu
Molokai
Maui
Lanai
Kahoolawe
Hawaii
LOCATION MAP
159°30'
159°40'
159°20'
22°10'
22°
0 2 4 MILES
0 2 4 KILOMETERS
Hanalei Bay
Hanalei
Wainiha
Kalihiwai
Kilauea
Moloaa Bay
Anahola Bay
Anahola
Kealia
Kapaa
Wailua
Waialeale
Hanamaulu
Hanamaulu Bay
Lihue
Puhi
Nawiliwili
Nawiliwili Bay
Waita Reservoir
Koloa
Omao
Lawai
Kukuiula
Poipu
Kalaheo
Hanapepe
Kaumakani
Waimea
Kekaha
Mana
WAIMEA CANYON
1140
1042
1030
0979
0975
1080
0100
0190
0880
0910
0850
0770
0845
0800
0790
0812
0715
0690
0680
0620
0612
0735
0710
0718
0600
1300
0310
0360
0380
0490
0550
0520
0525
EXPLANATION
STREAMFLOW-GAGING STATION
STREAMFLOW-GAGING AND WATER-QUALITY (CHEMICAL, TEMPERATURE) PARTIAL-RECORD STATION
STREAMFLOW-GAGING AND WATER-QUALITY (CHEMICAL, TEMPERATURE, SEDIMENT, MICROBIOLOGICAL) PARTIAL-RECORD STATION
CREST-STAGE GAGING STATION
0360 ABBREVIATED STATION NUMBER (COMPLETE NUMBER IS 16036000)

B

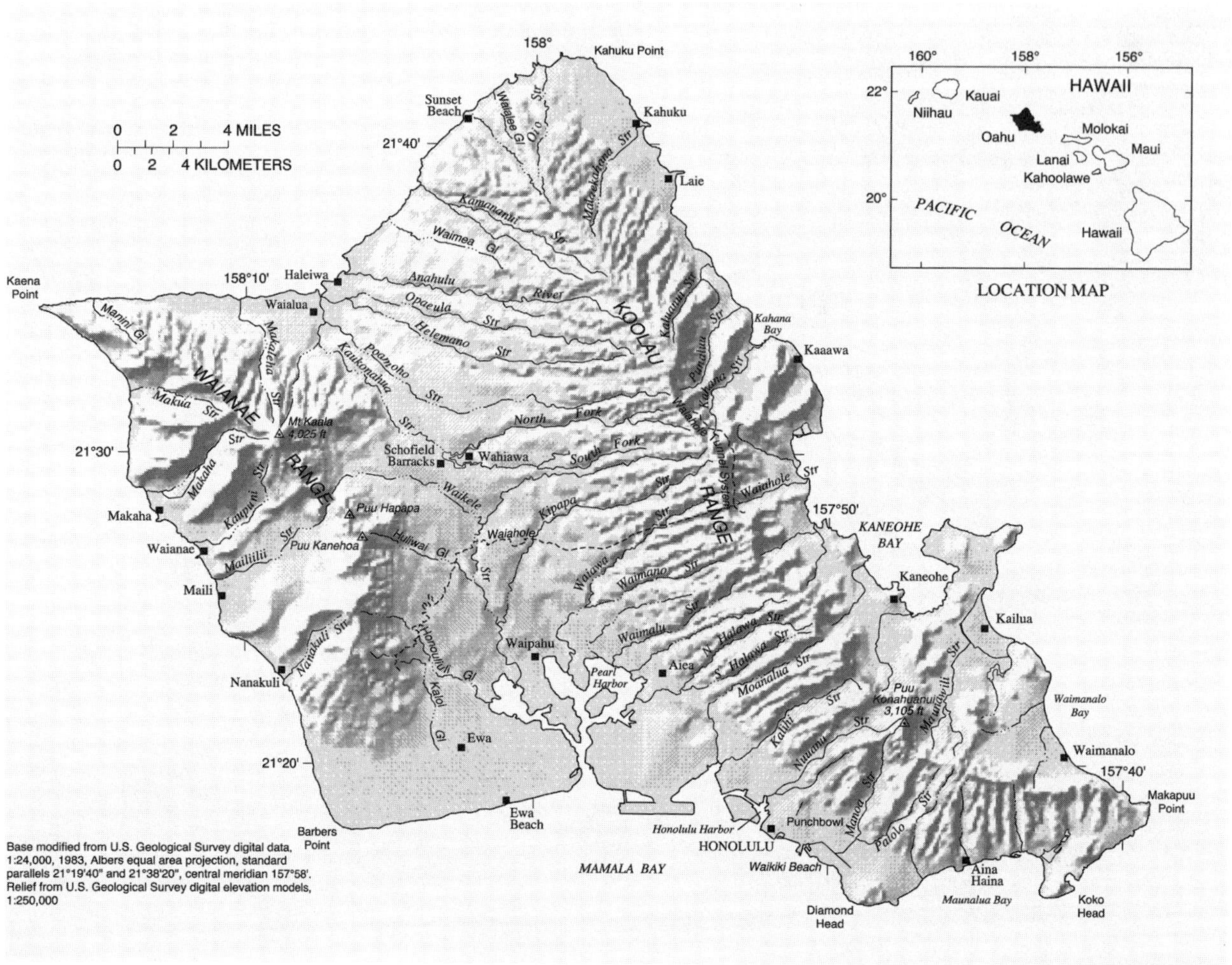

158°
Kahuku Point
Sunset Beach
Kahuku
Laie
0 2 4 MILES
0 2 4 KILOMETERS
21°40'
Waialee Gl
Oio Str
Kamananui Str
Malaekahana Str
Waimea Gl
HAWAII
160°
158°
156°
22°
20°
Kauai
Niihau
Oahu
Molokai
Lanai
Maui
Kahoolawe
PACIFIC OCEAN
Hawaii
LOCATION MAP
Kaena Point
158°10'
Haleiwa
Waialua
Anahulu River
Opaeula Str
Helemano Str
KOOLAU
Kaluanui Str
Punaluu Str
Kahana Bay
Kahana Str
Kaaawa
Manini Gl
Mokuleia Str
Poamoho Str
Kaukonahua Str
WAIANAE
Makua Str
Mt Kaala 4,025 ft
North Fork
South Fork
21°30'
Makaha Str
Kaupuni Str
RANGE
Schofield Barracks
Wahiawa
Waikele
Waiahole
Waiahole Str
157°50'
KANEOHE BAY
Makaha
Puu Hapapa
Kipapa
Waianae
Puu Kanehoa
Mailiili Str
Huliwai Gl
Waiahole
RANGE
Waiawa
Waimano Str
Kaneohe
Maili
Kailua
Waimalu Str
N. Halawa Str
S. Halawa Str
Nanakuli Str
Honouliuli Gl
Waipahu
Pearl Harbor
Aiea
Moanalua Str
Nanakuli
Kaloi Gl
Kalihi Str
Puu Konahuanui 3,105 ft
Maunawili Str
Waimanalo Bay
Nuuanu Str
Ewa
21°20'
Waimanalo
157°40'
Manoa Str
Palolo Str
Makapuu Point
Ewa Beach
Honolulu Harbor
Punchbowl
Barbers Point
HONOLULU
Base modified from U.S. Geological Survey digital data, 1:24,000, 1983, Albers equal area projection, standard parallels 21°19'40" and 21°38'20", central meridian 157°58'. Relief from U.S. Geological Survey digital elevation models, 1:250,000
MAMALA BAY
Waikiki Beach
Aina Haina
Maunalua Bay
Diamond Head
Koko Head


C

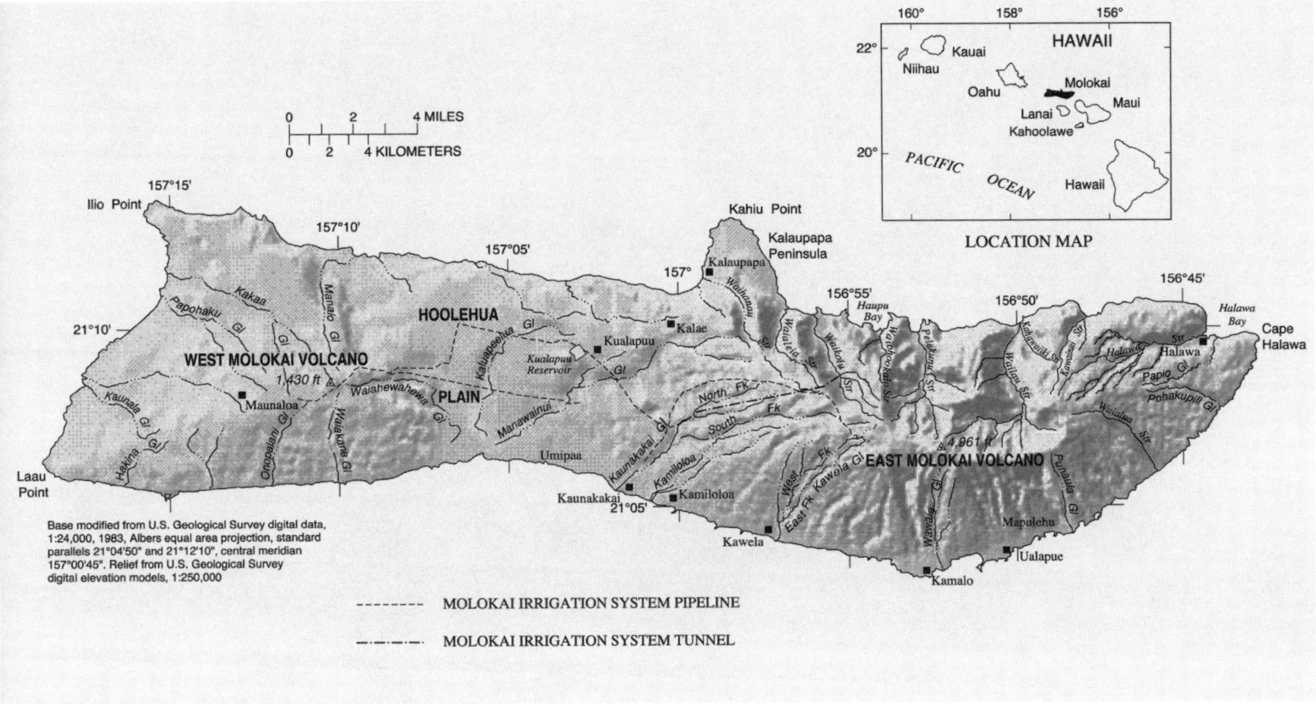
LOCATION MAP
HAWAII
160°
158°
156°
22°
20°
Kauai
Niihau
Oahu
Molokai
Maui
Lanai
Kahoolawe
Hawaii
PACIFIC OCEAN
0 2 4 MILES
0 2 4 KILOMETERS
157°15'
157°10'
157°05'
157°
156°55'
156°50'
156°45'
21°10'
21°05'
Ilio Point
Kahiu Point
Kalaupapa Peninsula
Kalaupapa
Laau Point
Cape Halawa
Halawa Bay
Haupu Bay
HOOLEHUA
PLAIN
WEST MOLOKAI VOLCANO
EAST MOLOKAI VOLCANO
1,430 ft
4,961 ft
Maunaloa
Kualapuu
Kualapuu Reservoir
Kalae
Kaunakakai
Kamiloloa
Umipaa
Kawela
Kamalo
Ualapue
Mapulehu
Halawa
Base modified from U.S. Geological Survey digital data, 1:24,000, 1983, Albers equal area projection, standard parallels 21°04'50" and 21°12'10", central meridian 157°00'45". Relief from U.S. Geological Survey digital elevation models, 1:250,000
MOLOKAI IRRIGATION SYSTEM PIPELINE
MOLOKAI IRRIGATION SYSTEM TUNNEL

D

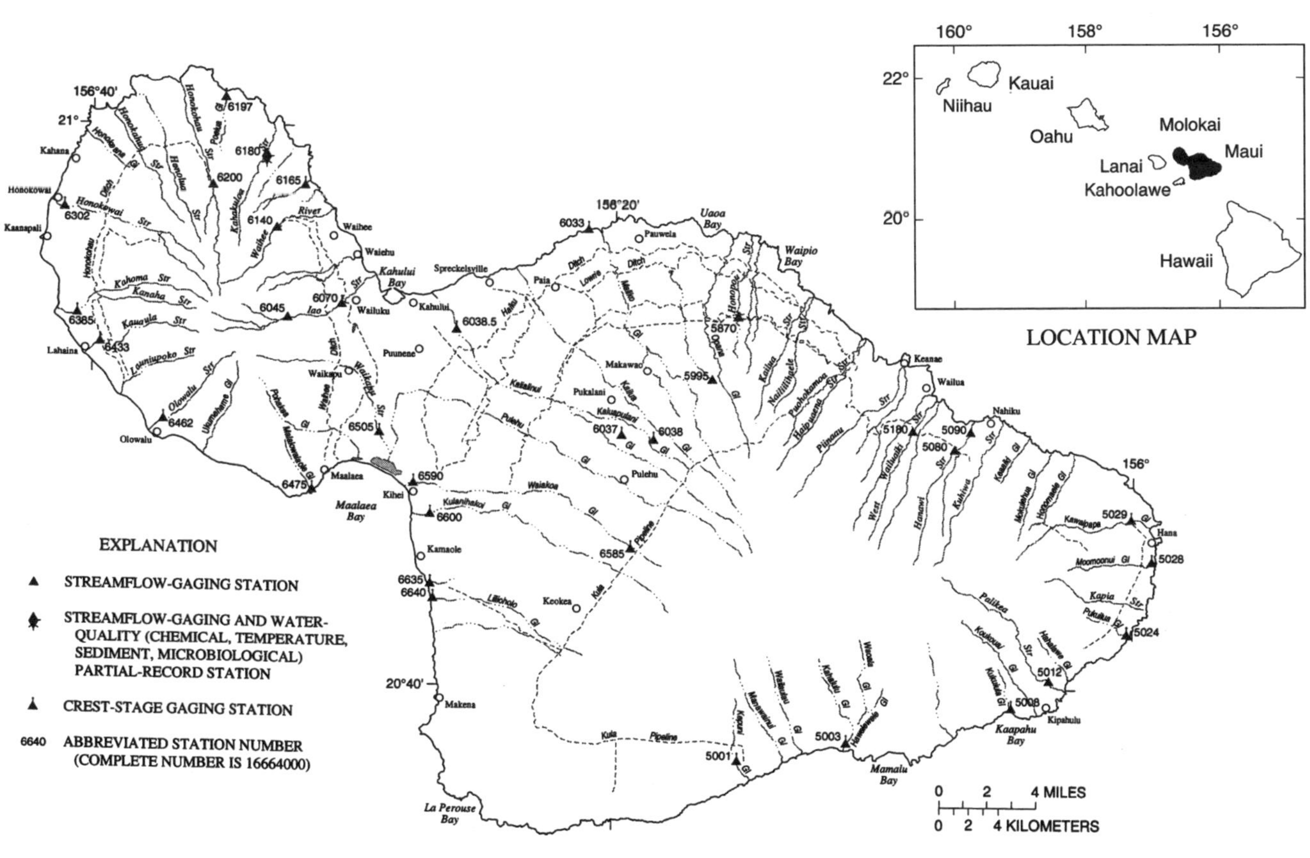
LOCATION MAP
160°
158°
156°
22°
20°
Kauai
Niihau
Oahu
Molokai
Maui
Lanai
Kahoolawe
Hawaii
156°40'
156°20'
156°
21°
20°40'
Kahana
Honokowai
Kaanapali
Lahaina
Olowalu
Maalaea
Maalaea Bay
Kihei
Kamaole
Makena
La Perouse Bay
Waihee
Waiehu
Wailuku
Kahului
Kahului Bay
Spreckelsville
Paia
Puunene
Waikapu
Makawao
Pukalani
Pulehu
Keokea
Pauwela
Uaoa Bay
Waipio Bay
Keanae
Wailua
Nahiku
Hana
Kipahulu
Kaapahu Bay
Mamalu Bay
Kula Pipeline
6197
6180
6200
6165
6140
6302
6070
6045
6385
6433
6462
6505
6475
6590
6600
6585
6635
6640
6033
6038.5
6037
6038
5995
5870
5180
5090
5080
5029
5028
5024
5012
5008
5003
5001
EXPLANATION
STREAMFLOW-GAGING STATION
STREAMFLOW-GAGING AND WATER-QUALITY (CHEMICAL, TEMPERATURE, SEDIMENT, MICROBIOLOGICAL) PARTIAL-RECORD STATION
CREST-STAGE GAGING STATION
6640 ABBREVIATED STATION NUMBER (COMPLETE NUMBER IS 16664000)
0 2 4 MILES
0 2 4 KILOMETERS

E

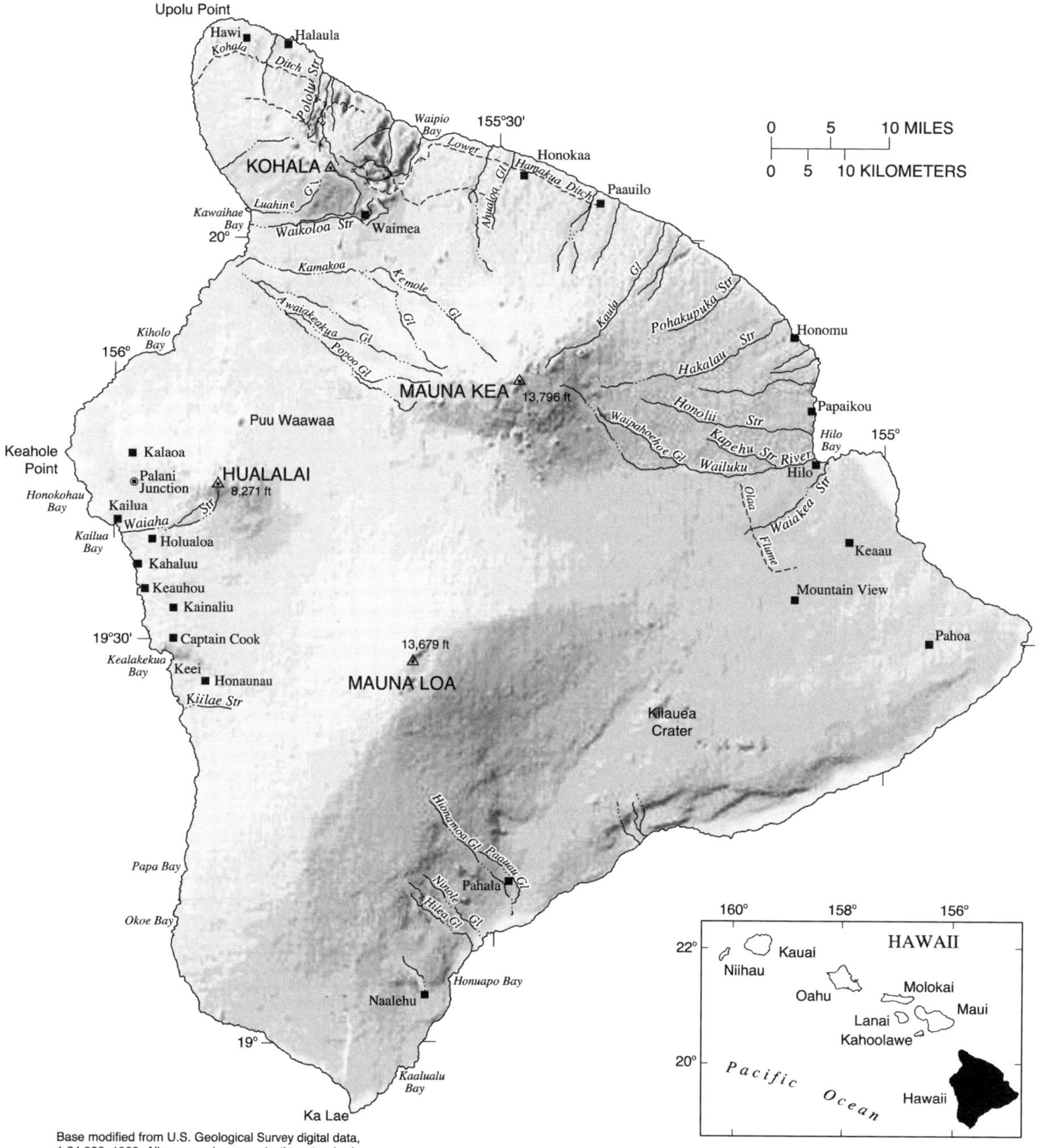


Base modified from U.S. Geological Survey digital data, 1:24,000, 1983, Albers equal area projection, standard parallels 19°08'30" and 20°02'30", central meridian 155°26'30". Relief from U.S. Geological Survey digital elevation models, 1:250,000

F

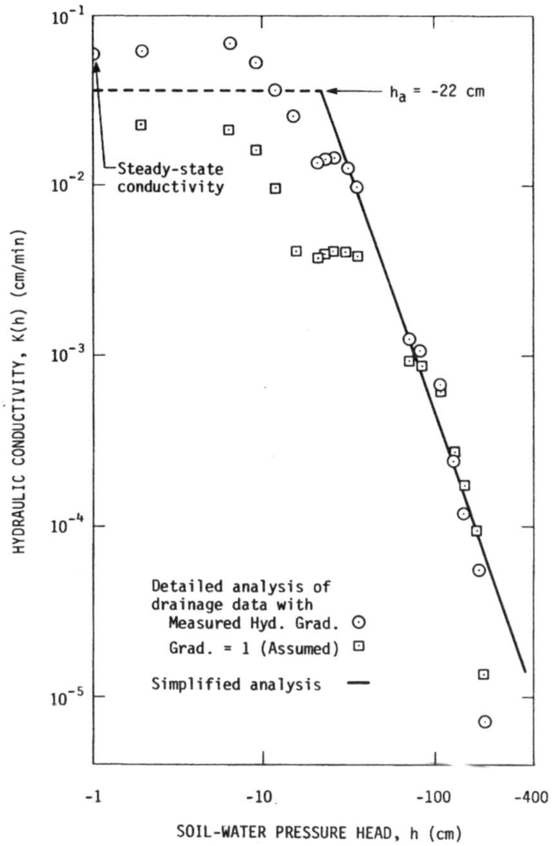


Figure 5.11. Unsaturated soil permeability for Moloka'i silty clay series. 1 cm/min = 47.2 ft/d = 14.4 m/d. (Reprinted from Chong et al. 1981 with permission from American Geophysical Union)

sufficient mass loading of TCE and to detect the contaminant in groundwater in a time frame of 10 to 20 years, spanning from the first use to its appearance in the underlying deep groundwater. Only intensive rain events that are known to occur in this area of low average annual rainfall can provide sufficient water to saturate the fractures, keep them open, and transport the TCE.

Lake-bottom permeability has been reported from laboratory tests of borings from Wahiawā reservoir, O'ahu (Harding Lawson Associates 1996). The hydraulic conductivity of the elastic silts is 0.02 m/d (0.07 ft/d), one order of magnitude less than that of clayey silt saprolite of blocky texture.

## Infiltration

Most Hawai'i surface soils absorb water readily. The steady-state infiltration capacity values, which were determined with infiltrometer cylinders surrounded by an earth berm, range from 0.3 to 1.5 m/d (1 to 5 ft/d) for the Wahiawā soil series and the Kāne'ohe soils series, 3.6 to 6.1 m/d (12 to 20 ft/d) for the Pōhākupu soil series and the Aloeloa soil series, and in excess of 6.1 m/d (20 ft/d) for the Waikāne soil series and the Mānana soil series (Willocks et al. 1961). Values between 0.3 and 1.2 m/d (1 and 4 ft/d) are reported for forest sites at the Lyon Arboretum (Ahuja and El-Swaify 1979). The lower value is for a large plot without ponding. These ranges have been encountered in other studies.

Reduction in infiltration as a result of various land uses is reported in comparative studies with double-ring infiltrometers on O'ahu, Hawai'i, Kaua'i, and Maui (Wood 1971, 1977). At the Honolua forest site on Maui, the steady-state infiltration capacity value for pineapple land is about four times less than that for the adjacent forest soil, 2.2 m/d (7.3 ft/d) versus 9.5 m/d (31.3 ft/d). Test values are less than 0.6 m/d (2 ft/d) at sugarcane sites and less than 1.1 m/d (3.5 ft/d) at pasture sites.

Urban effects on Oxisols (Wahiawā series) are appreciable, as revealed in measurements with double-ring infiltrometers following each step in urban land preparation (Murabayashi and Fok 1979). At the Mililani sites on O'ahu, the steady-state infiltration capacity values were 4.3 m/d (14.1 ft/d) for postpineapple conditions, 1.8 m/d (6.0 ft/d) after grubbing, 0.3 m/d (1.0 ft/d) after lot shaping, and 0.7 m/d (2.4 ft/d) after lawns were put in place. The overall loss of steady-state infiltration capacity was 83 percent.

The effect of antecedent moisture condition on infiltration was assessed in field tests for Oxisols on O'ahu (Green et al. 1982). The steady-state infiltration capacity values ranged from 0.9 to 3.7 m/d (3 to 12 ft/d) for dry antecedent conditions and only about one-half of that for wet conditions. This dry-wet difference may account for some of the variations reported in other studies.

Sorptivity, which is used in the Philip and Talsma-Parlange infiltration equations and is usually expressed using the International System of Units, was measured in Oxisols on Oʻahu (Green et al. 1982). The values for the three soil series — Molokaʻi, Lahaina, and Wahiawā — fall within a narrow range: 1.6 to 3.2 cm/min$^{1/2}$ for dry conditions and 1.3 to 1.5 cm/min$^{1/2}$ for wet conditions. Lower values are expected for wet soils. Sorptivity measurements made on Kahoʻolawe yield an average value of 1.1 cm/min$^{1/2}$ (Loague et al. 1996).

Suction head at the wetting front is used in the Green-Ampt equation to predict infiltration. Numerous suction-head measurements were made during infiltration experiments of Oxisols on Oʻahu (Green et al. 1982). Reference values available in the general literature vary from 5 cm for sands to 32 cm for clays (Rawls et al. 1993).

Surface crusting and sealing are not common in Hawaiʻi soils. The degree of these phenomena is minimal, as noted in erosion studies (Ahuja et al. 1976a).

## Overland Flow

The assertion that vegetal cover can reduce overland flow has been proved convincingly for Hawaiʻi agricultural soils. Other permeable Hawaiʻi soils, even though bare, can absorb plenty of rainwater before overland flow appears. Curve numbers have been evaluated for agricultural land for rainfall-runoff events in Hawaiʻi (Cooley and Lane 1982). These natural events were measured in two small watersheds (2 to 3 ha [4 to 7 ac]) cultivated with pineapple in Mililani (Wahiawā silty clay) and Kunia (Kolekole silty clay loam) from 1972 to 1979. Runoff volume was greatly reduced — by about 40 curve number units — in watersheds with complete vegetal cover as compared with bare conditions at the Mililani and Kunia sites. Further, the curve number values derived with Hawaiʻi data are much smaller, by 31 units, than those listed in the U.S. Soil Conservation Service handbook (1970) that had been determined previously for Hawaiʻi on the basis of mainland conditions. The former handbook values overestimate surface runoff volume. Similar findings apply in sugarcane-cultivated soils on Oʻahu (Paaloa silty clay in Waialua) and on the island of Hawaiʻi (Kūkaʻiau silty clay loam in Honokaʻa; Kaiwiki silty clay loam in Laupāhoehoe). Initiation of overland flow over bare soils was investigated with simulated rain (at 6.35 cm/h [2.5 in./h]) over a 22.8-m (75-ft) surface in ten different soil series on both islands (Ahuja et al. 1976a). Results show that the runoff initiation time varies linearly with antecedent saturation deficit (final minus initial soil water content) as predicted by the Green-Ampt equation. For wet soil with a deficit of less than 5 percent, the initiation times were just a few minutes. However, exceptionally long initiation times have also been observed: as long as 60 to 80 minutes, and 82 minutes for Waikāne, Oʻahu. These long times suggest non-Hortonian runoff generated by saturation from below the soil layers rather than limitation by the infiltration capacity of the soils. Additional rainfall-runoff relationships are discussed in Chapter 7.

## Transport

Subsurface transport in Hawaiʻi can be conveniently separated into two domains: those in the surface soils and saprolites (this chapter) and those in the regional groundwater zone (Chapter 6). Leachate from the top domain will experience some additional natural attenuative processes in the intermediate vadose zone, but the diminished concentrations may still be a potential source of contamination of the deep groundwater. Hydrologic transport in the top domain is intimately related to several concerns or problems in water contamination in Hawaiʻi. Trace amounts of volatile organics in deep potable groundwater originating at the surface have continued to be a principal water problem in Hawaiʻi, as attested in the State Health Department's assessment (Miike 1996). Although not a single case has been detected, concerns always exist about human enteric viruses gaining access to potable groundwater sources from land-treatment systems. Leakage and spills of petroleum hydrocarbons have created some localized contamination of surface soils and brackish groundwater. Gradual rising but subregulatory concen-

Table 5.4. Leachate Quality (1972–1975) of Bermuda Grass Sod–Covered Lahaina Series Soil from Mililani, O'ahu

| Water-Quality Parameters | Applied Water[a] Median | Applied Water[a] No. of Samples | Lysimeter Leachate[b] Median | Lysimeter Leachate[b] No. of Samples | Change Through Lysimeter (%) |
|---|---|---|---|---|---|
| Total suspended solids (mg/l) | 22.0 | 23 | 2.00 | 9 | -90.9 |
| Total dissolved solids (mg/l) | 332.0 | 140 | 248.00 | 117 | -25.3 |
| Total organic carbon (mg/l) | 25.0 | 39 | 1.50 | 29 | -94.0 |
| Total nitrogen (mg/l) | 18.4 | 217 | 0.02 | 190 | -99.9 |
| Total Kjeldahl nitrogen (mg/l) | 16.8 | 240 | 0.01 | 196 | -99.9 |
| Nitrite and nitrate nitrogen (mg/l) | 0.7 | 223 | 0.01 | 228 | -98.6 |
| Phosphate phosphorus (mg/l) | 10.7 | 196 | 0.02 | 151 | -99.9 |
| Potassium (mg/l) | 9.8 | 233 | 0.07 | 219 | -92.9 |
| Chloride (mg/l) | 50.0 | 225 | 50.00 | 179 | 0.0 |
| Electrical conductivity (µmhos/cm) | 430.0 | 223 | 375.00 | 183 | -12.8 |

SOURCE: Dugan and Lau 1981.

[a] Secondary wastewater effluent treated by activated sludge process.

[b] A hydraulic lysimeter packed with 1.52 m (5 ft) of soil and loaded with 262 cm of rainfall and 716 cm of effluent from July 1972 to February 1975 (32 months).

trations of nitrate in some potable groundwater sources of O'ahu may be associated with leaching of the residue of nitrogen chemicals used as agricultural fertilizers.

Managed vegetal soil systems are effective in removing solids and microbes from infiltrating waters. Bacteria and viruses are inactivated in surface soils as a combined effect of filtration, desiccation, adsorption, ultraviolet radiation, heat, and predation (Lau et al. 1975, 1980). Turbidity and suspended solids are readily removed by filtration. Nitrogen, phosphorus, and potassium (NPK) are removed by vegetation as nutrients. Strong evidence has been provided by the changes that take place in concentrations of water-quality parameters passing through a 1.52-m (5-ft)-deep lysimeter packed with Lahaina series soil, sodded with Bermuda grass, and loaded with wastewater effluent for a period of 32 months (see Table 5.4). Biodegradable organic matter expressed as biochemical oxygen demand is effectively decomposed in an intermittent wetting and drying operation that supplies the needed dissolved oxygen to sustain the aerobic condition. Loading of solids, biochemical oxygen demand, and NPK may not be a limiting factor in a well-managed system. The percolate from a land-treatment system is invariably of higher quality than the applied water. Experience with some systems in Hawai'i is presented in Appendix 5.3.

Vapor-phase transport may occur but is usually overlooked in a multiphase system. Vapor transport of TCE is minimal in the thick vadose zone at Schofield Barracks, as determined by simulated modeling of physical processes including advection, dispersion, and retardation, but not including biodegradation and hydrolysis (Harding Lawson Associates 1996). The advective process is far more dominant than the diffusive process in the liquid-phase transport of TCE in the Schofield Barracks vadose zone.

Classical advective-dispersive theory is found applicable in saturated soils (Moloka'i and Wahiawā series) in laboratory columns with chloride as the conservative solute. Dispersivity is unaffected by velocity, which is consistent with theory (Cagauan et al. 1968). The value of dispersivity ranges from 5.5 to 6.3 cm (2.2 to 2.5 in.). Similar results are obtained using potassium. The advective-dispersive theory is unsatisfactory for phosphates, suggesting that sorption or other processes are involved. Both soils, especially Wahiawā soils, are phosphorus sorptive. For unsaturated soils, field tests were conducted in a carved and partially sealed ring infiltrometer (30 cm [12 in.] in height) in Moloka'i silty clay soil with a nitrate solution. The coefficient of dispersion, *D*, was determined to be 0.0013 $m^2$/d (0.014 $ft^2$/d), for an average velocity of 0.25 m/d (0.82 ft/d). Contrary to theory, *D* was assumed to be independent of velocity (Khan 1979). This value of *D* was used in modeling for both soils and subsoils at a Mililani site (Loague et al. 1989) and at a Waiawa site (Loague et al. 1995).

Although rudimentary by today's precision, a laboratory study was conducted of the transport of two pesticides, DDT and Lindane, in percolating water through Wahiawā and Lahaina soils (Eto et al. 1967). Findings, including breakthrough in solution and residues in soil columns, were explained in terms of adsorption, volatilization, and presence of organic matter. Results of batch equilibration tests are expressed by Freundlich adsorption isotherms for these two as well as three other soils: Lolekaa, Mānana, and Makiki. In a historical context, there were no set limits for these chemicals in drinking-water standards in 1967. In fact, Rachel Carson's *Silent Spring* was published in 1962 when transport theories in porous media were just being formulated.

The values of chemical properties of the contaminants that are relevant to adsorption and volatility should be soil specific and, if necessary, evaluated by experiments. Experiments have been done for DBCP and 1,3-D (Telone). DBCP is found strongly adsorbed to the Ha'ikū series (Maui) and Leilehua series (O'ahu) of the Utisols order in laboratory experiments (Buxton and Green 1992). A typical value of the sorption coefficient, $K_d$, is about 2 l/kg, and the results fit the Freundlich isotherm. The DBCP sorbed to the soils that are high in organic matter is extremely difficult to desorb; the desorption coefficient is 100 to 1,000 times greater than the adsorption coefficient. These properties are the deciding factor in determining the need or the lack of need for remediating contaminated soils. The liquid-vapor diffusion coefficient of DBCP in Wahiawā silty clay was evaluated to be $6.1 \times 10^{-4}$ $cm^2$/s (0.057 $ft^2$/d) for a dry soil ($\theta = 2.3$ percent) and $2.6 \times 10^{-3}$ $cm^2$/s (0.242 $ft^2$/d) for a moist soil ($\theta = 31.6$ percent), based on laboratory experiments (Pringle et al. 1984). Telone is volatile and highly water soluble, and it has a relatively short half-life. Field leaching and evaporation experiments conducted at a Wahiawā silty clay site (Schneider et al. 1995) were necessary to quantify the parameters needed for simulation modeling (Lin 1993; Liu et al. 1993; Lin et al. 1995).

Biotic transformation of organic chemicals in the subsurface is measured by the time rate of the process for specific soils and for several chemicals that are used as nematocides in Hawai'i, including DBCP, Telone II, and fenamiphos and its metabolite, f. sulfoxide. DBCP biodegrades slowly, and the rate decreases with depth in the Hāmākuapoko silty clay series of the Utisols order: 0.0038/d at 45 cm (18 in.), 0.0034/d at 90 cm (35 in.), 0.0014/d at 135 cm (53 in.), and <0.0014/d at >135 cm (Green et al. 1986; Liu et al. 1987). This set of field data was used in transport modeling (Lin 1993; Liu et al. 1993) and later reworked into decay coefficients for use in another model (Orr and Lau 1987). The set of coefficients decreases with time after application: 0.0088/d in 6 months, 0.0046/d in 3.5 years, and 0.0001/d in 30 years, reflecting different processes at work in an overlapping way. Telone remains in soils longer than expected, even though it is supposed to be a fast-degrading chemical. This finding is based on laboratory tests, a field test in Kunia (Schneider et al. 1995), and modeling (Lin et al. 1995). Monod kinetics fits the data better than first-order kinetics. A pseudo first-order biodegradation constant was devised for ease of modeling work; however, the value of the constant is dependent on initial concentrations. Fenamiphos is a nonvolatile and noncarcinogenic chemical, but the human health concern is associated with its metabolite. The rate of biodegradation for fenamiphos is faster than that given by first-order kinetics (Lee et al. 1986).

## Hawai'i Data and Models

### Data

The database on hydrologic parameters of Hawai'i soils is scattered in journals and technical reports such as those cited. Hydrologic characterization has been focused principally on Oxisols and, to a lesser extent, Inceptisols, Histosols, and Andisols (Wood 1971, 1977; Ahuja et al. 1976a; Cooley and Lane 1982; Green et al. 1982; Loague et al. 1990). Landslides on urbanized land draw attention to Vertisols.

Spatial variability on a large scale in Hawai'i is addressed with geostatistical methods (Bresler and Green 1982; Yost et al. 1982). The basic concept in geostatistics as advanced by Matheron in 1965 is that hydrogeological variables and parameters are functions of space. Although highly varying, they are not purely random: if measurements are made at two different locations, the nearer the locations, the closer are the measured values. In other words, there exists some kind of statistical correlation in the unknown spatial distribution of the variable or parameter. Kriging is a method for optimizing estimation of the variable and parameters (de Marsily 1986). In Bresler and Green's work, a soil sampling scheme was proposed using mean, variance, and variogram for determining the unsaturated hydraulic conductivity of the heterogeneous soils in the watershed. For Oxisols in a 400-km$^2$ (154-mi$^2$) area in the Pearl Harbor drainage basin, a minimum of 30 sampling sites is necessary for estimating the value of this parameter for the purpose of stochastic modeling. In Yost et al.'s work, a semivariogram was used in the study of spatial dependence of soil chemical properties (pH, exchangeable cations, extractable silica, and phosphorus) at eighty sites spaced at 1- to 2-km (0.6- to 1.2-mi) intervals on the island of Hawai'i. In their work, spatial dependence was confirmed and rainfall was identified to impose uniformity on surface soil properties. Grouping soils with semivariograms is proposed for delineating regions of uniform soil chemical properties.

## Flow Models

Flow modeling concerning the unsaturated zone in Hawai'i ranges from inquiry into fundamentals to field-scale applications. These pursuits parallel and contribute to contemporary work done elsewhere.

### Hydraulics

Soils in Hawai'i can be treated as a continuum under drainage condition (Ahuja et al. 1980; Green et al. 1982; El-Kadi 1993). Darcy's Law and Richards equation are assumed to be applicable in Hawai'i soils (Chen et al. 1993; El-Kadi 1993).

A method for simplifying the fieldwork and data analysis was proposed for determining unsaturated hydraulic conductivity and water-retentive properties (Ahuja et al. 1980; Chong et al. 1981) and calibrated for the Lahaina, Moloka'i, and Wahiawā series of the Oxisol order. The work is based on the unit-hydraulic gradient, which is valid in the intermediate stage in drainage tests. Time variations of water content and negative pressure are fitted with power functions. The method has been adopted as the standard method by the American Society of Agronomy and the Soil Science Society of America (1986).

A two-domain model is proposed to deal with macroporosity and preferential flow (Chen et al. 1993). Assumptions are made that macropores dominate early drainage and that matrix pores dominate subsequent drainage after the macropores are emptied. The model is calibrated with the results of field experiments conducted in Oxisols at Wahiawā.

The Philip and Green-Ampt models are favored in predicting infiltration variations with time. However, the Talsma-Parlange equation fits accurately the data for longer times for the Moloka'i soil series at Kunia (Green et al. 1982). Initially, various empirical infiltration equations were used to fit field data (Willocks et al. 1961; Green and Guernsey 1981).

The U.S. Soil Conservation Service runoff curve number model is assumed to be applicable in Hawai'i drainage basins. The model is calibrated with Hawai'i natural rainfall-runoff events (Cooley and Lane 1982). Modeling the initiation time of overland flow, which is based on the Green-Ampt equation, has been validated with a simulated-rain field test in Hawai'i (Ahuja et al. 1976a).

Vertical heterogeneity in deep soils and saprolite can reduce percolation and groundwater recharge because the poorly permeable layers can induce saturation from above, lateral subsurface flow, and even discharge at seepage faces. This hypothesis was tested with a one-dimensional physically based model at the Mililani and Waipi'o sites over the 5-year period from 1975 to 1979 (Green et al. 1993). Values of the highly varying permeability in deep (30 m [100 ft]) sections were calculated from water-retention curves of the soil cores. Richards equation was solved numerically with VS2D, a code formulated by Lappala et al. (1987). Seasonal saturation was revealed at the expected depths, but the volume of lateral seepage was not quantifiable.

## Groundwater Recharge by Deep Percolation

Conservation of water volume of a soil-vegetation system can be written as follows:

$$P + IRR = SRO + AET + RCHG + SM \tag{5.12}$$

where P is precipitation, IRR is irrigation and other waters applied on land, SRO is surface-water runoff, AET is actual evapotranspiration, RCHG is groundwater recharge, and SM is the soil water in the soil-vegetation system. AET is a function of soil moisture content (see Chapter 4). A bookkeeping method is applied to the equation to determine the soil moisture content. Recharge is produced whenever the soil moisture content exceeds the field capacity of the soil.

Land is subdivided into small parcels so that, within each parcel, water and land traits involved can be considered approximately homogeneous. The traits include land use, rain, cloudiness, soils, surface-water runoff, and irrigation. A second but seldom addressed refinement for improving the accuracy is discretization of the time interval. The soil moisture accounting performed at monthly intervals tends to overestimate AET and thus underestimate recharge (Mather 1978). Soil moisture accounting was done on a daily basis in a water-balance study of the Pearl Harbor–Honolulu area (Giambelluca 1983).

Use of the rooting zone system and both spatial and time discretizations in mass balance accounting is relatively recent in Hawai'i. Perhaps the first but coarse treatment of spatial distribution was done for Kohala along the west coast of Hawai'i (Kanehiro and Peterson 1977). A greater refinement addressing both time and spatial distributions is Giambelluca's work (1983). The southern O'ahu area was discretized into 258 water-balance zones of homogeneity, and a 360-month groundwater recharge time series was computed using a daily time interval for each zone. The results of percolation contributing to groundwater recharge are presented as an annual time series and as spatial distribution of the annual averages in Figures 5.12 and 5.13, respectively. The detailed approach is not likely to be warranted for most practical water projects.

## Transport Models

Several approaches are used in Hawai'i to model leaching of chemicals (mostly pesticides and solvents) in surface soils and saprolite. The standard physically based models of solute transport are the Pesticide Root Zone Model (PRZM) and the Leaching Estimate and Chemistry Model Pesticide (LEACHP). Also employed are leaching indices RF and AF, and some innovative proposals.

The advection-dispersion model was applied to organics including DBCP and 1,3-dichloropropene (1,3-D, Telone II), f. sulfoxide, fenamiphos, and picloram and to inorganics including chloride, potassium, and phosphate. The models were numerical in most applications. An example is a two-dimensional simulation of both vapor and water transport of DBCP (Liu et al. 1993). The model, which was calibrated with data from field experiments conducted in East Maui, was used to test management options, including timing and method of field application, mulching, depth of chemical injection, and duration

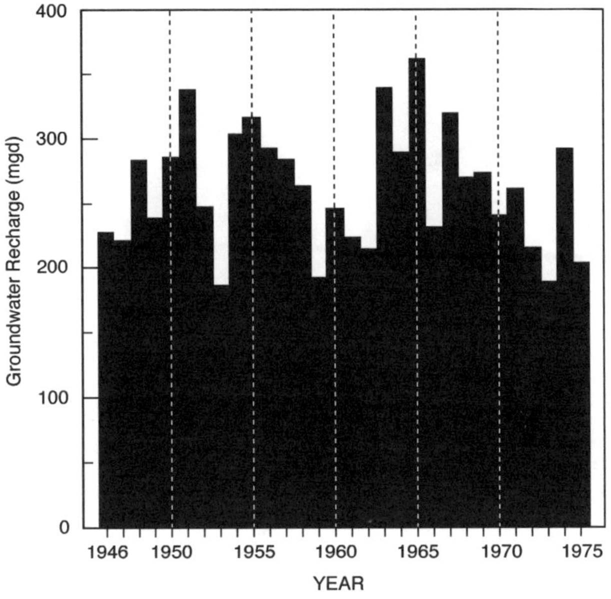


Figure 5.12. Computed annual groundwater recharge from overland sources, noncaprock area, Pearl Harbor region, O'ahu, Hawai'i. 1 mgd = 0.04381 $m^3$/s. (Adapted from Giambelluca 1983)

of water application. The results predict partitioning of DBCP as percentages escaping to the atmosphere, decaying in soil, remaining in the topsoil, and percolating.

Several conceptual models of the saprolite formation in the unsaturated zone at Schofield Barracks were tested with the multiphase transport code TOUGH2/T2VOC, which was developed and validated by researchers at the Lawrence Berkeley Laboratory (Pruess 1991; Falta et al. 1995). The work is a part of the Superfund remedial investigation of TCE contamination of the high-level groundwater in the Wahiawā aquifer (Harding Lawson Associates 1996). TCE, which was used as a cleaning solvent after 1945, was discovered in 1985 to exceed the U.S. maximum contaminant level of 5 μg/l. The thick (168 to 198 m [550 to 650 ft]) vadose zone consists of shallow soil (2 to 3 m [5 to 10 ft]) underlain by saprolite (15 to 55 m [50 to 180 ft]) overlying unweathered basalt. The modeling work was performed to help determine which conceptual model can fit the field data.

PRZM was developed by the U.S. Environmental Protection Agency as an operational procedure for leaching assessment. PRZM provides an approximate solution of the water-balance equation and the advection-dispersion equation. The model was calibrated with data from shallow soil experiments and applied to a deep (up to 240 m [787 ft]) vadose zone in Waiawa to test various soil- and water-management options for a proposed urban land development. The urban chemicals tested include chlorpyrifos (termiticide), diazinon (insecticide), metribuzin (herbicide), and nitrate. The simulation results indicate that potential contamination of the deep basal groundwater can be limited to a level far below maximum allowable contamination levels by selection and proper use of the chemicals coupled with simple soil and water management. A major concern about this modeling is the use of PRZM beyond its original intent (Oki et al. 1991; Loague et al. 1995).

The LEACHV model was formulated in Hawai'i by adding gas transport to the extant LEACHM model (Chen et al. 1995). LEACHV addresses air flow driven by barometric pressure changes at the soil surface, thus explicitly accounting for vapor-phase advection. Validation of the model was performed with California data for the volatile 1,3-D.

A lumped parameter model has been devised for the assessment of nitrate leaching (Ling and El-Kadi 1998). The model deals with nitrogen transformation and different modes of irrigation and fertilizer applications. The analytic solution requires climate data and some soil data. Simulation results compare favorably with those by the numerical model LEACHM and with the Hawai'i field data at the bottom of the root zone.

The retardation factor, RF, is an index for rating chemicals by their leaching potential in soils on the basis of adsorptivity and volatility of the chemicals. RF maps of several chemicals have been made using a geographic information system and soil properties as presented in soil taxonomic categories (Khan et al. 1986). The uncertainty of these RF values was assessed with the first-order uncertainty analysis FOUA (Loague et al. 1989). FOUA is a technique for estimating the uncertainty of a deterministic model that is due to the uncertainty of the individual

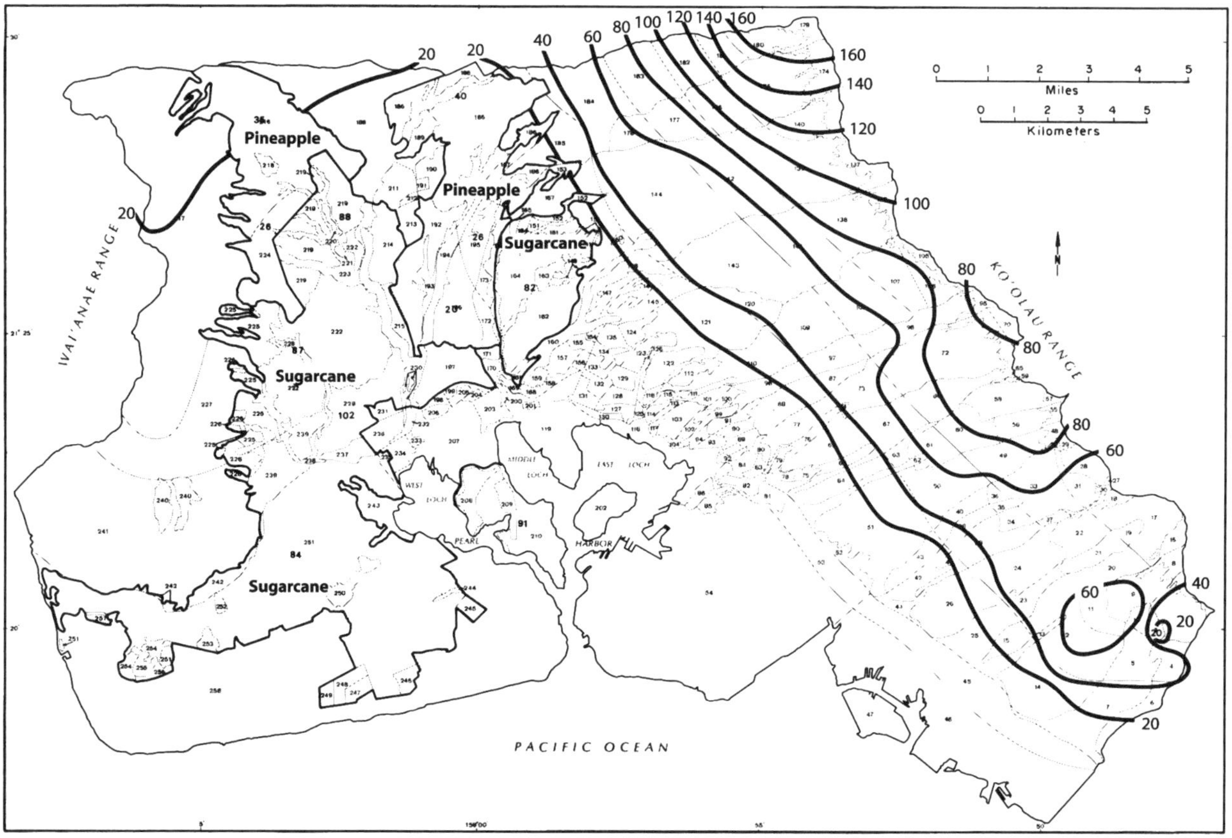


Figure 5.13. Distribution of computed annual groundwater recharge from overland sources, noncaprock area, Pearl Harbor–Honolulu, Oʻahu, Hawaiʻi. Point values shown for sugarcane and pineapple areas. Recharge measurements provided in inches. (Adapted from Giambelluca 1983)

parameters used in the model (Cornell 1972). This exercise indicates that reduction of uncertainty is possible by using the lowest soil taxonomic category.

The attenuation factor, AF, is an extension of RF by incorporating depth to the water and annual recharge rate. A test was performed on diuron and atrazine used in the Pearl Harbor sector (Loague et al. 1990). The results indicate a significant amount of uncertainty and caution in the use of AF for regulatory purposes.

Other transport models proposed for or used in Hawaiʻi include systems and stochastic approaches. In the former, all of the individual flow and transport processes involved are lumped together in a system response function that is calibrated with observed data and then used to predict an output for a given input. First introduced in subsurface solute transport by Jury in 1982, this approach was applied to simulate transport of fenamiphos and other chemicals. The response function was assumed to be a gamma distribution function whose two parameters were correlated by regression with soil and chemical properties (Liu et al. 1994). The model is judged best suited for steady-state flow and low-volatile organics.

An uncommon model is proposed for macrodispersion in an unsaturated heterogeneous soil (Liu et al. 1991b). The soil column is created by random field turning bands, the flow satisfies the Richards equation, and the solute plume is computed with the finite element method. The dispersive plume is composed of three phases: linear, nonlinear, and asymptotic to a constant. The growth of the plume approximates that by Dagan's formula. The first phase appears as advective with minimal lateral mixing, as reported earlier by Liu (1988). The travel depth, 1 m (3.28 ft), is shorter than the required 5 m (16.4 ft) to reach a symptotic constant according to the Fickian state. The value of dispersivity used in the study ranges from 1 to 3 cm (0.39 to 1.18 in.) for the top 1 m (3.28 ft) of soil.

A solute transport model is proposed to simulate preferential flow on a field scale (Soicher 1996). The model consists of parallel columns. Whenever saturated, a column becomes a preferential pathway and conducts fluid with increased velocity for the rest of the event. A two-dimensional model (101 columns each with 101 nodes) is created with a random field generator and tested for hypothetical infiltration events with a conservative solute, assuming that the Richards and advection-dispersion equations are applicable. Test results mimic features of observed preferential flows in reported field studies.

## Appendix 5.1. Instruments and Basic Computations

A tensiometer is a device for measuring the negative pressure (matrix potential) in soil water. It consists of a porous ceramic cup attached to the lower end of a rigid tube. The tube is sealed at the upper end with a removable cap, which allows filling the tensiometer with water and purging air accumulations. A vacuum gage is installed near the top part of the water-filled tube to register the negative pressure of the water in the tensiometer (Figure 5.14). When the tensiometer is inserted into the soil, the soil imbibes water from the tensiometer, and, as this occurs, the water tension in the tensiometer increases until it reaches equilibrium with the soil water outside the cup. Typical use of tensiometer measurements is illustrated in Example 5.1.

Piezometers are holes bored in the ground and, if the hole is likely to cave in, fitted with a rigid tube. The bottom of the tube, which is open, is the point of measurement. The water level in the tube indicates the positive water pressure at the point of measurement. The water pressure at the base of a 30.5-cm (1-ft) water column is 2.986 kPa (0.433 psi).

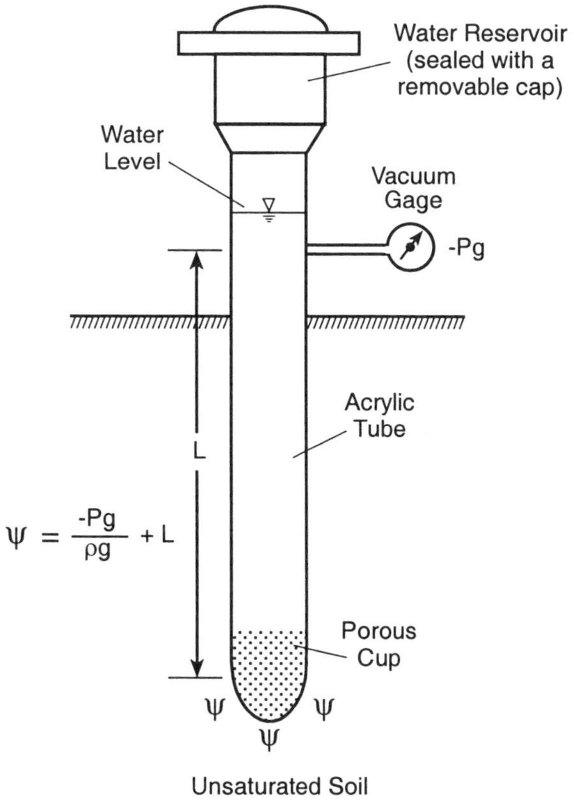


Figure 5.14. A tensiometer. ψ = soil-water tension, L is a correction. (Adapted from Stephens 1996)

The double-ring, ponded-water infiltrometer is a typical device for measuring infiltration. It consists of a small ring and a large ring, the latter known as a buffer ring, both driven into the ground in a concentric way. During a test, water is applied inside the inner ring while the water level is kept low and constant. Water is also ponded in the annulus space between the two rings to minimize lateral spreading of water in the subsurface, which would exaggerate the infiltration in the surface area of the small ring. In a highly instrumented field study conducted at Lyon Arboretum, Oʻahu, the lateral flow was quantified to be negligible when a buffer ring with a 60-cm (24-in.) diameter was used. A field method that involves insertion of two multidepth tensiometers, one inside the inner ring and the other in the annulus, is available for determining simultaneously vertical and radial hydraulic conductivities (Ahuja et al. 1976b).

### Example 5.1

Two tensiometers are positioned 100 cm (39.4 in.) apart along a vertical line. The negative pressure head is –60 cm (–23.6 in.) at the top tensiometer (A) and –100 cm (–39.4 in.) at the bottom tensiometer (B). The hydraulic conductivity at A and B is $2 \times 10^{-3}$ cm/min ($0.8 \times 10^{-3}$ in./min) and $4 \times 10^{-4}$ cm/min ($1.6 \times 10^{-4}$ in./min), respectively.

Compute the hydraulic gradient *J*, flow direction, specific discharge *q*, and flow rate *Q* per 10,000-ft² surface area.

Solution:

Choose an arbitrary datum at 210 cm (82.6 in.) below B and set up a table as shown here:

| Location | Pressure Head (cm) | Position Head (cm) | Total Head (cm) |
|---|---|---|---|
| A | -60 | 310 | 250 |
| B | -100 | 210 | 110 |

$$J = -\frac{\partial h}{\partial z} = -\frac{110 - 250}{310 - 210} = \frac{140}{100} = 1.4$$

Flow direction A to B downward

$$\text{Average } K = \frac{(2 \times 10^{-3}) + (4 \times 10^{-4})}{2} = 1.2 \times 10^{-3} \text{ cm/min}$$

$$q = KJ = (1.2 \times 10^{-3})\,(1.4) = 1.68 \times 10^{-3} \text{ cm/min}$$
$$= 7.94 \times 10^{-2} \text{ ft/d}$$
$$Q = Aq = (10{,}000 \text{ ft}^2)\,(7.94 \times 10^{-2} \text{ ft/d})\,(7.48 \text{ gal/ft}^3)$$
$$= 5{,}900 \text{ gal/d per } 10{,}000\text{-ft}^2 \text{ surface area}$$
$$= 4.1 \text{ gal/min per } 10{,}000\text{-ft}^2 \text{ surface area}$$

# Appendix 5.2. Runoff Curve Number Method

The volume of rainwater $P$ of a rain event may be separated into two portions: direct surface runoff $Q$ and abstractions caused by infiltration and other retentive processes. A fraction is formulated by dividing $Q$ by $P - I_a$ where $I_a$ is the initial abstraction until ponding occurs. After direct runoff has begun, the additional volume of water $F$, which is retained primarily by infiltration in the watershed, is expressed as a fraction of some potential maximum retentive volume $S$. The basic assumption of the U.S. Soil Conservation Service (1969) runoff curve number method is that these two fractions are equal. This equality is combined with the equation for conservation of water volume of the rain event, $P = Q + I_a + F$. By eliminating $F$, $Q$ for the rain event in the watershed is obtained as follows:

$$Q = \frac{(P - I_a)^2}{P - I_a + S} \tag{5.13}$$

$I_a$ is determined to be 0.2$S$ based on an analysis of rainfall-runoff data of numerous small watersheds. Curve number is defined as 1,000/(10 + $S$). Equation (5.13) is reduced to

$$Q = \frac{(P - 0.2S)^2}{P + 0.8S} \tag{5.14}$$

A family of curves $Q(P)$ is plotted in Figure 5.15 using curve number as a parameter (Chow et al. 1988).

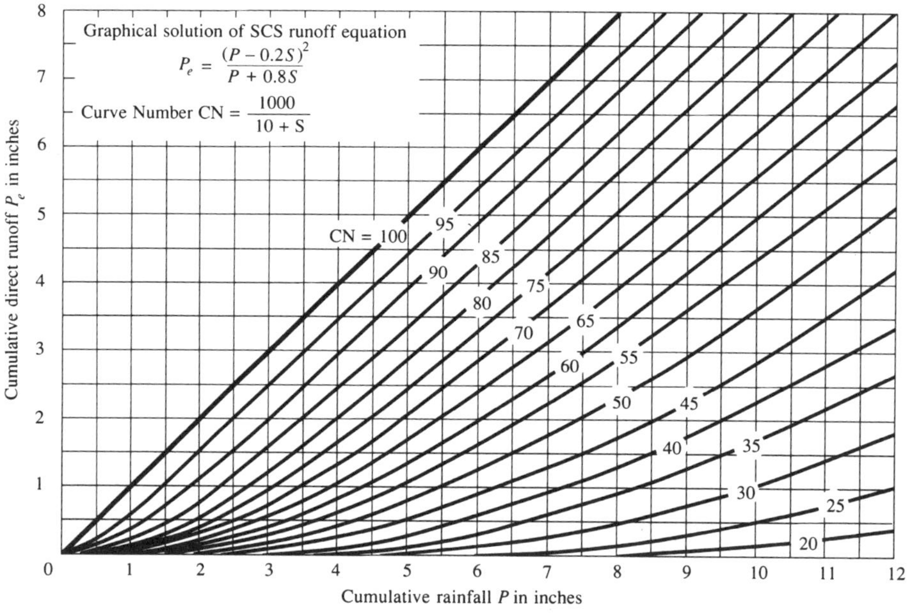


Figure 5.15. Rainfall and direct runoff relationship. (Based on U.S. Soil Conservation Service 1969 and reprinted from Chow et al. 1988 with permission from McGraw-Hill Companies)

## Appendix 5.3. Land-Treatment Systems

Land treatment is controlled application of wastewater to vegetated or bare soils to achieve reduction and alteration of the constituents in the wastewater. Land-treatment systems use the natural physical, chemical, and biological processes in the soil-water vegetation matrix (Reed et al. 1995).

The rapid infiltration system is most suited for Hawai'i and has been successfully demonstrated with vegetation including Bermuda grass, California grass (a fodder for cattle feed), and sugarcane. Efficacy of this system is evidenced using field lysimeters and full-scale demonstration prospects (Lau et al. 1975, 1980, 1989). Permeable soils and year-round vegetal growth are the intrinsic natural attributes of the system. The hydraulic loading used was high and intentional to test the capacity of the system. The most accurate method used for the assessment of flow and transport was the 1.52-m (5-ft)-deep Bermuda grass sod–covered hydraulic lysimeter packed with Lahaina series soil and irrigated with secondary wastewater effluent at a site in Mililani, O'ahu. Under a water application rate (irrigation plus rainfall) averaging 10 mm/d (0.4 in./d) over 960 days, the evapotranspiration averaged 4.4 mm (0.17 in./d) (Dugan and Lau 1981). Under irrigation rates as high as 98 mm/d (3.8 in./d) for a 5-day week over 510 days, evapotranspiration of California grass averaged 4.6 mm/d (0.18 in./d) in percolate-style lysimeters (Handley and Ekern 1984). The growth of grasses, especially California grass, was robust. The ample percolation is recognized as a potential water source for groundwater recharge. This consideration is enhanced by the high quality of the percolate (see Table 5.4).

In a demonstration project in the 'Ewa plain, O'ahu, California grass and sugarcane were grown in 0.2- to 0.4-ha (0.5- to 1.0-ac) plots, in fine, mixed isohyperthermic soils of the Mollisols order (Lau et al. 1989; Lau 1994). The primary wastewater effluent from the Honouliuli Wastewater Treatment Plant was used as irrigation water and applied intermittently at an average rate as high as 508 mm/wk (20 in./wk) by the border-flood method for California grass and by the flood-drip method for sugarcane. Plots were free of standing water a few hours after water applications. Based on monitoring with piezometer wells, the underlying shallow groundwater was freshened to 245 mg/l in chloride concentration. The applied nitrogen was substantially stripped by the system with a resultant nitrate-nitrogen concentration of 4.8 mg/l in the groundwater (maximum contaminant level being 10 mg/l), bacterial quality by monitoring was satisfactory, and toxics (pesticides and heavy metals) analyzed were all below applicable maximum contaminant levels. No adverse environmental concerns—including surface runoff, insect infestation, and groundwater contamination—were identified. Soil clogging was not evident because of primary treatment of the wastewater and fill-and-draw operations. The biomass produced was of usable quality and adequate quantity.

A scale-up facility was not realized because of land-use economics. The land requirement is approximately 134 ha/m$^3$/s (14.6 ac/mgd) of recharge water reaching the water table. The 'Ewa plain region is under intense urban development.

Current regulations in Hawai'i require tertiary treatment of the wastewater before it can be applied on land. For instance, spray irrigation is permitted only for R-1 water. The treatment processes for R-1 water include oxidation, filtration, and disinfection. The fecal coliforms in the treated water must not exceed 2.2 CFU/100 ml for the preceding 7 days, 23 CFU/100 ml in more than one sample in a 30-day period, and 200 CFU/100 ml in any sample. The turbidity is limited to less than 2.0 NTU at any time. Besides its use in spray irrigation, R-1 water can be used for firefighting, commercial laundering, drinking water by farm animals, and direct-contact irrigation of vegetables/fruits that are eaten raw (Hawai'i Department of Health 2002).

CHAPTER SIX

# Groundwater

The mystique of groundwater arose when humans first encountered water gushing from springs and primitive artesian wells. This water must have seemed inexhaustible and to have originated from mysterious sources. Eventually humankind discovered and appreciated the enormous subsurface reservoirs that are available naturally at no cost and that contain water usually suitable for drinking without treatment. Groundwater has subsequently become the universal premier drinking water supply source. However, because most groundwater is invisible and the subsurface is not easily accessible, accurate assessment of groundwater behavior remains a continuing challenge.

## Fundamentals of Groundwater

### Aquifers

Aquifers are geologic formations from which substantial amounts of water can be extracted. An aquifer may be composed of any type of permeable and porous rock, including unconsolidated sand and gravel (see Chapter 5). Poorly permeable formations in which the flow of water is highly restrained are called aquitards, and those in which the flow of water is virtually precluded are called aquicludes. Aquicludes serve as impermeable boundaries of aquifers. Exploitable aquifers usually are areally extensive and vertically thick.

Following rainfall some of the infiltrated water may pass through the shallow soil and root zone into an intermediate vadose zone that retains some water but often dries up. Perched water bodies may occur in the vadose zone on areally limited and poorly permeable strata such as clay lenses, ash beds, and buried soils. Beneath the intermediate vadose zone lies a zone of saturation in which the main groundwater resource is found. It usually begins less than 30 m (100 ft) below the ground surface but may begin much deeper in many Hawaiian terrains.

The top of the saturated zone is called the water table and the aquifer is known as unconfined. At the unconfined water table is a capillary fringe created by surface tension that is not fully saturated and is generally thin (see Figure 5.5*a*).

The energy possessed by a mass of groundwater at any point is termed head, also known as piezometric or hydraulic head. Head is indicated by the water elevation above a datum, as observed in a piezometer. This instrument is an open-ended tube with its lower end inserted at the point where head is measured. Groundwater head consists of only two components (see Figure 5.5*b*): pressure head, $p/\rho g$, where $p$ is water pressure, $\rho$ is water density, and $g$ is gravitational acceleration; and position head, $z$. Pressure in groundwater is always positive (equal to or greater than atmospheric pressure). The datum is an arbitrarily chosen elevation where $z$ is set as zero. In Hawai'i the datum is commonly sea level. The velocity head of groundwater flow is too small to matter in energy analysis.

A confined aquifer is bound at the top and theoretically at the bottom by either aquitards or aquicludes (Figure 6.1). The confined aquifer water pressure is high enough to force the head to reach above the lower surface

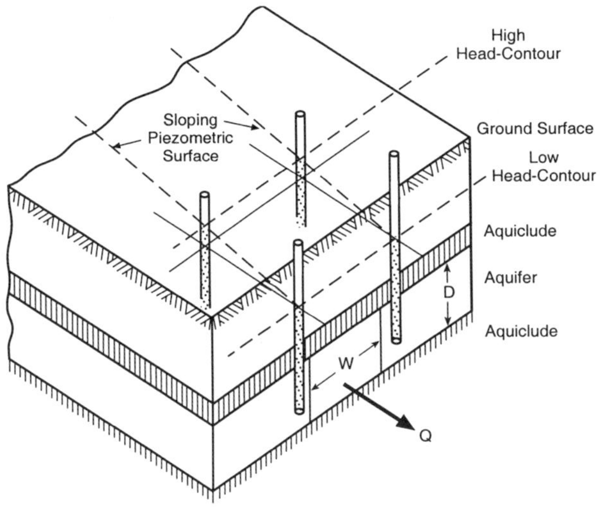


Figure 6.1. Groundwater flow in a confined aquifer: discharge *Q* of groundwater flow through a cross-sectional area of width *W* and thickness *D*.

of the top confining layer, which is also the top of the saturated zone of the aquifer (see Figure 5.5*a*). When the water level rises above the ground, the aquifer is called artesian. Confined aquifers are recharged from infiltrated rainwater occurring in areas beyond the confining layer and/or from underflow, that is, groundwater flow from adjacent connecting aquifers. Hydraulically, the water pressure everywhere in a confined aquifer is greater than atmospheric pressure.

Unconfined aquifers are vulnerable to contamination when the overlying surface becomes contaminated. On the other hand, confined aquifers are naturally protected from direct surface contamination by the upper confining layer; the only possible exception is when downward leakage occurs through the confining layer.

## Flow and Transport

### Flow

Groundwater is in a continuous state of slow and diffusive motion in the void network of a rock formation. Typically, the average velocity is on the order of 0.3 m/d (1.0 ft/d). By comparison, the average velocity of dry-weather streamflow is ordinarily about 1.0 m/s (3.3 ft/s), which is faster than groundwater velocity by an order of nearly five magnitudes. The distributive pattern of flow is represented by a velocity vector field in which flow lines are drawn in such a way that the tangent of the line at a point coincides with the velocity vector at that point. The field indicates convergent, divergent, and other variable directions of flow.

The physics of groundwater flow is described by Darcy's Law:

$$q = KJ \tag{6.1}$$

where $q$ is specific discharge (length/time), which is converted to discharge, $Q$, per unit cross-sectional area $A$, $K$ is hydraulic conductivity (length/time), and $J$ is hydraulic gradient (dimensionless), or the slope of the water table or piezometric surface (see Chapter 5). Darcian flow is laminar, and its energy loss results from friction alone.

Actual groundwater velocity $V$ averaged over cross-sectional area $A$ is $q/n$, where $n$ is porosity. The magnitude of the hydraulic gradient is the difference of groundwater head between two points along a flow line divided by the distance between them. The amount (volumetric rate) of groundwater flow within a width $W$ between two adjacent flow lines is computed with Darcy's Law as $Q = K(DW)J$, where $D$ is the aquifer thickness (Figure 6.1).

Hydraulic conductivity $K$ varies tremendously among rock types from $10^{-10}$ to $10^{+7}$ m/d ($3 \times 10^{-10}$ to $3 \times 10^{+7}$ ft/d) (Smith and Wheatcraft 1993). Generally speaking, $K = 1$ m/d (3 ft/d) is considered sufficiently permeable to allow economic exploitation. $K$ varies with soil texture but much less widely than it does for rocks (see Chapter 5, Figure 5.3).

Homogeneity (identical properties at every point) and isotropy (identical properties in all directions) are the usual initial assumptions of the hydraulic properties of the rock formations in flow and transport analysis. In reality, layered geologic media tend to be anisotropic, favoring horizontal over vertical flow.

Groundwater flows under mild hydraulic gradients on

the order of 0.01 to 0.0001. Flatter gradients are characteristic of more permeable aquifers.

The change in volume of water stored in the saturated zone of an aquifer responding to a change in head is measured by its storage coefficient or storativity. It is defined as the volume released from or taken into storage per unit change in groundwater head per unit surface area of the aquifer. For an unconfined aquifer it is equal to specific yield, or effective porosity, which is the ratio of the water volume drained, after saturation, under gravity to the bulk volume. The storage coefficient in most unconfined aquifers ranges from a few percent to 30 percent. When applied to confined aquifers, storativity reflects the compressibilities of the water and the rock. Its value is very small — only on the order of $5 \times 10^{-3}$ to $5 \times 10^{-5}$.

Regional groundwater flow tends to conform to the patterns of aquifer formations, which commonly are dominated by horizontally extensive layers. Groundwater generally flows along the path of least resistance from recharge areas to discharge locations. Beyond these generalities, many factors modify the flow directions, including geologic and hydrologic boundaries and methods of groundwater extraction.

The pattern of circulation and distribution of groundwater flow may be estimated by constructing a flow net as a first approximation. The contours are lines of equal head that are often influenced by surface water features and geologic boundaries. Flow lines are then drawn perpendicular to the contours, assuming aquifer isotropy. Flow lines and contours together form a flow net, a simple yet powerful tool for a general evaluation of groundwater flow patterns and volumes.

Realistic modeling of groundwater flow is possible if sufficient data are available. Modeling involves solving the governing flow equations for given initial and boundary conditions constrained by reasonable assumptions. The resulting head and velocity patterns depict the circulation of groundwater flow.

Flow in unconfined aquifers involves a free surface (water table), which renders the partial differential equations nonlinear and extremely difficult to solve analytically. Fortunately most regional groundwater flows are driven by low gradients and have water table configurations that permit application of the Dupuit approximation, which assumes horizontal flow vectors and vertical equipotential lines. This simplification permits analytical solutions to many problems of the flow field.

Complete discussions of groundwater hydrology and hydraulics are provided in standard textbooks such as Todd (1980) and Bear (1979).

## Transport

Considerations of transport of solutes in the subsurface water were initiated when Slichter reported on his salt experiments in 1905. By the 1950s, the phenomenon of hydrodynamic dispersion in groundwater began to receive attention, especially where seawater intrusion affected freshwater aquifers. But it was not until groundwater contamination was recognized as a major environmental concern in the 1980s that transport phenomena finally received the degree of attention formerly reserved exclusively for flow phenomena.

The nature of groundwater transport pertains to the interaction between the subsurface environment and contaminants — mostly dissolved chemicals but occasionally microbes as well. The characteristics of most concern are the persistence, mobility, and potency of the contaminants. Potency or strength is measured by concentration, persistency by the length of time required for breakdown or disappearance, and mobility by how easily the contaminants are transported. Conservative, nonreactive inorganic chemicals, such as chlorides, are among the most persistent and mobile.

Contaminants dissolved in water are transported as solutes. The transport of other contaminants such as gasoline and oil, a group known as nonaqueous-phase liquid, is more complicated. Their migration is influenced by many properties, including surface tension, density, solubility, and volatility.

The nature of groundwater transport is presented in detail in Chapter 5. Everything discussed there about the transport processes is applicable to the discussion that follows. Many of the processes that decompose contaminants become much less effective below the relatively thin

(1 m [3 ft]) mantle of "living" filter of soil and vegetation. Beneath the filter, many of nature's attenuative processes decrease sharply. On the other hand, the long residence time of water in the saturated zone favors dilution and slow-rate processes such as hydrolysis.

Groundwater transport of solute contaminants can be depicted as millions of dissolved particles introduced into the ambient flow. If inserted in a single dose, they spread and travel as an isolated plume; if fed continuously, they form a continuous plume. In either case, they disperse much more along the flow direction (longitudinal) than across it (transverse). In the case where a continuous discharge of contaminated water is strong relative to the specific discharge of the ambient groundwater flow, the ultimate transport is essentially one of displacement, with dispersion occurring only at the fringes.

A full understanding of the natural processes involved and an accurate description of the concentration patterns are tasks still in progress. However, methodologies for approximate practical solution are available; the choice of methods depends on the importance of the problem, data available, and acceptable assumptions and approximations. For example, nonreactive (conservative) contaminants travel with and disperse in water. The arrival pattern of conservative contaminants is determined by solving coupled partial differential equations of flow and mass transport of the contaminant using a priori knowledge of the aquifer dispersivity and initial and boundary conditions. This approach is not adequate for nonconservative contaminants that may be sorbed or that undergo biochemical transformation. Evaluation of sorption and biodegradation requires parameters such as the sorption distribution coefficient and the decay coefficient. See Chapter 5 for additional discussion.

## Aquifer Heterogeneity and Scale Dependence

A porous medium is traditionally treated as a continuum, in which fluid flow is deterministic (Bear 1979). The continuum-deterministic approach overcomes the difficulties inherent to the heterogeneity of geological media, which is impossible to fully characterize by measurements and to deal with by mathematical analysis.

In the 1980s subsurface hydrology began to undergo examination as a stochastic process. The problem focuses on the stochastic properties of the geological media (i.e., the spatial distribution of the hydraulic properties of the heterogeneous media) that could be applied to large-scale flow and transport problems. According to Gelhar (1993), the mean behavior of flow and transport phenomena based on stochastic approach is similar to classical deterministic descriptions in many cases. As a result, there are no absolute choices as to which approach is more appropriate to use. Methods for stochastic solutions to several types of seawater intrusion are presented according to which outcome of uncertainty is quantified in terms of the mean and variance of the inputs (Cheng and Ouazar 1999; Dagan and Zeitoun 1999).

Aquifer parameters of interest to contamination problems can take on a totally different order of magnitude, depending on the scale (linear dimension) of the problem. Scale dependence reflects aquifer heterogeneity. Dispersivity and hydraulic conductivity, which are controlled solely by the solid structure of the rock formations, are the basic parameters of concern (Lau et al. 1958). A collection of field dispersivity values indicates clearly scale dependence as noted in Figure 6.2 (Gelhar et al. 1992). Neuman (1990) suggested universal scaling of hydraulic conductivities and dispersion in geologic media.

Scale dependence of hydraulic conductivity, $K$, in heterogeneous media has been formulated with the equation $K = CV^m$, where $C$ is a parameter characteristic of the geological medium, $V$ is the volume of the medium used as a measure of scale, and $m$ is the exponent of the relationship (i.e., slope of the line on a log-log plot [Schulze-Makuch et al. 1999]). Based on thirty-nine geological media, the value of the exponent is shown to depend on whether the flow is in porous media (0.5), double-porosity media (0.5 to 1.0), or fracture and conduit media (1.0). The relationship is valid to an upper bound value beyond which $K$ remains constant with scale. No variation of $K$ with scale was observed for homogeneous media.

## Seawater Intrusion

Many fresh groundwater bodies in coastal regions either adjoin or overlie saltwater bodies. The intrusion of seawater into freshwater limits the usefulness of aquifer water as a source for drinking and other beneficial uses.

The physics of buoyancy in which less-dense freshwater floats on denser saltwater results in the following formula:

$$\zeta = -\delta h_f \qquad (6.2)$$

where $\zeta$ is the depth to the freshwater-saltwater interface below sea level, $h_f$ is the freshwater head above sea level, and $\delta = \rho_f/(\rho_s - \rho_f)$, where $\rho_f$ and $\rho_s$ are the density of freshwater and seawater, respectively. For a freshwater density of 1.000 and a seawater density of 1.025, $\delta$ equals $40 = 1/(1.025 - 1)$. Thus, a freshwater head rising to 1 ft above sea level is balanced by 40 ft of freshwater extending below sea level, assuming static condition. In other words, a 1-ft freshwater head can suppress the interface to 40 ft below sea level. The formulation is commonly credited to W. Badon-Ghyben (1888/1889) of Holland and B. Herzberg (1901) of Germany. $\delta$ is sometimes known as the Ghyben-Herzberg ratio.

The Ghyben-Herzberg relationship assumes static equilibrium and is based on hydrostatic pressure distribution in the freshwater region, stationary seawater, and an immiscible interface. Hubbert (1940) presented a generalized expression applicable at the interface as $\zeta = (1 + \delta)\, h_s - \delta h_f$, where $h_s$ is seawater head. This equa-

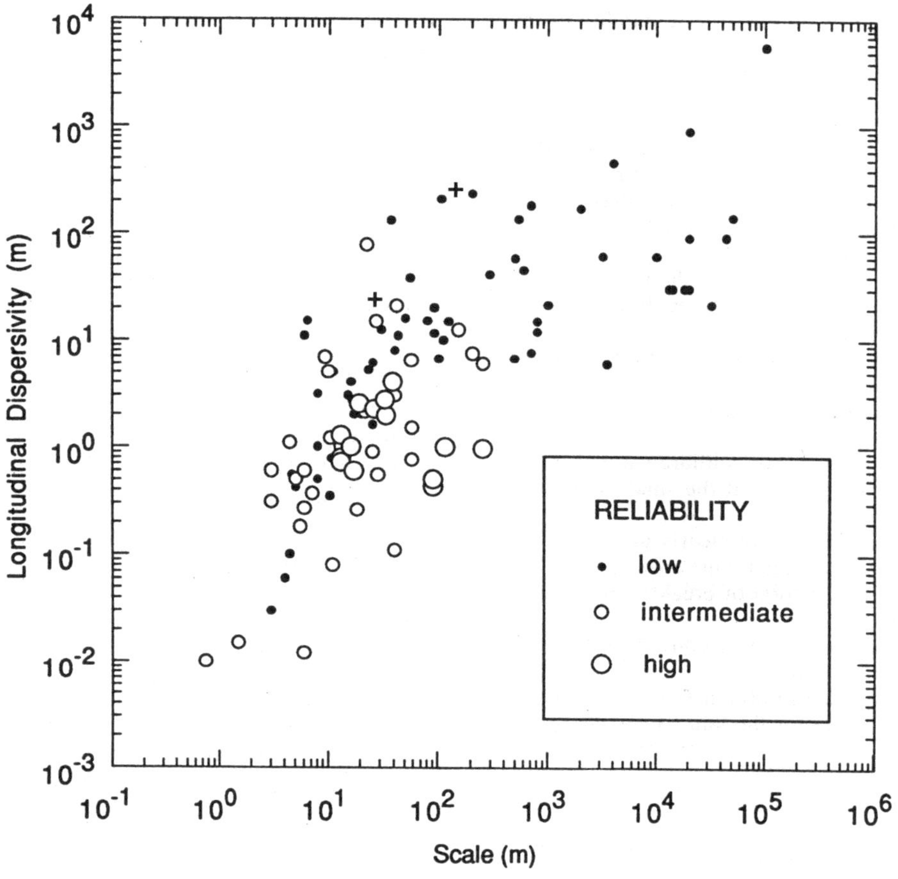


Figure 6.2. Field longitudinal dispersivity values versus the scale of measurements and reliability of data. Data are based on fifty-nine field sites of widely different geologic materials. (Reprinted from Gelhar et al. 1992 with permission from American Geophysical Union); 1 m = 3.28 ft. Two added points (+) are based on field helium experiments in Ko'olau flank lava. (Reprinted by permission from Gupta et al. 1990)

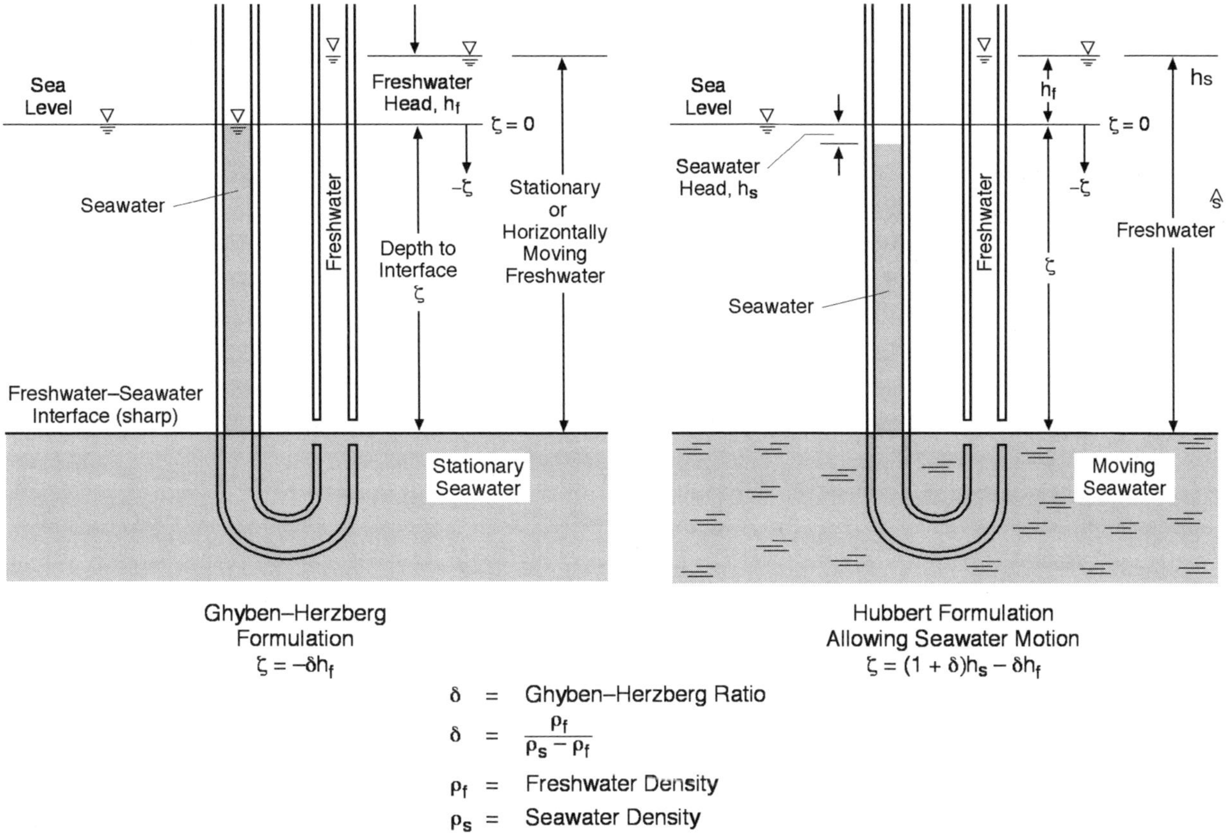


Figure 6.3. Ghyben-Herzberg relationship and Hubbert formulation. Not to vertical scale.

tion is reduced to the Ghyben-Herzberg formula when the seawater is stationary (i.e., $h_s = 0$ [see Figure 6.3]).

The Ghyben-Herzberg relationship is valid for dynamic equilibrium between steady horizontal freshwater flow and stationary seawater separated by a sharp interface (Bear 1979). The validity is based on Dupuit's assumption for horizontal flow. According to Dupuit (1863), if the hydraulic gradient is small enough ($10^{-2}$ to $10^{-3}$) to be approximated by its horizontal component, the flow direction is virtually horizontal and the equipotential lines are virtually vertical lines that have hydrostatic pressure distribution.

The Ghyben-Herzberg ratio is temperature dependent because the density of each type of water is inversely proportional to temperature. The density of freshwater is 1.000 at 5°C (41°F), and that of seawater is 1.025 at 15°C (59°F) (de Marsily 1986).

The approach of coupling sharp interface and Dupuit's horizontal flow in using the Ghyben-Herzberg relationship has enabled analytic solutions to many problems in seawater intrusion (Bear 1979). For example, the seaward freshwater steady flow $Q_o$ per unit width in unconfined aquifers without overland recharge is related to Ghyben-Herzberg head as follows:

$$Q_0 = \frac{K}{2}\left(\frac{1+\delta}{\delta^2}\right)\zeta^2\frac{1}{L} \tag{6.3}$$

where $\zeta$ is the depth to the interface and equals $\delta h_o$ and $h_o$ is the freshwater head, and $L$ is the distance from the discharge front at the coast. A simpler but identical form of the equation is $Q = 41Kh_o^2/2L$ in which $(1 + \delta)$ is replaced by 41 and $\zeta^2/\delta^2$ is replaced by $h_o^2$. A similar equation has been derived for aquifers confined from above; it differs from equation (6.3) only by replacing the bracketed term with $1/\delta$ and replacing $\zeta$ with the depth to the interface measured from the base of the confining layer (Harder et al. 1953; Bear 1979).

Seawater motion beneath the freshwater has long been recognized, but freshwater motion is typically the dominant flow. Thus, seawater stationarity is considered a reasonable assumption. Theoretically, saltwater flow can affect freshwater head and flow (Hubbert 1940; Bear 1979).

A transition zone instead of a sharp interface exists because freshwater and seawater are miscible. The zone contains a mixture of freshwater and seawater and varies in thickness. Along a vertical, the concentration of chlorides or salinity (the most common indicators of seawater) increases sigmoidally with depth in conformance with the equation of hydrodynamic dispersion. Typically, the transition zone is thin at a distance far from the coast, then thickens gradually seaward (i.e., along the direction of flow under natural conditions). The transition zone manifests the effect of variable density and may be described by a pattern of isochlors (line of equal chlorides). The Ghyben-Herzberg ratio applies theoretically to the midpoint of the transition zone, provided the flow is horizontal and the salinity-depth distribution is symmetrical about the midpoint (Appendix 6.1).

The mass balance of solutes involving variable density was mathematically formulated by Bear (1979), but solution of the equations had to await numerical techniques such as the model SUTRA (Voss and Souza 1987). A sharp interface is an acceptable assumption if the transition zone is thin compared with the thickness of the freshwater lens.

Upconing phenomenon is the bulging of the interface directly beneath the intake of a groundwater-extraction facility. The rise can become unstable and discharge saltwater into the intake when robust pumping exceeds certain limits. Semiempirical criteria have been developed for salt-free operation. For example, one such criterion is that the well bottom be set four times higher than the crest (maximum rise) of the sharp-interface mount, $m_o$, above the original interface at steady state (Schmorak and Mercado 1969). The crest may be determined using a modified theory by Dagan and Bear (1968) as follows:

$$m_0(t) = \frac{Q_p}{2\pi \frac{1}{\delta} K_x d} \left( 1 - \frac{2nd}{2nd + \frac{1}{\delta} K_z t} \right) \tag{6.4}$$

where $Q_p$ is the pumping rate; $d$ is the depth to the original interface below the bottom of the well; $K_x$ and $K_z$ are longitudinal and transverse hydraulic conductivities, respectively; $n$ is porosity; and $t$ is time since pumping began. Additional work has been done to explain field-related observations (Reilly et al. 1987).

Bear et al. (1999) presented current knowledge of seawater intrusion in coastal aquifers — concepts, methods, and practices. The theoretical aspect was handled with a rigorous mathematical presentation of the physical and conceptual models, including three-dimensional sharp interface and three-dimensional miscible transport, followed by an elementary coverage of analytical solutions and stochastic solutions. A series of numerical models including SUTRA and SHARP (see Models section later in this chapter) and a comprehensive survey of computer codes were detailed. Optimal exploration of fresh groundwater and management aspects were dealt with analytically for prevention and remediation of seawater intrusion. Updated information was provided for geophysical, geochemical, and isotopic aspects.

## Groundwater in Hawai‘i

Although hardly recognized in a formal sense as one of the most precious and outstanding natural resources in the Hawaiian Archipelago until the explosion of economic development following the granting of statehood, groundwater has been a vital component of the culture and economy in Hawai'i since the arrival of the first settlers more than 10 centuries ago as well as a controlling element of the natural environment and an esthetic adornment. The importance of water is deeply imbedded in the Hawaiian culture, and most of the water sources that have fashioned that culture originate as groundwater. The perennial streams, springs, and wetlands relied upon so profoundly by the early Hawaiians are groundwater discharges.

Except in regions of persistent and high rainfall, every perennial stream in Hawai'i is sustained by groundwater flow. In mountainous areas this flow drains mostly from high-level dike aquifers and less commonly from high-level perched aquifers. Drainage may be focused as springs or diffused as seepage into stream channels. Among large perennial streams sustained by high-level groundwater are the following drainages: from Kailua to Punalu'u in windward O'ahu; Waihe'e, 'Īao, and Honokōhau in West Maui; Waikolu, Pelekunu, and Wailau in Northeast Moloka'i; Waipi'o and its tributaries in the Kohala mountains of Hawai'i; and virtually every drainage in the island of Kaua'i.

In the lower reaches of the volcanic slopes where the free groundwater table lies above the ground elevation, seepages generate streams, coastal springs, and wetlands. The voluminous low-level springs along the margin of Pearl Harbor in southern O'ahu are dramatic manifestation of the escape of basal groundwater. Even more voluminous basal groundwater springs drain into Hilo Bay and the smaller bays to the east of it. An unrecognized value of groundwater in the sediments not far below the ground surface in coastal plains is its accessibility by tree roots. Rainfall on the Honolulu plain is insufficient to ensure the survivability of trees; the roots must thrust through the sediments to the water table for the trees to thrive.

## Size of Groundwater Resources

The largest basal aquifer in Hawai'i lies in Mauna Loa flank lavas between the Hilo coast and the high-rainfall area to about the 1,520-m (5,000-ft) elevation. An enormous volume of cool, fresh groundwater moves through the aquifer to discharge freely at the coast, unimpeded by a caprock. Discharged as a spring, it would be among the most voluminous in the world.

Perhaps no other island on the planet has the same high ratio of groundwater resources to land area and population as does O'ahu. The environmental features created by nature that have resulted in O'ahu's vast groundwater supply and its exploitability could hardly be improved upon had they been designed by technocrats.

Groundwater is the preferred source for the domestic water supply throughout the state. The large population on O'ahu depends almost exclusively on groundwater. The groundwater resources of West Maui are on the verge of being fully developed, but in East Maui a vast supply remains in its pristine state. The fresh groundwater resources of Moloka'i and Lāna'i, although far less voluminous per unit of land area than those of the larger islands, are crucial to the welfare of the islands' inhabitants. In Kaua'i, the most geologically complex of all the islands, groundwater resources are universal but most often occur in smaller, less-easily-developed aquifers than those in the other major islands of Hawai'i, Maui, and O'ahu.

## Occurrence of Groundwater

Aquifers in Hawai'i consist of either volcanic rock or sedimentary rock. Volcanic aquifers are much larger and more extensive than sedimentary aquifers and, in fact, constitute the only aquifers capable of supplying potable water. The yield of sedimentary aquifers is almost always brackish water, and usage is restricted to irrigation.

"High level" and "basal" are the two fundamental varieties of groundwater in the islands. High-level groundwater is either isolated from or beyond the reach of seawater intrusion; basal groundwater rests on and is hydraulically continuous with underlying seawater. Major types of groundwater were recognized early in the twentieth century in a succession of discoveries (D. C. Cox 1981). The Ghyben-Herzberg ratio was also discovered independently by Andrews in Honolulu in 1909.

## High-Level Groundwater

High-level groundwater either is trapped in lava aquifers between quasi-vertical dikes or occurs in lavas underlain by quasi-horizontal low-permeability strata. Dikes are the remnant conduits through which lavas were extruded from volcanic rift zones. Commonly less than 1.5 m (5 ft) wide, they are very poorly permeable and often virtually impermeable. The horizontal strata that perch groundwater include low-permeability ash and tuff beds, unusually massive and dense lava flows, and soils and erosional debris associated with unconformities.

Dike aquifers discharge by subterranean flow through fractures in the dikes or by overflow into stream channels where dikes have been truncated by erosion. In the dike complex, where the proportion of dike rock to total rock mass is greater than 5 to 10 percent, the individual aquifers are small, but in the marginal dike zones, where the proportion of dike rock to total rock mass is less than 5 to 10 percent, they can be extensive. Exploitable groundwater resources in dike zones are substantial, even though they are volumetrically meager in comparison with those in basal aquifers. However, their usual elevation of occurrence, their ease and low cost of development in areas where they are accessible, and their freedom from the threat of seawater intrusion make them a highly valuable resource. A comprehensive discussion of dike-impounded groundwater in O'ahu is given in Takasaki and Mink (1981).

In contrast to high-level dike water whose movement is constrained by vertical barriers, the movement of high-level perched water is controlled by horizontal barriers. Perched aquifers are small and tend to be discontinuous. They generally are not large enough to serve as water supply sources for more than a small number of enterprises or a few households. Most perched water bodies are found within the volcanic rock succession, but very small aquifers may exist within sedimentary sequences.

## Basal Groundwater

The greatest and most voluminous aquifers saturate flank lavas seaward of rift zones. They were named "basal aquifers" by the U.S. Geological Survey during early investigations of groundwater in Hawai'i, presumably because the elevation of the water table of a resource is lower than that of high-level water tables. The highest confined basal water table in Hawai'i was encountered at 12.8 m (42 ft) above sea level in the Honolulu coastal plain after well drilling first started in 1879.

Basal groundwater rests on seawater, and its depth below sea level is controlled by the difference in density between the two types of water. The surrogate for density is salinity of the waters. Theoretically, groundwater resting on seawater or brackish water, which is less dense (that is, having a chloride content of less than 19,000 mg/l), is termed "basal." For the condition of pure freshwater at 5°C resting immiscibly on seawater at 15°C, the respective densities are 1.000 and 1.025, and thus by the buoyancy principle the depth of freshwater below sea level is forty times its elevation above sea level. This ideal case, which is never truly achieved, is referred to as the Ghyben-Herzberg ratio, and bodies of freshwater floating on seawater in porous media are called either basal or Ghyben-Herzberg lenses.

The term "basal water" has spread from Hawai'i throughout the Pacific islands and has become the standard nomenclature for referring to freshwater overlying seawater. A companion term, "parabasal," refers to groundwater hydraulically continuous with basal water but resting on an impermeable stratum rather than seawater. The term was first used to describe common groundwater occurrences on the island of Guam where the impermeable basement on which the limestone aquifers lie is highly irregular. The basal water aquifers in Hawai'i basalts lack a bottom aquiclude, but parabasal groundwater may occur in narrow zones above low-permeability uncomformities.

The volumetrically and dimensionally greatest groundwater resources are in volcanic rock aquifers, principally in the flank lavas of the primal shield volcanoes in each island. Less extensive basal aquifers occur in lavas that originated with later volcanic eruptions. Smaller, and generally fully brackish (slightly salty) in contrast to low-salinity freshwater that dominates the large volcanic basal

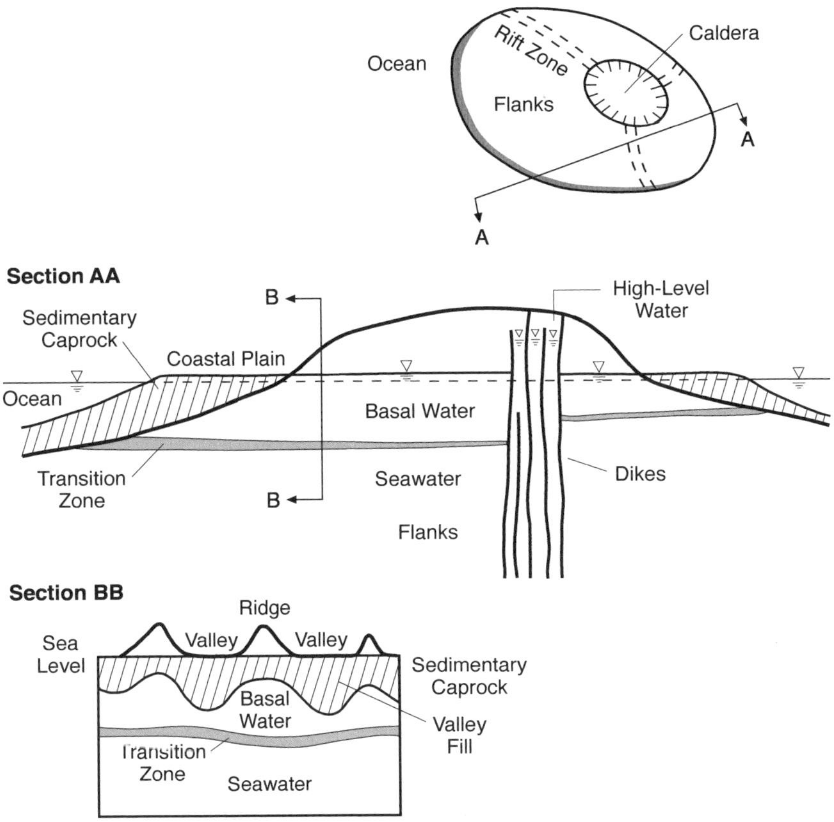


Figure 6.4. Schematics of basal and high-level groundwaters in basaltic aquifer with coastal sedimentary caprock. Vertical dimension is highly exaggerated. (The top figure is redrawn from Macdonald and Abbott 1970)

aquifers, are the basal aquifers in sedimentary sequences forming coastal plains. These sedimentary successions compose the caprock, which impedes discharge from volcanic basal aquifers into the sea because at the base of and within the sequences are very low-permeability strata. Nevertheless, the top coralline stratum of the sedimentary succession usually carries a brackish basal lens. Brackish basal lenses also occur in arid to semiarid regions where groundwater flow is not restrained by a caprock or other subsurface impediment.

The natural flow gradient in large basal aquifers is typically very small, on the order of 0.2 m/km (1 ft/mi) because of the remarkably high regional permeability of basalts. Extraction of groundwater from large volcanic basal aquifers is not usually accompanied by large drawdowns in the laminar flow zone of the aquifer a few meters (feet) away from the well. Unusual drawdown is induced by turbulent flow caused by an increase in velocity of the water as it enters the well bore.

Figure 6.4 depicts schematically the principal groundwater bodies and some related geological features in the Hawaiian Islands.

## Groundwater Quality

The subsurface journey of rainwater in Hawai'i begins as infiltration, principally in the rain forest in the mountains and secondarily in leeward slopes and plains, and ends at the coast. Groundwater that accumulates in the mountain conservation zone is superb in quality and needs no treatment before being used as drinking water. Typically, it has only 10 to 20 mg/l of chloride, 35 to 45 mg/l of silica (a measure of dissolution of basalt), and less

Table 6.1. Chemical Composition of Rainwater, Dike Water, and Uncontaminated Basal Water on the Islands of Hawai'i and O'ahu, Hawai'i

| Water-Quality Parameter | Rainwater Mean | | Dike Water Ko'olau Range Median | Uncontaminated Basal Water Southern O'ahu Mode | Seawater |
|---|---|---|---|---|---|
| | Hilo[a] | Mauna Kea[b] | | | |
| $SiO_2$ (mg/l) | — | — | 15 | 36 | 1 |
| $Ca^{+2}$ (mg/l) | 1.11 | 0.39 | 7.2 | 8.0[c] | 400 |
| $Mg^{+2}$ (mg/l) | 1.54 | 0.45 | 4.6 | 6.0 | 1,272 |
| $Na^{+1}$ (mg/l) | 6.4 | 1.62 | 11 | 20 | 10,561 |
| $K^{+1}$ (mg/l) | 0.66 | 0.11 | 1.4 | 2.0 | 380 |
| $HCO_3^{-1}$(mg/l) | — | — | 50 | 65 | 140 |
| $SO_4^{-2}$(mg/l) | 2.66 | 0.89 | 3.5 | 5.5 | 2,649 |
| $Cl^{-1}$ (mg/l) | 11.0 | 3.11 | 16 | 22 | 18,980 |
| $NO_3^{-1}$(mg/l) | 0.61 | 0.39 | 0.55 | 1.1 | 0.05 |
| $PO_4^{-3}$(mg/l) | — | — | — | 0.20 | 0.01 |
| pH | 5.4 | 5.1 | 7.9 | 7.9 | — |
| $F^{-1}$ (mg/l) | — | — | 0.07 | 0.07 | 1.3 |
| Specific conductance (μmho at 25°C) | — | — | 155 | 205 | — |
| Total dissolved solids, calculated (mg/l) | — | — | 111 | 165 | 34,380 |

SOURCE: Visher and Mink 1964.

[a]Elevation 9.2 m (20 ft), distance from coast 1 km (0.6 mi).

[b]Elevation 404–802 m (1,325–2,630 ft), distance from coast 6.3–12.2 km (3.9–7.6 mi).

[c]Mean value.

than 1 mg/l of nitrate as $NO_3^{-1}$. Other constituents including fluorides occur naturally and in concentrations that are far below recommended maximum drinking water limits (Table 6.1).

Meteoric water in Hawai'i is virtually pure, but within the sphere of influence sea spray adds trace amounts of salts and nitrogen, and volcanic emission lowers the pH and adds some sulfate (see Chapter 3).

During infiltration, rainwater dissolves a considerable amount of carbon as $HCO_3^{-1}$ from the $CO_2$-rich soil gases in the mantle of vegetation and soils. During percolation in the unsaturated zone and flow in the saturated zone, a substantial amount of silica is dissolved from basaltic rocks. Leaching of basalt adds alkalies and alkaline earth metals. On the other hand, basalt contains very small amounts of chlorides that can be easily leached.

Gradation in salt content has been indicated in different dike waters. $Cl^-$ in Waiāhole Tunnel water is 12 mg/l, which is low compared with the 18 mg/l in Waimānalo water and the 29 mg/l in Wai'anae Range water. Such gradation has been attributed to variations in rainfall abundance and evaporative climate.

The rain-recharged groundwater forms the uncontaminated freshwater core of the basal lens. The quality of basal-lens water is affected by dissolved matter generated by surface activities from above and by the encroachment of saltwater from below. Historically, the infiltration of excess irrigation water was a primary source of introduced dissolved constituents, carrying with it salinities higher than those occurring in the pristine water, along with residue of chemicals applied as fertilizers and pesticides. In recent years, urban use of organic chemicals as solvents and fuels has contributed to contamination from the surface.

Seawater intrusion, indicated by salinity, is induced by natural perturbation to some extent (e.g., tidal fluctuation and seasonal variations in recharge) but mostly by stresses induced by the extraction of groundwater by pumping, especially with high-capacity pumps in deep wells. Since groundwater development began over a century ago, freshwater levels have lowered and seawater encroachment has advanced. These phenomena, however, are controllable to an acceptable state when correct management practices are employed.

Groundwater temperature is an important variable in the understanding of the structure and dynamics of Hawai'i's groundwater systems and the system's communication with the ocean.

Rainwater becomes heated after infiltration. On O'ahu, the heat-generating processes include (1) equilibration in the soil temperature, (2) frictional heating during groundwater movement, (3) mixing with warmer return irrigation water, and (4) geothermal heating at the base of the lens. The recharge water in the central Ko'olau Range of southern O'ahu is about 19.7°C (67.5°F). The pumped basal water from wells and infiltration galleries is warmer, averaging 21.3°C (70.3°F) in the Honolulu aquifers and 20.6°C (69.1°F) in the central Pearl Harbor aquifers. The lower temperature in the Pearl Harbor aquifers is probably due to a greater portion of the recharge water originating from higher altitudes where soil temperatures are cooler. The warming effect of return irrigation water is noticeable in the spring discharges along the inland margin of Pearl Harbor. The average springwater temperature is 21.2°C (70.2°F), which is 0.6°C (1.1°F) warmer than the water in the deeper portion of the basal lens (Mink 1964).

At a specific location along the subsurface journey of the groundwater, the vertical profiles of water temperature indicate the layering of water in a thick lens. The Waipi'o monitoring well in the Pearl Harbor aquifer is an example as shown in Figure 6.5 (Voss and Wood 1993). The consequences of return irrigation on temperature have been moderated since heavily irrigated sugarcane cultivation has ceased.

Groundwater that leaks into the ocean is considerably colder than the warm surface coastal and ocean water by as much as 6°C (10.8°F). This phenomenon has been detected by infrared sensing near points of concentrated discharge along the coasts (see Chapter 8).

Water-quality data for understanding circulation at greater depths than previously reported were provided from a 1,056-m-deep (3,468-ft-deep) borehole drilled near the shoreline of Hilo Bay (Thomas et al. 1996). At a depth of 700 m (2,297 ft) saline water occurred at a temperature of 5°C (41°F), which is nearly identical to that in the open ocean at the same depth. The radiocarbon $^{14}C$ age of the deep saline groundwater was about 6,000 to 7,000 years before present. The rate of fluid flow advancing from the ocean was inferred to be 2.2 m/yr (7.2 ft/yr). In the same borehole, a 200-m-thick (656-ft-thick) zone of freshwater (100 mg/l $Cl^{-1}$ by calculation) was discovered at a depth of 310 m (1,017 ft) in the Mauna Kea basalts below a soil horizon that occurs at the contact of Mauna Loa and Mauna Kea formations. Thomas et al. suggested that the freshwater originates from rainfall on Mauna Kea at the 2,000-m (6,562-ft) elevation level and discharges as deep submarine springs in Puna Canyon at a depth of 600 to 800 m (1,968 to 2,625 ft) below sea level.

In the Hilo deep borehole, there is little evidence of ion exchange with formation clays or sediments. This is at variance with previous reports indicating depletion of sodium and potassium with corresponding enrichment of calcium and magnesium in subbasal lens waters on O'ahu (Visher and Mink 1964; Voss and Wood 1993). The absence of a thick blanket of marine and terrestrial sediments off the coast of Hawai'i is responsible for the lack of base exchange.

Dating a groundwater sample is not simple. In principle, it is similar to dating a relic bone or wood by mea-

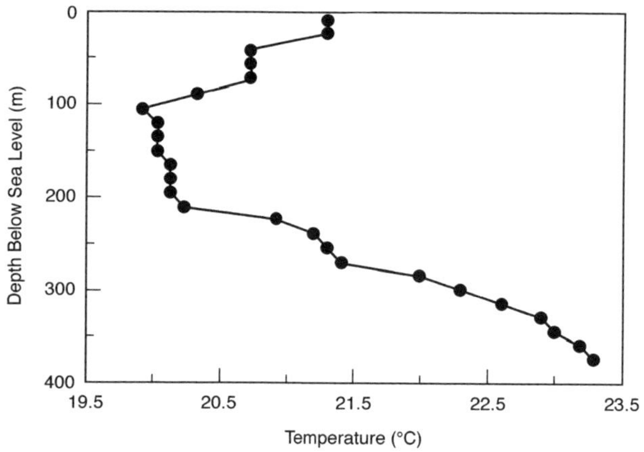


Figure 6.5. Depth profile of water temperature in the Waipi'o monitoring well (1988), O'ahu, Hawai'i. 1 m = 3.28 ft. (Adapted from Voss and Wood 1993)

suring its radiocarbon activity and employing the radioactive decay law. Weak activity means old age. Tritium, a radioactive isotope of hydrogen, is also commonly used in groundwater dating.

Both isotopes exist naturally in the atmosphere as a result of cosmic ray bombardment. Their activity exists at very low levels, and the decay rate is assumed to be constant for thousands of years. Biological dynamic equilibrium is maintained between these isotopes and all living organisms. The high-altitude testings of nuclear bombs in the 1950s spiked the activity to very high levels, distinguishing clearly the pre- from postbomb era.

After entering the water table, the water loses contact with the atmosphere, and the radioisotopes in the water disintegrate without replenishment from the atmosphere. The rate of decay is precise but varies with each element. The half-life is the time period during which half of the original activity is disintegrated. Tritium's half-life is 12.27 years, and radiocarbon's is 5,730 years.

Measurement of radiocarbon activity is made on the carbon content, which occurs predominantly as $HCO_3^{-1}$ for the pH range of 7.0 to 9.5. A large volume (210 to 420 l [55 to 110 gal]) of water is required for accurate measurement. Measurement of tritium activity is made directly on the water sample (usually 500 ml [0.1 gal]) after enrichment by electrolysis.

W. F. Libby, a Nobel laureate recognized for his discovery of radiocarbon in 1947, made the first measurement of Hawaiian groundwaters in 1961 (Libby 1962). He interpreted the spring water in a high-level Moloka'i tunnel as "young," only about 5 or 6 years old. The aging process commences the moment the infiltrated water enters the water table. Age is the residence time of the groundwater in the aquifer. Long residence time or old age may imply that the water has advanced far away from its origin as meteoric recharge or has been isolated from circulation. In any event, "old" groundwater, if extracted from the aquifer, may not be easily replenished.

The first survey of the radioactivity of carbon and hydrogen in waters was made on O'ahu in 1970–1973 (Table 6.2). The results indicate that bomb tritium was definitely present, as expected, in precipitation and surface waters. It was also present in perched groundwater and some dike waters. Marginal amounts were present in some basal waters. Bomb radiocarbon was obviously present in surface waters, but its presence in groundwaters was very limited. On the other hand, low to extremely low radiocarbon was associated with basal waters in the Honolulu area, subbasal saltwaters, and saltwaters in deep confined carbonate aquifers. These data indicate that perched waters (e.g., springs in Tantalus) have the shortest residence times, followed by dike water, basal water, and subbasal saltwaters specifically in a sequence of increasing residence times (Hufen et al. 1974).

The early 1970s high tritium content in natural waters served unambiguously as qualitative identification of recent recharge. The low radiocarbon associated with basal waters has prompted subsequent studies.

Radioisotopic age has been used to infer regional groundwater flow velocity and residence time in Hawai'i. Experience indicates that isotopic data must not be used alone to draw conclusions; instead, they should be used in conjunction with hydrologic and chemical information. In a study of the Waipi'o-Waipahu area of the Waipahu Aquifer System in the Pearl Harbor Aquifer Sector, the $^{14}C$ age of the Waipahu well water was found to be younger than that of Waipi'o well water, which is further inland and upgradient (Voss and Wood 1993).

Table 6.2. Isotopic and Chemical Quality of Natural Waters, O'ahu, Hawai'i, 1970–1973

| Type of Water | Tritium (TU) | Radiocarbon (% modern) | $^{13}C$ (δ‰ PDB) | $Cl^{-1}$ (mg/l) | $HCO_3^{-1}$ (mg/l) |
|---|---|---|---|---|---|
| Precipitation | 15–25 | — | — | 5 | — |
| Surface water | 12–23 | 130–140 | — | 12 | 20–40 |
| Perched groundwater | 18–24 | 100–108 | -16.9[a] | 21 | 125 |
| Dike water (Ko'olau Range) | 1–14 | 99–100 | -18.8 | 15 | 45 |
| Basal water | | | | | |
| Pearl Harbor | 0–3 | 94–102 | — | 40–90 | 60–90 |
| Honolulu | 0–1 | 90–96 | -18.0 | 45–100 | 60–80 |
| Saltwaters | | | | | |
| Subbasal lens basalt aquifer | 0 | 33[b] | — | ~18,000 | 100 |
| Confined carbonate aquifer | 0 | 13 | -3.8 | ~10,000 | 108 |

SOURCE: Hufen et al. 1974.

NOTE: TU, tritium unit (1 TU = 0.0072 dpm/ml); % modern, percentage deviation from modern standard (95% NBS oxalic acid); δ‰ PDB, per mil deviation from PDB standard.

[a]Single measurement.

[b]Average of two samples: one taken at 400 m (1,312 ft) below mean sea level in Kalihi Aquifer System and the other at 336 m (1,102 ft) below mean sea level in Pearl Harbor Aquifer Sector.

This implies incorrectly that groundwater could become younger in the direction of flow. To resolve this incongruity, the high activity value (young age) for Waipahu water is attributed to the effect of return irrigation water. The phenomenon of return irrigation water as a young water had been well established in a previous isotopic study (Hufen et al. 1980).

The regional velocity of the subbasal saltwater in the Waipi'o-Waipahu area was isotopically determined to be 0.006 m/d (0.02 ft/d). The radiocarbon age differs by 1,600 to 2,800 years for these two locations spaced 4,500 m (14,765 ft apart). This value is over 200 times less than the freshwater flow velocity. The saltwater body is thus virtually stationary, a state commonly assumed in dynamics studies. The saltwater flow direction is determined by isotopic concentrations to be landward.

Determining regional freshwater flow velocity with $^{14}C$ has also been attempted. For the Wahiawā Aquifer System, O'ahu (Harding Lawson Associates 1996), the values are in the range of 0.04 to 0.08 m/d (0.14 to 0.28 ft/d). For the East Mauna Kea Aquifer System, Hawai'i (Thomas et al. 1996), the value is 0.04 m/d (0.12 ft/d). These values are much lower than the range commonly encountered in basal lenses elsewhere in Hawai'i.

Residence time, $T$, is the time required to displace by advection the volume of groundwater storage, $V$, in an aquifer. The equation is $T = V/qA$, where $A$ is the cross-sectional area perpendicular to specific discharge $q$. Residence time has been assessed for the Honolulu and Pearl Harbor aquifers to be on the order of a few decades. On the other hand, the radiocarbon residence time of the freshwater core in the basal lenses is on the order of hundreds of years. The reason for the inconcordance has been hypothesized but not proven. One hypothesis is that certain unexplored dike water other than that intercepted by high-level tunnels can be very old. Discharge into and mixing with the lens water would result in an isotopically old water (Voss and Wood 1993). Another hypothesis is that age stratification is vertical within the freshwater core (Hufen 1974). Still another hypothesis is that some calcite is dissolved in the path of groundwater flow. The dissolution of calcite has been repeatedly rejected in the studied basal lenses (Hufen 1974; Voss and Wood 1993; Harding Lawson Associates 1996). The only exception is the Wilder Station wells in Honolulu (Hufen 1974). The radiocarbon age was adjusted to 290 years from an apparent age of 1,330 years because of the exceptionally high bicarbonate concentration of 110 mg/l.

The isotopic residence time of groundwater is in the range of 500 to 800 years for Wahiawā Aquifer System and 2,200 years for East Mauna Kea Aquifer Sector.

Deuterium and $^{18}O$ are stable isotopes of hydrogen and oxygen, respectively. When water vapor condenses, the rain that forms has higher deuterium and $^{18}O$ concentrations than the remaining water vapor. In the case of trade-wind precipitation on Oʻahu, the Koʻolau Range intercepts the trade-wind rain clouds before the Waiʻanae Range does. The groundwater originating in the Koʻolau Range is isotopically heavier than the groundwater originating in the Waiʻanae Range. The hypothesis has been substantiated in the high-level groundwater in the Wahiawā Aquifer System (Figure 6.6*a*). This information has been used to determine groundwater flow patterns in the aquifer (Figure 6.6*b*).

Geothermal anomalies have long been suggested as occurring in four Hawaiian islands, but only the Kīlauea east rift, island of Hawaiʻi, has a proven geothermal reservoir (Hawaiʻi Department of Planning and Economic Development 1982). The HGP-A well completed in Puna in 1976 penetrated an extremely hot (358°C [676°F]) reservoir at a depth of 1,768 m (5,800 ft) below sea level.

The discharge fluid from the HGP-A well is a mixture of water (~70 percent steam) and small amounts of other gases consisting mostly of $CO_2$, $H_2S$, $N_2$, and $H_2$. The fluid is only slightly saline (925 mg/l $Cl^{-1}$). The water is nearly depleted in magnesium (1 mg/l) but contains high concentrations of silica (420 mg/l as $SiO_2$) and sulfide (100 mg/l). Geochemical thermometer readings based on the data of Fournier and others are in excellent agreement with the measured downhole temperature of 275°C (527°F) (Kroopnick et al. 1980).

Mass balance calculations involving carbon, its isotopes $^{14}C$ and $^{13}C$, and $^{18}O$ indicate that 10 percent of the water comes from the ocean, 66 percent from groundwater recharged by local rainwater, and 24 percent from a hydrothermal source (Thomas 1977). The radiocarbon activity of 20 percent of modern $CO_2$ is attributed to the mixing of carbon from groundwater and shallow hydrothermal waters with dead magmatic carbon. The helium concentration and isotopic values indicate that the fluids recharging this well are most likely in contact with a deep-seated magmatic source.

In 1980, the HGP-A well was producing fluids at 5,800 kg/h (6.4 tons/h) with a wellhead pressure of 1,075 kPa (156 psi) and a wellhead temperature of 190°C (374°F) (equivalent to 3.5 megawatts of electrical energy). In 2000, a total of 30 megawatts from three Puna wells supplied about 25 percent of the island of Hawaiʻi's electrical needs.

## Aquifer Classification

On each island numerous aquifers have been identified, ranging from small to very large and containing brackish water to freshwater. The boundaries and general features of a few of the aquifers, especially those in Oʻahu, have been described. Several aquifers have been the subject of analytical and numerical modeling, but generally the actual hydraulic boundaries of aquifers are not known in detail.

The boundaries of the continuous flow of intermediate-depth and deep groundwater do not necessarily correspond with land surface features. An aquifer may underlie many surface drainage basins. The study of the occurrence and behavior of groundwater in complex systems is constrained by many assumptions. Narrowing such constraints on understanding requires continual investigations and the application of new knowledge to traditional and innovative methods of analyses.

Many investigations of groundwater resources have been made, but aquifer nomenclature had never been

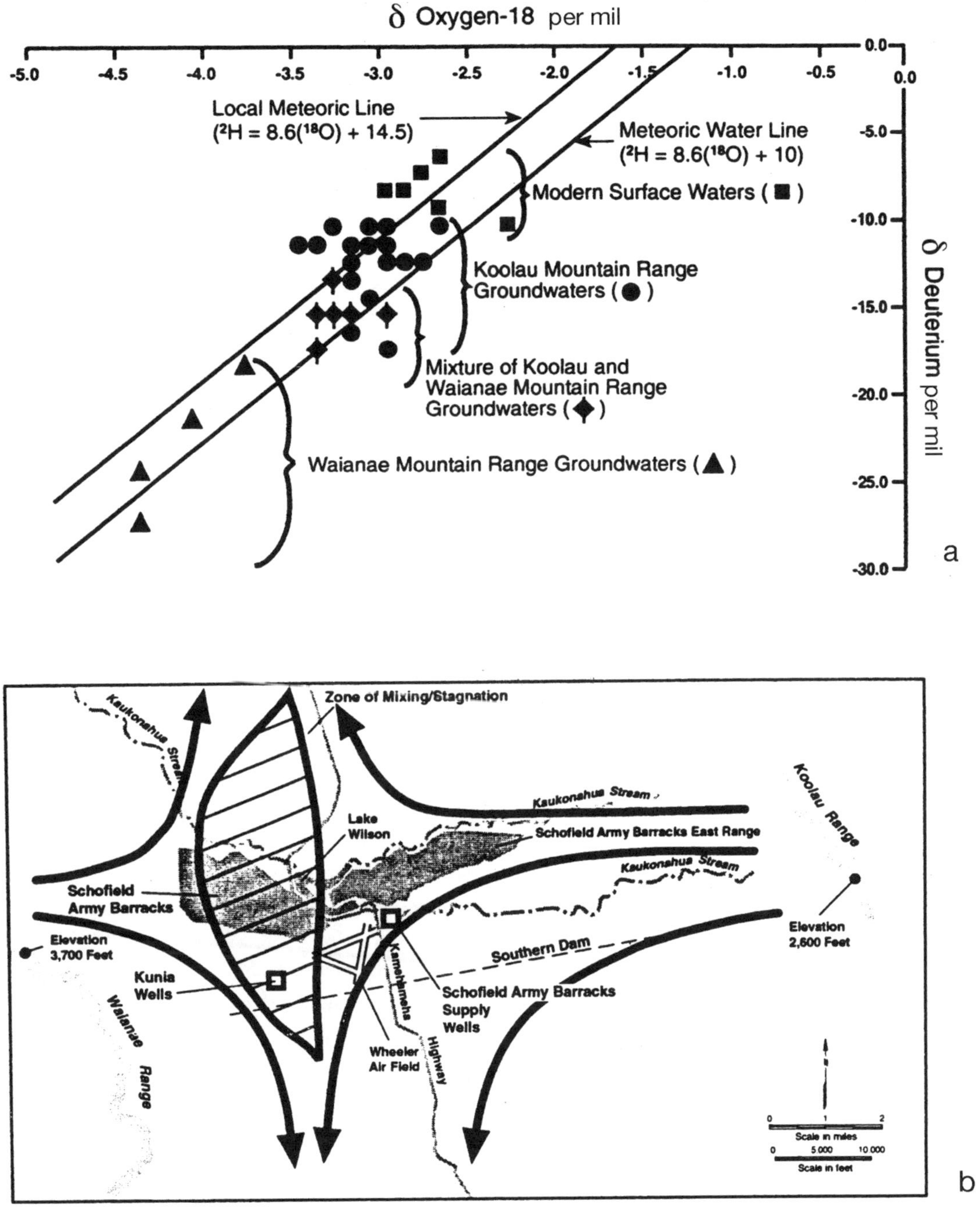


Figure 6.6. Groundwater flow patterns estimated from geochemical data, central O'ahu, Hawai'i: *a*, deuterium and oxygen 18 values in central O'ahu groundwaters; *b*, flow pattern based partially on heavy isotopic data. (Adapted from Harding Lawson Associates 1996)

standardized until the State Commission on Water Resource Management was created. A variety of terms had been created for identical aquifers, causing confusion among planners and regulators as well as engineers and scientists. Without a set of rules for naming aquifers and for assigning fundamental criteria by which to regulate development, rational planning often was misunderstood. Both the Hawai'i Department of Health and the Commission on Water Resource Management addressed the state of confusion by proposing a system of aquifer classification that would serve as the template for the management and protection of groundwater resources, even though for the great majority of cases knowledge of the resources, as well as of their capacity and extent, is poorly known. The development of the classification system has standardized nomenclature and has assigned sustainable yields. The boundaries and sustainable yield of the aquifers can be redefined as new information is gathered and evaluated.

The first classification was proposed for O'ahu by Mink and Sumida (1984), and then for O'ahu and all other islands by Mink and Lau (1990a, 1990b, 1992a, 1992b, 1993a, 1993b). These classifications are currently employed by the Commission on Water Resource Management and the Department of Health for regulating and protecting the groundwater resources throughout the state.

The aquifer classification system comprises a hierarchical scheme starting with the largest division, the aquifer sector, and then, in descending order, aquifer systems, aquifer types, and aquifer units. The sectors, systems, and types have already been identified and are in use by State agencies, but the units, the smallest divisions, await identification.

An aquifer sector embraces a large region based on geographic, judicial, and historic boundaries as modified by hydrogeological factors. A sector usually includes more than one aquifer system, with the systems hydraulically connected to a discernible degree. The sector attempts to retain boundaries that have historically appeared in groundwater reports and discussions.

Aquifer systems are the focus of the classification. Systems may include a number of aquifer types. The hydrogeological conditions in an aquifer system may vary, but groundwater is hydraulically continuous. Aquifer types are differentiated by the same distinctive features of hydrology and geology. Regulatory and protective measures employ the aquifer system as the unit of control.

The division of southern O'ahu into sectors, systems, and types illustrates the utility of the classification. An explanation of the classification codes is given in Table 6.3, and a map with the location of the sectors and systems is shown in Figure 6.7.

Boundaries are consistent with historical divisions and hydrogeologic features and also with the boundaries proposed by the two investigators who contributed most to comprehension of the water resources of the island, C. K. Wentworth and H. T. Stearns. The Honolulu Aquifer Sector extends from Makapu'u Point to the ridgeline separating Moanalua Valley from south Hālawa Valley, which is a synthetic rather than a real hydrogeologic boundary. The sector includes five aquifer systems whose boundaries agree with those employed in the original literature of Wentworth as well as by subsequent investigators. The Pālolo Aquifer System is the same as area 1 of Wentworth; the Nu'uanu Aquifer System, the same as area 2; the Kalihi Aquifer System, the same as area 3; and the Moanalua Aquifer System, the same as area 4. These four systems are hydrogeologically separated by alluviated valleys but are strongly hydraulically connected. The highest basal water heads occur in the Nu'uanu system, with heads descending both eastward and westward. East of the Diamond Head rift zone, the Wai'alae Aquifer System, equivalent to Wentworth's area 5, is poorly connected to the Pālolo system to the west.

In addition to the large basal aquifers in each system, substantial high-level groundwater resources occur. Sediments of the coastal plain are saturated with brackish basal water.

West of the Red Hill interfluve, the Pearl Harbor Aquifer Sector extends for 24 km (15 mi) to the crest of the Wai'anae Mountains, northwest along the Ko'olau crest, and west again along the boundary of the high-level Wahiawā Aquifer System to the Wai'anae crest. Included in the sector are the Waimalu Aquifer System, the Waipahu-Waiawa Aquifer System (originally two

Table 6.3. Aquifer Classification for Hawai'i

Aquifer and Status Codes[a]

Aquifer Code = Island
+ Aquifer Sector
+ Aquifer System
+ Aquifer Type

Thus, 30104111 = Aquifer Code
where 3 = O'ahu
01 = Honolulu
04 = Moanalua
1 = basal
1 = unconfined
1 = flank

and
(11111) = Status Code
where 1 = currently used
1 = drinking
1 = fresh, <250 mg/l $Cl^-$
1 = irreplaceable
1 = high vulnerability to contamination

Status Code (Groundwater)

Development Stage
1 Currently used
2 Potential use
3 No potential use

Utility
1 Drinking
2 Ecologically important
3 Neither

Salinity (mg/l $Cl^-$)
1 Fresh (<250)
2 Low (250–1,000)
3 Moderate (1,000–5,000)
4 High (5,000–15,000)
5 Seawater (>15,000)

Uniqueness
1 Irreplaceable
2 Replaceable

Vulnerability to Contamination
1 High
2 Moderate
3 Low
4 None

| Aquifer Type | Hydrology[b] |
|---|---|
| 1 Basal | Freshwater in contact with seawater |
| 2 High level | Freshwater not in contact with seawater |
| 1 Unconfined | Where water table is upper surface of saturated aquifer |
| 2 Confined | Where groundwater is confined by a poorly permeable stratum and is under greater than atmospheric pressure |
| 3 Confined or Unconfined | Where actual condition is uncertain |

| Aquifer Type | Geology[c] |
|---|---|
| 1 Flank | Horizontally extensive lavas |
| 2 Dike | Aquifers in dike compartments |
| 3 Flank/Dike | Indistinguishable |
| 4 Perched | Aquifer on an impermeable layer |
| 5 Dike/Perched | Indistinguishable |
| 6 Sedimentary | Nonvolcanic lithology |

SOURCE: Mink and Lau 1990a.

[a] Where sedimentary caprock aquifers rest on primary basalt aquifers, two aquifer and status codes separated by a slash indicate numerator code is upper aquifer and denominator is lower aquifer.

[b] First two digits from hydrologic descriptors (points 1, 2).

[c] Last digit from geologic descriptor.

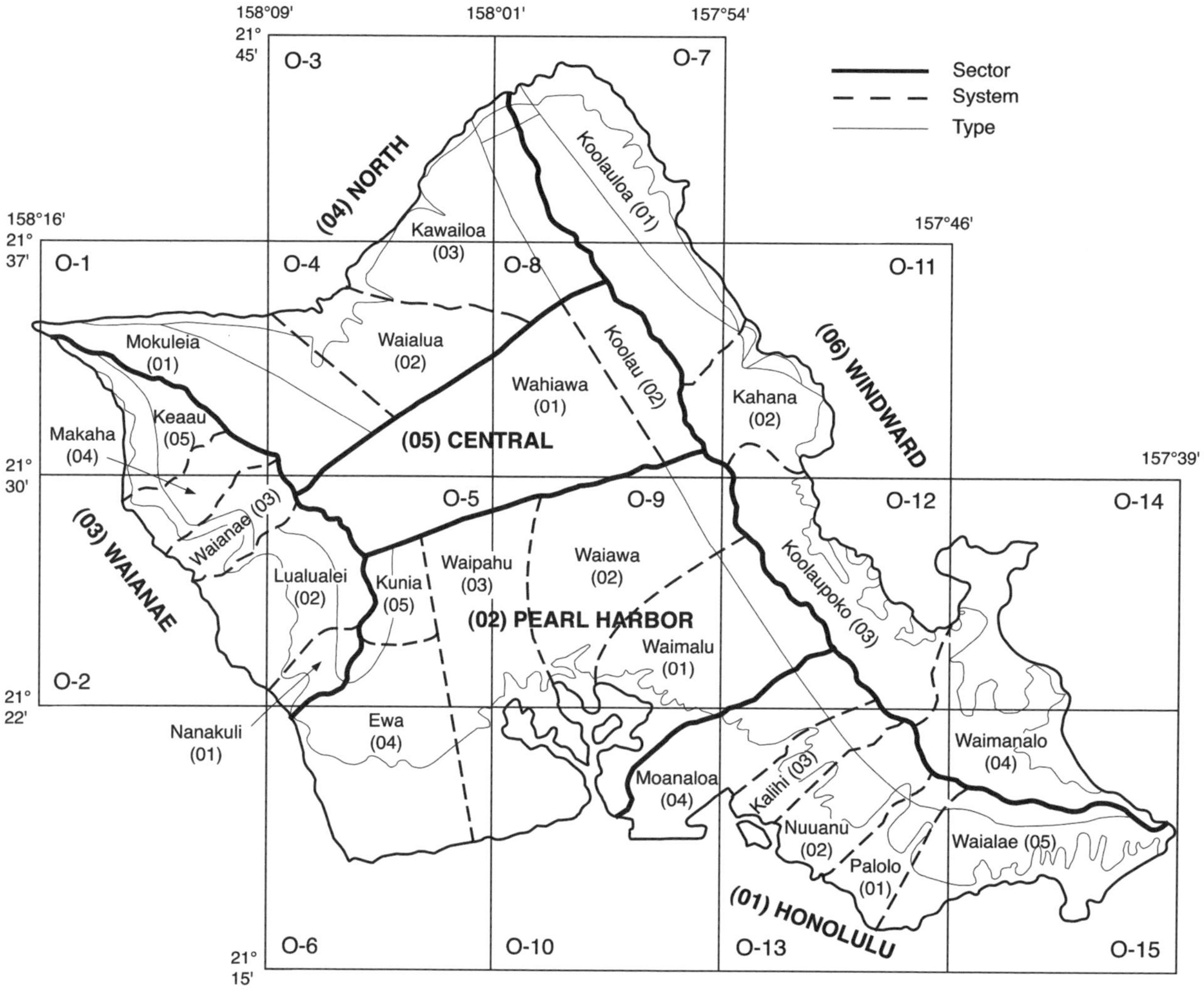


Figure 6.7. Layout of aquifer sectors, systems, and types for O'ahu, Hawai'i. (Adapted from Mink and Lau 1990a)

systems separated by the former demarcation between irrigated sugarcane land, which, at the time the classification was developed, was restricted to the Waipahu Aquifer System), and the 'Ewa-Kunia Aquifer System (also originally two systems with irrigated sugarcane land restricted to the 'Ewa system). Sugarcane is no longer grown on O'ahu.

All of the systems in the Pearl Harbor sector include basal and high-level water in the volcanics as aquifer types and basal water in sediments. The hydraulic connection between the Waimalu system and the Moanalua system of the Honolulu sector is continuous, as is the connection between the Waimalu system and the Waipahu-Waiawa system. The hydraulic connection between the Waipahu-Waiawa and 'Ewa-Kunia systems is controlled by the Ko'olau/Wai'anae unconformity, which suppresses but does not exclude groundwater transfer.

Using the classification system, all groundwater resources in each island can be located and categorized, even though understanding of most aquifer systems is weak. As investigations continue and new information is collected, regulatory and protective measures can be modified. Managing groundwater resources is work continually in progress.

Accompanying the aquifer code is a groundwater status code that identifies each aquifer by its development stage, utility, salinity, uniqueness, and vulnerability to contamination (see Table 6.3 for details). The aquifer classification codes for O'ahu are listed in Appendix 6.2.

## Behavior of Lenses

### Basic Flow Premises

Darcy's Law was first demonstrated to be valid for basaltic aquifers by Wentworth (1946). He showed that except for the approximately 1 m (few feet) closest to the wells, laminar flow prevailed in Honolulu basaltic aquifers pumped at high rates. He deduced a value of unity for the exponent $n$ in the equation $H = CQ^n$, where $H$ is drawdown, $Q$ is the pumping rate, and $C$ and $n$ are constants.

Flow of groundwater in Hawaiian basal lenses is essentially horizontal except near the discharge front. This is based on the Dupuit simplification, which becomes applicable because of the typical observed flatness ($<10^{-3}$) of the hydraulic gradient. In the Dupuit characterization, isopiestic lines are vertical and the Ghyben-Herzberg relation is applicable.

### Framework of Basal Lenses

Hawaiian freshwater lenses vary considerably in thickness. The depth to the 250 mg/l isochlor, the generally accepted maximum concentration chloride limit for potable water, reaches as low as 260 m (850 ft) in the Nu'uanu Aquifer System, O'ahu, the thickest known freshwater lens in Hawai'i. Where the freshwater and saltwater merge the salinity is invariably dispersed. A transition zone is formed, varying in thickness from a few to several hundred meters (10 ft to 1,200 ft).

Hawaiian basal lenses do not have a sharp interface, as a consequence of appreciable hydrodynamic dispersion. Salinity variation with depth is represented by an S-shaped curve in the transition zone symmetric about the 50 percent isochlor. Shallow lenses have relatively thin but not negligible transition zones (see Figure 6.8). The first investigation of thin lenses was conducted near Kahului, Maui, with electrical resistivity measurements and depth sampling (Swartz 1937). Much later, an exceptionally thick transition zone (Figure 6.9) was discovered during the first investigation of thick lenses in a deep well drilled near the copious Kalauao Springs on the shore of Pearl Harbor in southern O'ahu (Visher and Mink 1964). Depth variations of salinity and temperature in thick lenses are periodically measured in many deep monitor wells on O'ahu, Maui, and Hawai'i.

Water temperature and chemical composition define three principal layers constituting a lens. The middle layer is the core of the freshwater lens. It is the coolest, tends to be isothermal, and represents recharge from the wet mountains. Underlying it is an increasingly warmer layer, especially where the lens is thick due to the presence of caprock and where the water is virtually static and extends throughout the transition zone into the saltwater as geothermal warming takes effect. The top layer, if present, is a thin warmer layer reflecting local recharge, such as that generated from return water from sugarcane irrigation. The presence of a caprock limits ocean-island water exchange and keeps the ocean temperature gradient from influencing the lens water (Mink 1964), but in the absence of caprock and subsequent free movement of ocean water, the geothermal gradient is ineffective.

The chemical composition of saltwater underlying the lens where caprock occurs is essentially that of seawater, but the concentrations of several constituents ($Ca^{+2}$, $Mg^{+2}$, $Na^{+1}$, $K^{+1}$, $SO_4^{-2}$) altered as a result of ion exchange and sulfate reduction (Visher and Mink 1964).

Motion of the saltwater is induced by solute dispersion, convective water circulation, and tidal fluctuations. Tide-induced water movements are appreciable in highly permeable basalt and fossil coral limestone (Dale 1974).

### Heads in an Exploited Ghyben-Herzberg System

A Ghyben-Herzberg system is sustained by the buoyancy of freshwater relative to seawater. The difference

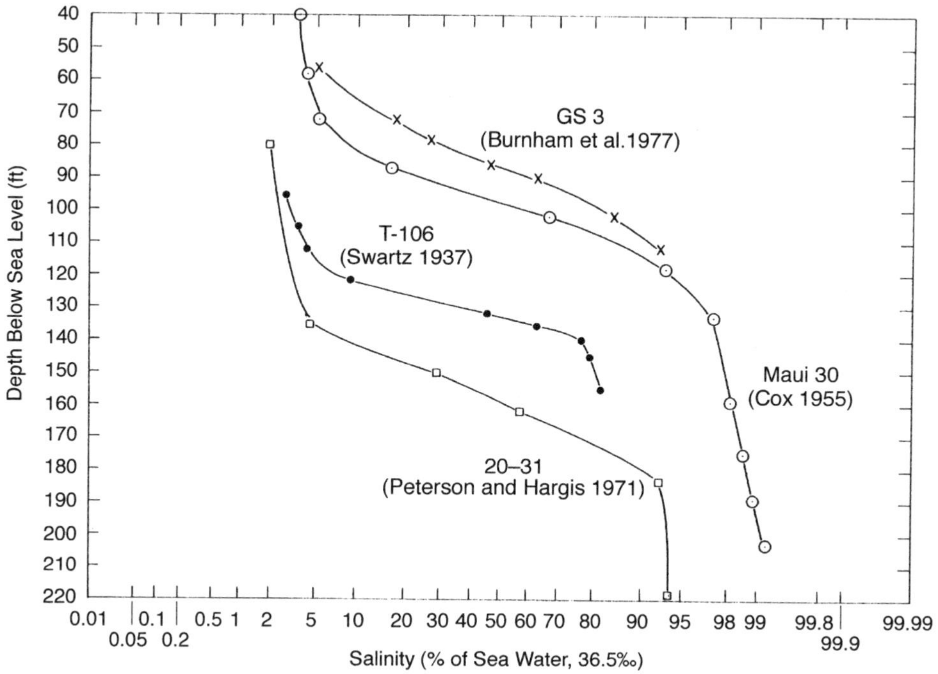


Figure 6.8. Salinity-depth distributions in thin lens, Kahului, Maui, Hawai'i. Based on Burnham et al. 1977; Cox 1955; Swartz 1937; Peterson and Hargis 1971. 1 ft = 0.305 m. (Adapted from Mink and Lau 1980)

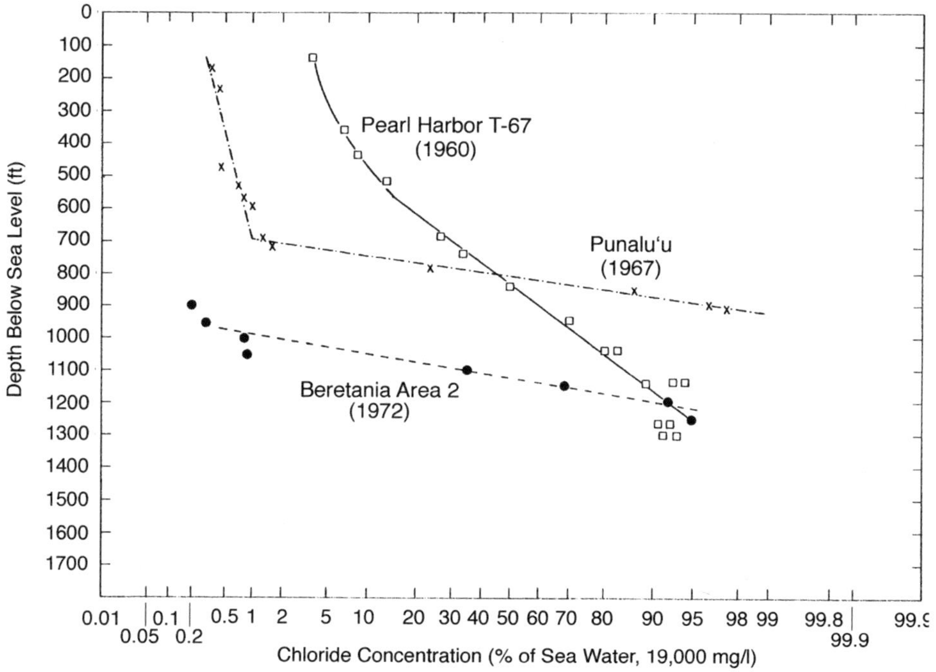


Figure 6.9. Chloride-depth distributions in several thick basal lenses, O'ahu, Hawai'i. 1 ft = 0.305 m. (Adapted from Mink and Lau 1980)

in densities between seawater (ρ = 1.025 as a standard) and freshwater (ρ = 1.000 as a standard) results in a depth of 40 feet of freshwater below sea level for every foot above. These standard densities, however, refer to freshwater at 5°C (41°F) and seawater at 15°C (59°F); at other temperatures the ratio differs from 40:1. Rarely, if ever, does a column of freshwater rest directly on seawater; the column will have a different average density from that of standard freshwater to achieve the ratio. The ideal density conditions refer to static waters in which neither dispersive mixing nor advective movement takes place. Even though these ideal conditions appear to be unusually restrictive—requiring, for instance, a sharp interface between the freshwater lens and the underlying saltwater—most Ghyben-Herzberg systems in islands and along coasts, where Dupuit conditions prevail inland of the coastal discharge boundary, are surprisingly accurately described by the 40:1 ratio. Bear (1979) suggested that these conditions may be assumed at a distance from the discharge front of 1.5 to 2.0 times the depth of flow of the lens.

A sharp interface is not necessary in the application of the 40:1 ratio. The zone of dispersion between seawater and the freshwater lens, commonly called the transition zone, is theoretically and typically symmetrical at about the 50 percent relative seawater concentration isochlor. This symmetry is typically very persistent, allowing the 40:1 ratio to be applied to the depth below sea level to the 50 percent isochlor (see Appendix 6.1). The freshwater lens commonly becomes stratified into a thick freshwater core, an upper limb of the transition zone between the freshwater core and the 50 percent seawater isochlor, and a lower limb of the transition zone between the 50 percent isochlor and seawater equal in thickness to the upper limb. These conditions prevail away from the discharge boundary.

The freshwater core does not necessarily contain potable or even agriculturally usable water. It refers to the upper portion of the lens beyond the symmetry of the transition zone. Not unusually, it consists of water with 1,000 to 2,000 mg/l chloride, or a 5 to 10 percent relative seawater concentration. Frequently, no variation in concentration occurs in the freshwater core, especially in thick lenses. Where concentrations are variable, the changes do not follow an equation of symmetry.

The transition zone, upper and lower limbs, follows the symmetry of the error function curve. The midpoint of the S-shaped curve is the 50 percent relative seawater concentration. The curve plots as a straight line on normal probability paper with depth below sea level on the arithmetic scale and concentration as percentage seawater on the probability scale. The behavior of a Ghyben-Herzberg system is clearly and directly encapsulated in such a symmetry plot. A variable is obtained by dividing the depth below sea level to the 50 percent concentration by the Ghyben-Herzberg ratio, δ, or 40. This gives the true Ghyben-Herzberg head in contrast to the transitory head measured at the water table. The Ghyben-Herzberg head, which may be called the storage head, measures the vertical size of the lens and the volume of water in storage (Mink 1980). It does not change easily; for every vertical foot of water that accumulates in storage above sea level, 40 feet must accumulate below sea level for the true Ghyben-Herzberg condition to prevail. An example of the slowness of change is suggested by the semimonthly data obtained at a deep monitoring well (457 m [1,500 ft]), probing the lens and below the lens in the Nu'uanu Aquifer System in Honolulu on O'ahu (Lao 1995). The depth below sea level to the 50 percent concentration rose 21.3 m (70 ft) from 335 m (1,100 ft) to 314 m (1,030 ft) over a period of 18 years.

Water table or piezometric surface (see Figure 5.5*a*) levels in a freshwater-seawater system are transitory phenomena that change too rapidly and erratically to reflect the Ghyben-Herzberg balance in storage. They are ambient levels subject to sporadic infiltration, pumpage, and other environmental influences. Near the coast tides cyclically affect ambient levels. The intervals over which the addition to or subtraction from lens storage take place are generally too short for the bottom of the lens to be measurably affected. The deeper storage responds over the long term, during which time accumulations and depletions have been averaged.

Water levels and storage head rarely coincide in exploited aquifers. For an aquifer at rest over a period of time sufficiently long for adjustments to accumulations and depletion of storage to take place, ambient water levels will approach storage head. The time period is shorter for permeable aquifers whose outflow is not impeded by caprock or other coastal lithologies.

Storage head is in balance with the Ghyben-Herzberg 40:1 requirement but is not necessarily an equilibrium head. The equilibrium head is the storage head at steady state, in which output (draft plus leakage) is exactly balanced by input (recharge). This is possible only where a system can be described rationally with averages of outputs and inputs. In a cyclic system, such as where a highly defined wet season is followed by a dry season, a single value of equilibrium head is not possible. However, when the cycle of outputs and inputs is uniform, a quasi-steady-state cycle of equilibrium heads will eventually set in.

In summary, Ghyben-Herzberg systems are defined by storage head, which is determined from the position of the 50 percent seawater concentration contour in the middle of the transition zone. Ambient water levels are transient phenomena and not a true measure of the corresponding changes in storage of a Ghyben-Herzberg system of substantial thickness.

## Storage and Leakage

Leakage is defined as the natural seaward freshwater flow that ultimately discharges into the ocean, either directly or through the caprock. The flow may be diminished by draft (extraction by pumping) after originating from recharge areas. For example, much of the leakage from the Pearl Harbor Aquifer Sector is by discharge of the Pearl Harbor springs, which historically has varied substantially over time.

Leakage is related to storage head, which, in addition to being the measure of storage in the lens, is also the head in Dupuit horizontal flow. Leakage continues as long as there is recharge. A mathematical functional relationship among recharge, draft, leakage, and head was derived by Mink (1980), then numerically deduced by Souza and Voss (1989). The results of these models are supported by long-term measurements in the Pearl Harbor aquifers.

It is essential to distinguish steady state (dynamic equilibrium) from the transient state, particularly in considering storage changes in thick lenses. Under steady state, the head and other dependent variables stay unchanged with time, whereas during a transient period they are time dependent.

In addition to hydraulic factors, storage is enhanced by poorly permeable caprock acting as a coastal aquiclude or aquitard. For example, the thick lens in the Nuʻuanu Aquifer System is capped by a wedge of sediments that makes up the Honolulu coastal plain. The wedge thickens seaward to coastal depths more than 213 m (700 ft) below sea level in the valley extensions and to about 122 m (400 ft) below sea level between valleys. Where the caprock behaves as a weak aquitard, thin lenses are expected. The Lahaina coastal alluvium, for example, is insufficient to act as a strong caprock. Where caprock is absent, as along the North Kona coast, brackish-water lenses reach far inland. Not only do coastal plain sediments act as caprock, but massive and dense lava flows also may impede outflow locally. For example, where it covers the Wailuku basalt at the coast, the Honolua basaltic andesite in West Maui behaves as a caprock.

The effects of saltwater on a freshwater lens arise from three factors: (1) saltwater density, (2) saltwater motion, and (3) saltwater-freshwater miscibility. The most effective of the three is the density factor, which results in the realization of the Ghyben-Herzberg system. The impact of all three factors combined was analyzed in a study of the Pearl Harbor Aquifer Sector with SUTRA, a numerical model (Voss and Souza 1987). The combined effects of just the density and motion factors have been analyzed with a numerical model known as SHARP, which assumes immiscibility and therefore a sharp interface (Essaid 1986).

When irrigated sugarcane dominated the land area of central Oʻahu, lowering of the freshwater water table in the Pearl Harbor aquifers during seasons of maximum pumping was considerable. The amount exceeded the decline that would have accounted for the volume of freshwater extracted plus leakage if the water table level actually had represented the storage head. However, storage head is measured at the interface, which is resistant to seasonal perturbations. If the Ghyben-Herzberg ratio had been applied to the measured water table oscillations, the freshwater volume released would have been enormous, far exceeding the extracted and natural leakage amounts combined. This phenomenon was first observed and analyzed by Wentworth (1942).

Wentworth's explanation was conceptual: the interface cannot rise by the amount called for by the Ghyben-Herzberg ratio within such a short time period. The interface would continue to rise for a much longer time before eventually reaching the new equilibrium level as predicted by the ratio if the pumping were continued.

Another explanation offered assesses aquifer boundary effects during pumping (Mink 1980). Much of the observed water table oscillations reflect dewatering and rewatering of the topmost portion of the lens, conforming to classical image well theory. Storage head at the interface responds only weakly to water table changes. Still another explanation offered assesses the influence on freshwater head by seawater flow (Essaid 1986).

Understanding the behavior of the transition zone in thick lenses remains incomplete. But the following appears clear: seasonal heads should be used with extreme caution to infer the storage of a thick lens. The correlation between the measured mid-depth of the transition zone (50 percent isochlor) and the Ghyben-Herzberg ratio has been explained by Mink and Lau (see Appendix 6.1).

## Managing Groundwater Resources: Sustainable Yield

Sustainable yield refers to the forced withdrawal (draft) of groundwater at a rate that could be sustained indefinitely without affecting either the quality of the pumped water or the volume rate of pumping. It depends upon the steady-state (dynamic equilibrium) storage head at a point selected as the minimum acceptable during continuous pumping. This storage head is the elevation of the unconfined water table or confined piezometric surface above sea level in balance with the depth of the lens, which is not necessarily coincident with the measured water level that is affected by pumping. There is not a unique value for sustainable yield; the value depends on the storage head that will preserve the integrity of the groundwater resource.

Sustainable yield is equal to a fraction of the recharge. In a basal lens the fraction is usually more than one-half and sometimes greater than three-fourths where initial heads are high. In high-level aquifers about three-fourths of the recharge can be taken as sustainable yield.

The simplest way to understand a groundwater system from the management point of view is to treat the system as a single unit exhibiting global rather than local response behavior. At the equilibrium state, global sustainable yield depends on the initial head, the rate of recharge, and the selected equilibrium head. A sustainable yield exists for each value of an equilibrium head between the limits of head greater than zero and less than the initial head. Sustainable yield is always less than recharge. If it were to equal or exceed recharge, head and storage would eventually become zero because outflow from the aquifer is a combination of natural leakage and draft, and the natural leakage would continue as long as the head is above sea level. For a sustainable yield to exist, a balance must be established among draft, recharge, and leakage, which is controlled by head.

Good estimates of sustainable yield need a reliable database. In most of the state of Hawaiʻi not enough is known about the extent and behavior of groundwater to allow more than a weak estimate of sustainable yields. Only in southern Oʻahu, Lānaʻi, and sections of East and West Maui, where many years of investigation have been devoted to unraveling the complexities of groundwater occurrence, can the sustainable yields be accepted with a reasonable degree of confidence.

Estimates of the sustainable yield of each aquifer system start with the simple predevelopment water balance equation. The balance is computed for each aquifer system using averages based on the data record. The averages for aquifers in volcanic rock incorporate high-level and basal aquifer regions but exclude caprock areas.

Estimates made for aquifer systems are limited to the basal aquifers, except where high-level groundwater is dominant or reaches the coast. The typical sequence of aquifers in Hawai'i is from a zone of high-level water in mountainous regions to a basal aquifer terminating either at the coast or beneath caprock some distance off the coast. Groundwater in high-level aquifers passes into basal aquifers. Where it is removed directly from a high-level source, the sustainable yield calculated for the basal aquifer must be reduced.

Estimates of sustainable yield are not exact and should be used with caution in making planning documents. The estimates are constrained not only by the scanty database but also by the fact that they do not consider the reality of feasible development methods. Where resource exploitation is already under way, hypothetical sustainable yield must be amended to "allowable" sustainable yield, which is equivalent to safely developable yield.

Considerations restricting the unqualified use of sustainable yield estimates are as follows:

1. The estimate is based on the water balance method for predevelopment conditions. Movement of groundwater from one aquifer system to another is not taken into account.
2. The sustainable yield is correlated with an equilibrium head chosen on the basis of experience, but the experience may not be relevant to a given aquifer. An equilibrium head higher than the one selected would result in lower sustainable yield, and the converse would be true for a smaller equilibrium head.
3. Assumptions about the initial state of an aquifer may be faulty.
4. Sustainable yield is calculated as the total supply developable under optimal means of exploitation. Where suboptimal methods are practiced, allowable sustainable yield should replace the hypothetical value.
5. The sustainable yield estimate should not be equated to feasibly developable water, either technically or economically.
6. Sustainable yield can be constrained by criteria not treated here, such as legal and environmental aspects of water development.

In view of these limitations, sustainable yield estimates should be regarded as a guide for planning rather than as inflexible rules.

## Behavior of High-Level Water

Groundwater that does not rest on seawater is classified in Hawai'i as "high-level water." It includes two principal members, perched water and dike-impounded water. Perched water mostly occurs in small, thin, and unconnected aquifers resting on poorly permeable substrates. Dike-impounded water occurs in aquifers trapped between quasi-vertical dikes of negligible permeability. Dikes usually are 0.1 to 1.5 m (0.5 to 5 ft) thick, and the exploitable aquifers between controlling dikes may range to a hundred or so meters (several hundred feet) in dimensions.

### Perched Water

Although they are the source of many small springs located high above sea level, individual perched aquifers are neither a copious nor a reliable source of water supply. However, poorly connected or unconnected perched aquifers may yield appreciable accumulated base flow in streams and collection ditches. This is especially true for the islands of Kaua'i and Maui but less so for the other islands. Perched water supplies of consequence are not found in O'ahu, Moloka'i, and Lāna'i. In Kaua'i perched water occurs in permeable basaltic layers lying on less-permeable layers within the Kōloa Volcanics.

In East Maui the controlling geological association for perched water is the low-permeability transition lithology of sediments and ash that separate the basement Honomanū basalt from the Kula formation and Hāna formation. Groundwater accumulates in the andesitic lavas of the Kula formation and basaltic lavas of the Hāna formation before draining to streams. In West Maui a similar relationship exists between the shield-building Wailuku basalt and the overlying Honolua Volcanics.

The volume of water draining to streams is greater in East Maui than in West Maui.

The Upper Hāmākua Ditch in the Kohala mountains of the island of Hawaiʻi is sustained by perched water. A substantial accumulation also occurs perched on ash beds in the Kaʻū region. In West Hawaiʻi the high levels of the water table suggest perching phenomena rather than dike-impounded groundwater because erosion has not been sufficiently deep to expose the core of the rift zones as in Oʻahu.

## Dike-Impounded Groundwater

The report by Takasaki and Mink (1981) discussed in detail the occurrences of groundwater in the dike zones of Oʻahu, but the conclusions are applicable to dike-zone aquifers in all of the other islands as well. Dikes are typically concentrated in linear rift zones in which the distribution density is greatest in the center of the zone and diminishes toward its boundaries. The central portion where dikes account for more than 10 percent of the rock mass is called the "dike complex," and the area where the dikes account for 5 percent or less of the rock mass is called the "marginal dike zone." The principal source of dike-impounded groundwater is the marginal dike zone.

Each island contains dike-impounded groundwater. For Kauaʻi the exploitable aquifers are in the Waimea Canyon Volcanics, the basement formation. For Oʻahu substantial aquifers occur in the rift zones of both the Waiʻanae and Koʻolau Ranges and perhaps in the Central Aquifer Sector. The largest developed source of dike water in West Maui originates in Honokōhau Valley, although dike aquifers are found in the upper reaches of each valley. The base flow of ʻĪao and Waiheʻe Streams is provided by dike water. The topography of East Maui is too young to have the rift zones exposed and the dike aquifers revealed. In central Lānaʻi, wells tap dike-impounded water. The base flow in the streams of East Molokaʻi is derived from dike water. Even Kahoʻolawe and Niʻihau have dike-impounded aquifers, but the volume and quality of the water on these islands have not been investigated. Dike-impounded aquifers in the Kohala mountains of Hawaiʻi yield groundwater to the Lower Hāmākua and Kohala Ditches. Dike groundwater may also occur in the Kaʻū District. The groundwater associated with the rift zone of the Koʻolau Range on Oʻahu, however, is the most notable because of the magnitude of the yield and the controversy surrounding its allocation.

In Oʻahu, tunnels have been driven to exploit groundwater in both the Waiʻanae and Koʻolau Mountains. The Waiʻanae tunnels were described in a report by Mink (1978). The tunnels in the Koʻolau Range have been discussed by many investigators (Stearns and Vaksvik, Wentworth, Board of Water Supply reports, U.S. Geological Survey reports), but the most comprehensive study is that by Takasaki and Mink (1981). The tunnels with outstanding yields are in Waiheʻe Valley and in the Waiāhole system, which collects water from the major valleys of Kahana, Waikāne, and Waiāhole. The Waiāhole system includes development or collection tunnels (Kahana, Waikāne I, Waikāne II, Uwau, and Waiāhole) that are driven nearly perpendicular to the trend of the linear rift zone, which approximately parallels the summit line; a transmission tunnel that directs the collected water to the Waiāhole portal; and the "main bore," which is a tunnel that connects the Waiāhole portal with the tunnel terminus in Waiawa Valley on the leeward side of the range. Beyond Waiawa, ditches, short tunnels, and siphons transmit the water to the fields of central Oʻahu.

Total average yield of the Waiāhole system has been 106 million liters per day (mld) (28 million gallons per day [mgd]), nearly all of which is provided by gravity drainage of dike-impounded groundwater. Construction of the main bore was started in 1913 and completed in 1916. An enormous volume of groundwater stored in the dike aquifers was lost by drainage during construction of the tunnels. The computed lost storage was 174,848 million liters (46,195 million gallons) (Takasaki and Mink 1981). Years after the loss, attempts were made to construct concrete bulkheads across tunnel cross-sections to simulate dikes, with the goal of allowing storage to accumulate. None was successful except in a minor way. The most

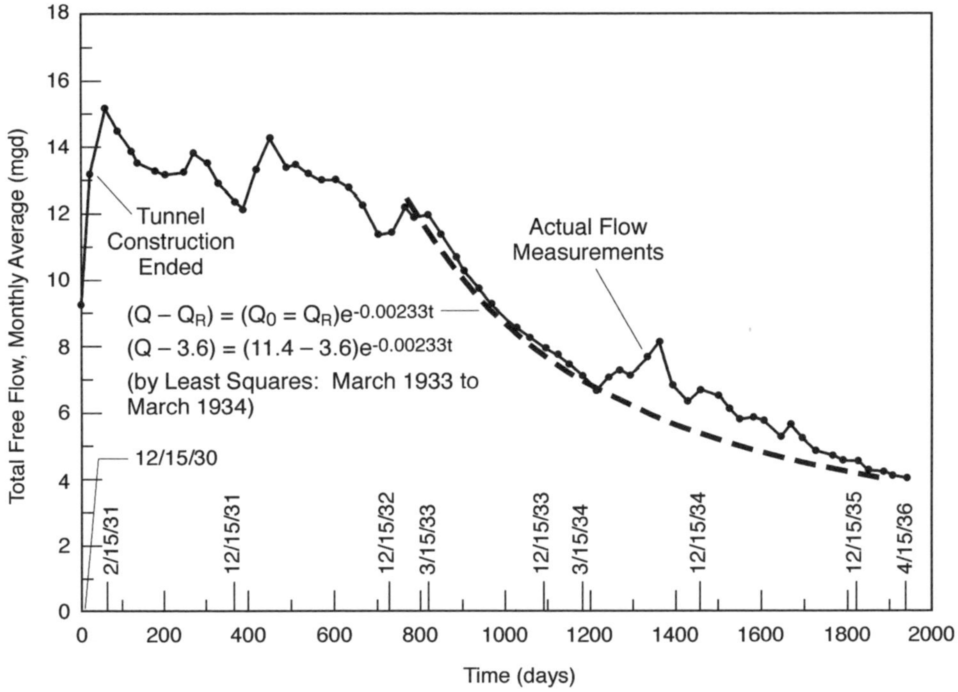


Figure 6.10. Free-flow decay at Kahana Tunnel, O'ahu, Hawai'i. 1 mgd = 43.8 l/s. (Adapted from Takasaki and Mink 1981)

recent attempt at bulk-heading was made in the Kahana Tunnel in 1990, but to what extent storage was gained has not yet been tested.

The hydraulics of drainage from storage and recharge in the collection tunnels are discussed in Takasaki and Mink (1981). Free drainage with constant recharge is expressed as the familiar exponential decay equation

$$(Q - Q_R) = (Q_0 - Q_R)\exp(-bt) \tag{6.5}$$

where $Q$ is total drainage at time $t$, $Q_R$ is constant recharge, $Q_o$ is initial drainage, and $b$ is the recession constant. The recession constant incorporates diffusivity, $T/S$, and length parameter ($T$ is transmissivity, $S$ is storage coefficient). The equation closely matches the decay data for Kahana Tunnel (Figure 6.10).

Where dike-impounded aquifers in rift zones have been exploited by means of tunnels, the nonstorage drainage has proven to be steady and reliable. The minimum drainage flow of the Waiāhole system has been approximately 72 mld (19 mgd), that of the Lower Hāmākua system also 72 mld, and that of the Honokōhau system of West Maui about 53 mld (14 mgd). These values are 60 to 70 percent of the average drainage yields.

Dike-impounded aquifers are also exploited with vertical wells. Because of the complexity of the aquifer boundaries, yields usually are moderate. Only in the marginal zone have wells been successful.

## Groundwater Pollution: Surface Contamination

### Contamination Sources

From 1979 to 1983 widespread contamination of basal groundwater sources in Hawai'i was identified and required instant remediation. Many drinking-water wells tapping the basal lens in the Pearl Harbor Aquifer Sector were tainted with trace amounts (low ng/l) of volatile organic chemicals such as 1,2-dibromo-3-chloropropane (DBCP), ethylene dibromide (EDB), and 1,2,3-trichloropropane (TCP), which had been certified for use as pesticides (nematocides) for pineapple agriculture during the preceding approximately 35 years. EDB had also been used as an additive to aviation gasoline. Military records show many spills (Figure 6.11). Since then, a total of seventeen organic chemicals of health concern has been discovered, some in concentrations requiring remedial actions, for all islands except Moloka'i and Lāna'i (Table 6.4). Another organic chemical, trichloroethylene (TCE), which is commonly used as a solvent, has been detected in the Wahiawā Aquifer System and in wells in the Pearl Harbor Aquifer Sector downgradient of Wahiawā. However, no other types of contaminants including pathogenic microorganisms (bacteria, viruses, and protozoans) have ever been detected and reported in potable groundwater sources.

Petroleum hydrocarbons are also surface contaminants of consequence, but their presence has been limited to soils and small pockets of brackish waters in sedimentary sequences of the coastal plains. Benzene, which is a dissolved volatile precursor of petroleum contamination, has not been detected in groundwater drinking sources. Petroleum as a free product is a potential for fire and explosion hazard. Like water, contaminated soils are subject to regulations.

Nitrates in concentrations above background level have been reported in basal and caprock water as a result of leaching of nitrogen from fertilizers applied to sugarcane fields (Mink 1962a; Tenorio et al. 1970; Lau et al. 1989). In all cases, the detected levels of nitrate have not exceeded 10 mg/l as N, the accepted maximum concentration level. Excessive concentration in drinking water presents a health risk, causing the blue-baby syndrome in infants.

## Surface Unsaturated Zone

All groundwater contaminants originating at the surface must pass through a zone composed of several layers. The shallow layers of this zone consist of about 30 cm (1 ft) of biologically active soil and approximately 60 cm (2 ft) of subsoil. Beneath the subsoil is a layer of saprolite, usually 15 to 30 m (50 to 100 ft) thick, overlying unaltered, unsaturated parent volcanic rock. Saprolite is the in-situ mantle of weathered volcanic rock. Its texture resembles the parent rock, but compositionally it consists of hydrates and clays derived from the original minerals. Its permeability is many magnitudes lower than that of its unaltered parent. In some areas the saprolite is saturated and serves as a temporary reservoir for percolating recharge, retarding the transport of and attenuating many contaminants, including organic chemicals, pathogenic microorganisms, and nitrates.

Numerous processes in the shallow zone participate in providing natural protection against contamination. They may be grouped into (1) surface geological media (surface soils and saprolites), (2) vegetation, and (3) atmospheric air and sunlight.

Surface soils provide filtration of suspended and colloidal solids and bacteria and serve as media for biological transformation. Oxisols enriched with soil organic carbon (2 to 4 percent) derived from decomposed vegetation have exceptionally strong sorption and resistance to desorption. DBCP desorption distribution coefficient $K'_D$ exceeds 1,500 l/kg and is 100 to 1,000 times greater than the sorption coefficient $K_D$. This means that DBCP residues at a level present in soils and saprolites (42 to 139 μg/kg) do not constitute a serious source for future groundwater contamination (Lau et al. 1987).

Saprolite is extremely inhomogeneous and marked by low hydraulic conductivity (0.003 to 0.305 m/d [0.5 to 1.0 ft/d]), which is grossly inferior to that of fresh rocks (Mink 1982). It is an effective aquitard, retarding pesticide transport in percolating water, but fracture zones can occur in thick saprolites to allow quick percolation as suggested in the Wahiawā Aquifer System (Harding Lawson Associates 1996).

Vegetated soils behave as a living filter. In tests using Bermuda grass grown in a 1.5-m (5-ft) Oxisol soil lysimeter and irrigated with secondary wastewater effluent, it was demonstrated that continuous and substantial removal of nutrients by plant uptake occurs. Both nitrogen and phosphorus were decreased by 99 percent and potassium by 93 percent. Much of the phosphorus probably became fixed in the soil. Aerobic conditions in the surface

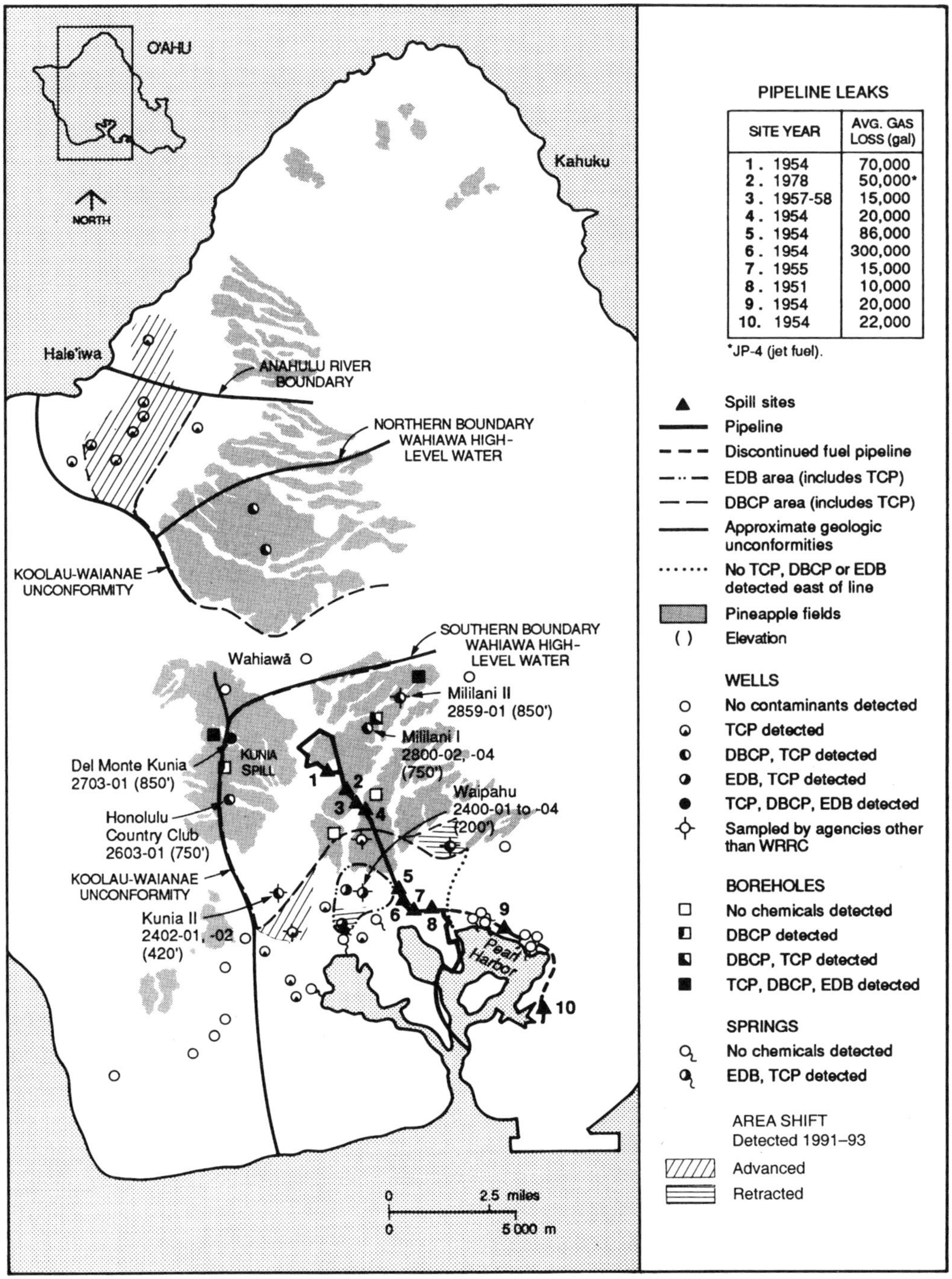


Figure 6.11. Spatial distribution of organic chemical contamination in southern, central, and northern O'ahu, Hawai'i. 1 gallon = 3.785 liters. (Adapted from Lau 1987)

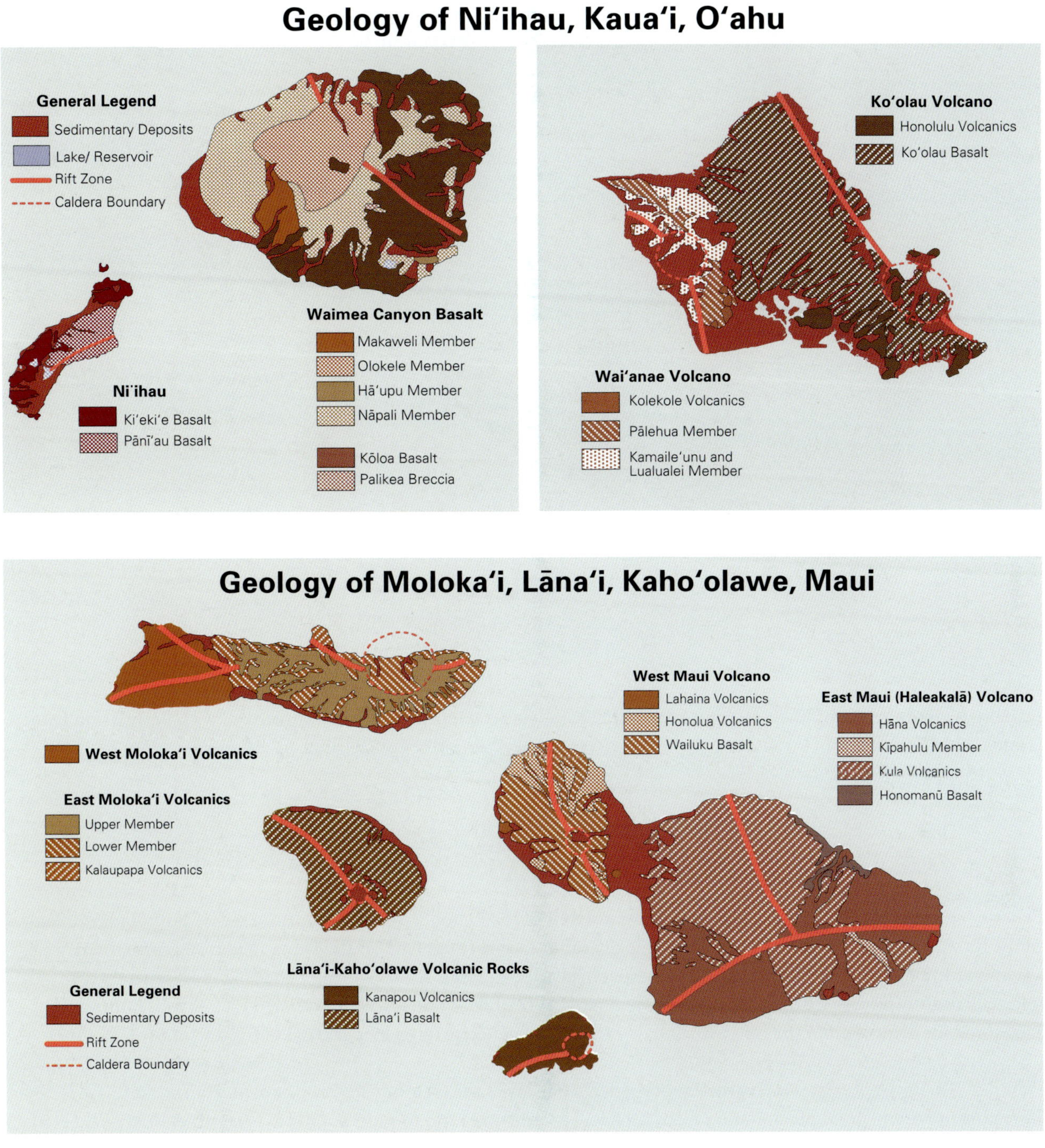


Plate 1.1. Simplified geologic maps of major Hawaiian Islands. (Reprinted from Clague 1998 with permission from J. Juvik and University of Hawai‘i Press)

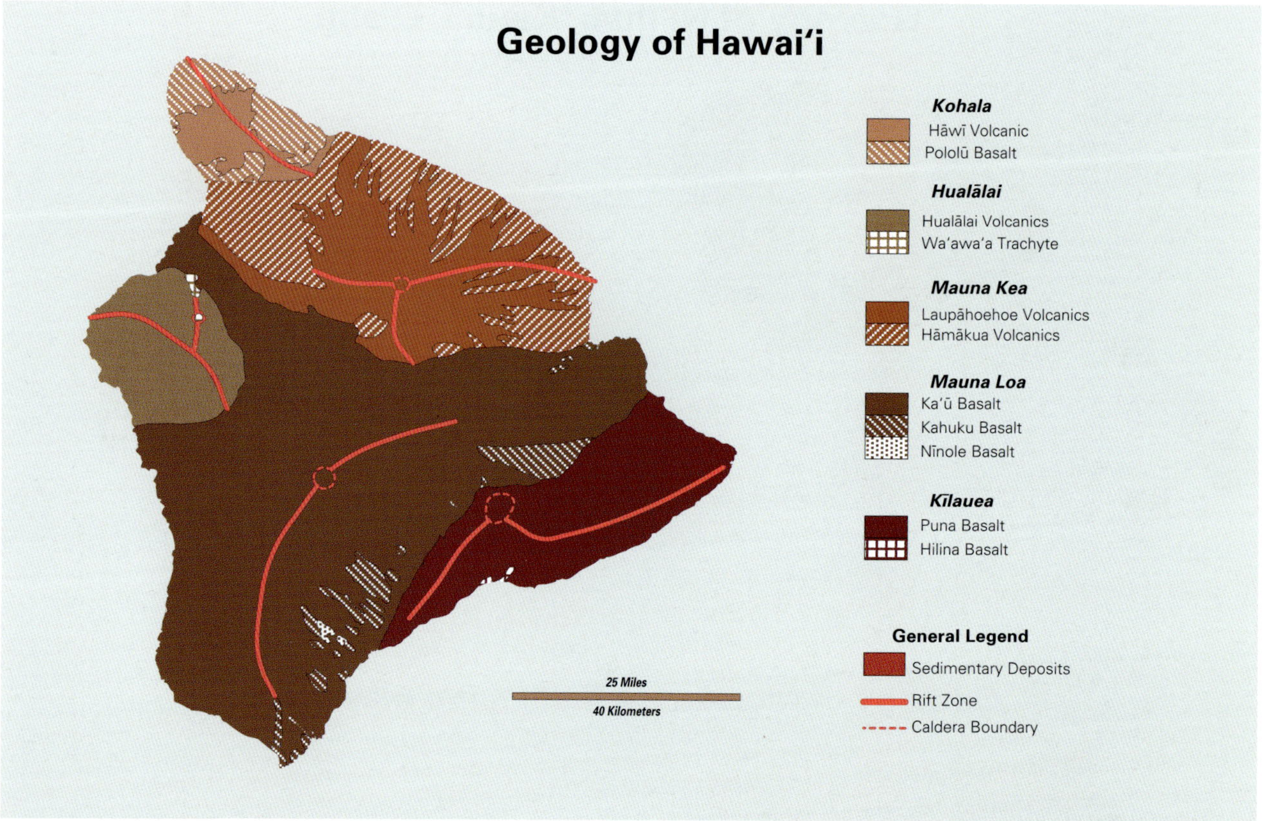


Plate 1.1. *(Continued)* Simplified geologic map of Hawaiʻi Island. (Reprinted from Clague 1998 with permission from J. Juvik and University of Hawaiʻi Press)

Plate 3.1. *(Facing page, top)* Hydrometeorological analysis of the October 29, 2000, storm, Hāna, Maui: Radar reflectivity. Kamuela radar (panels *a* and *b*) and Molokaʻi radar (panel *c*) focused on the Hāna region at approximately 2.5, 4.5, and 10 km (8,200, 14,800, and 32,800 ft), respectively, above mean sea level. Color shading represents 10-dBZ intervals beginning at 15 dBZ. These panels reveal most of the life of one cell that persisted for nearly 4 hours and accounted for 80 percent of the rain associated with the storm. (Reprinted by permission from Lyman et al. 2005)

Plate 3.2. *(Facing page, bottom)* Hydrometeorological analysis of the October 29, 2000, storm, Hāna, Maui: Total storm precipitation. Total precipitation from Molokaʻi radar from 1200 to 1900 hours HST based on Rosenfield et al. (1993) Z-R relationship with addition of a +3.5 dBZ for range correction; isohyets in 50-mm (2-in.) increments; topographic contours, 500 m (1,640 ft). Black dot is location of gage that received 686 mm (27 in.); black triangle is the Hāna airport gage that accumulated 249 mm (9.8 in.). (Reprinted by permission from Lyman et al. 2005)

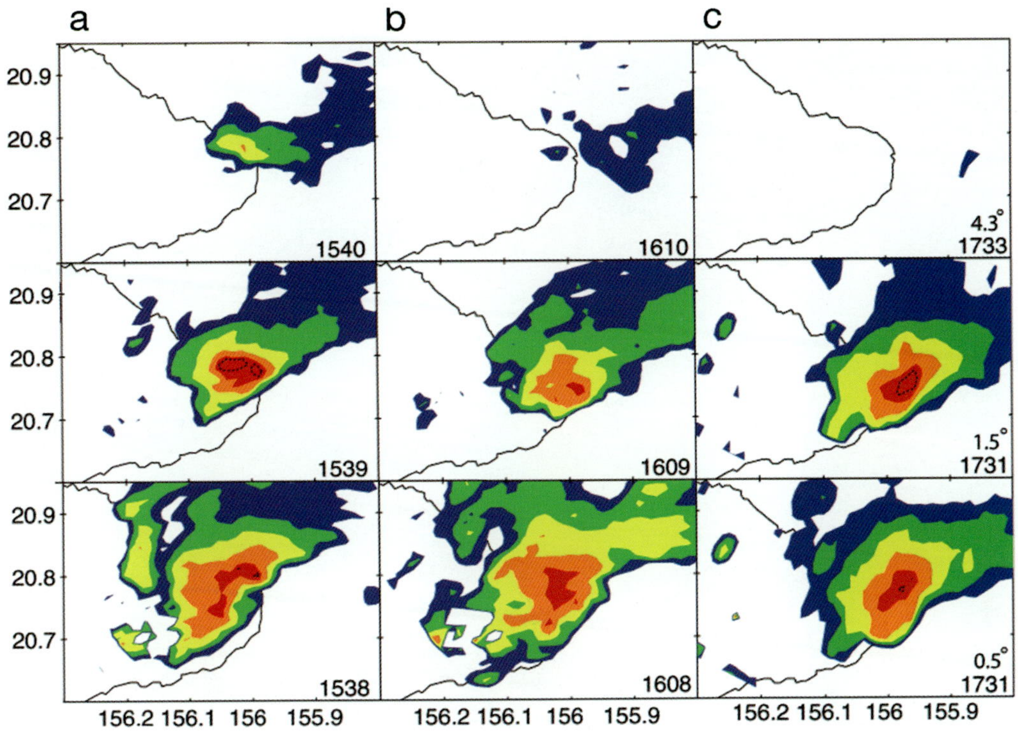
a
b
c
20.9
20.8
20.7
1540
1610
4.3°
1733
1539
1609
1.5°
1731
1538
1608
0.5°
1731
156.2 156.1 156 155.9

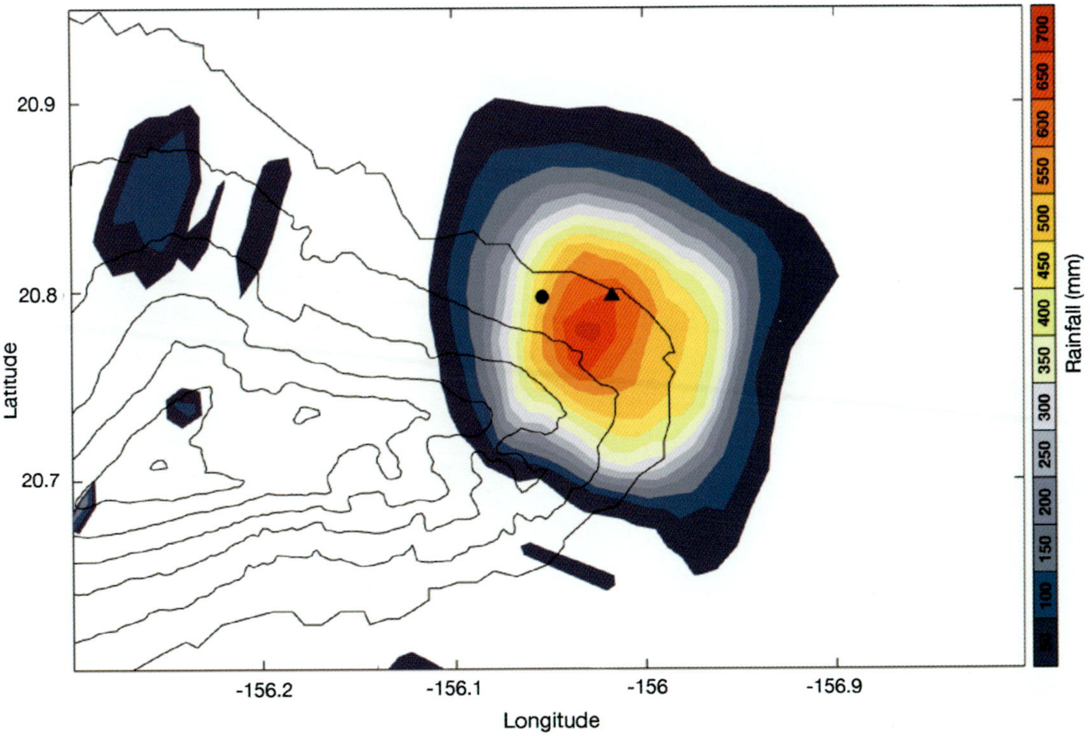
Latitude
20.9
20.8
20.7
-156.2
-156.1
-156
-156.9
Longitude
Rainfall (mm)
100
150
200
250
300
350
400
450
500
550
600
650
700

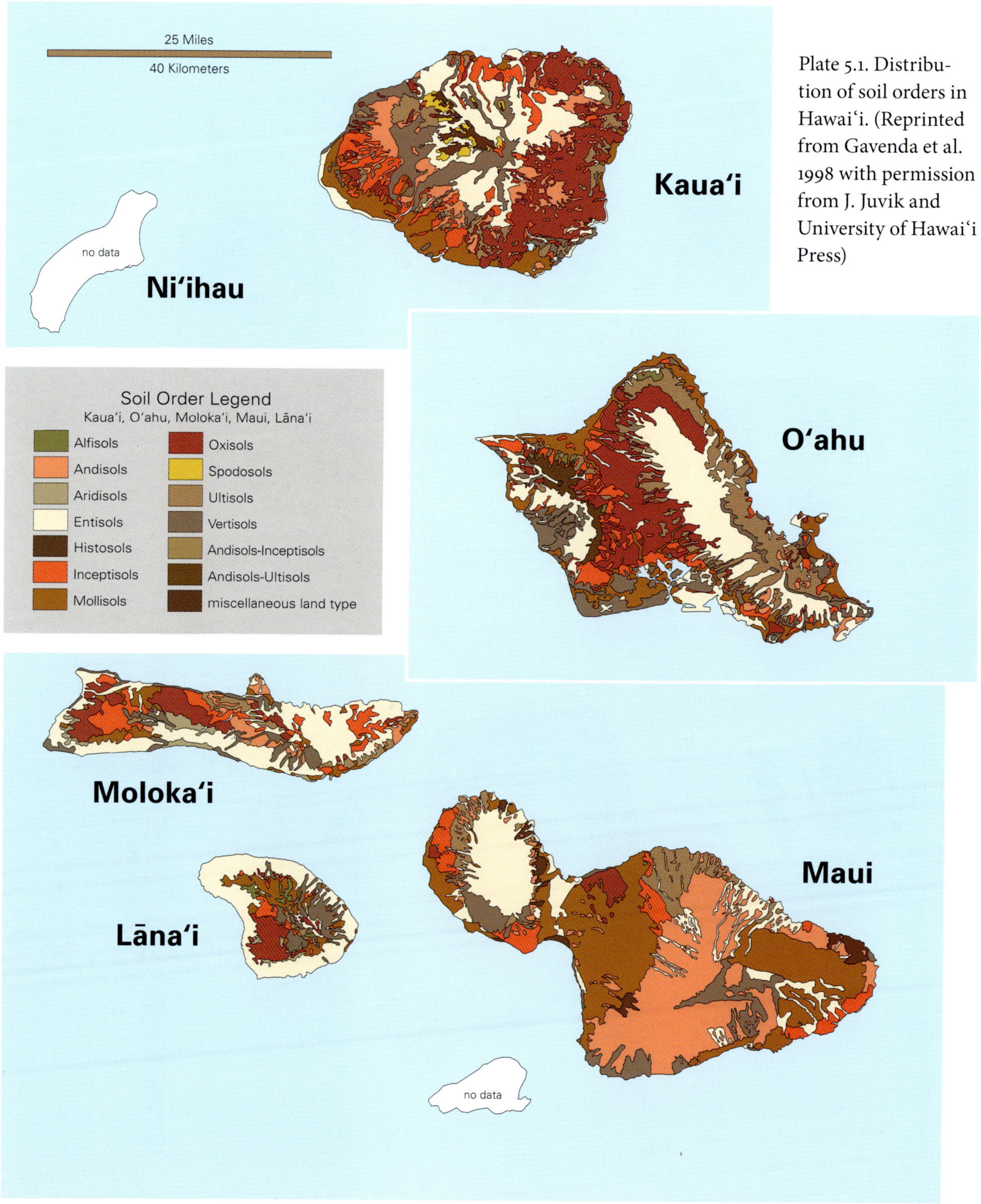


Plate 5.1. Distribution of soil orders in Hawai'i. (Reprinted from Gavenda et al. 1998 with permission from J. Juvik and University of Hawai'i Press)

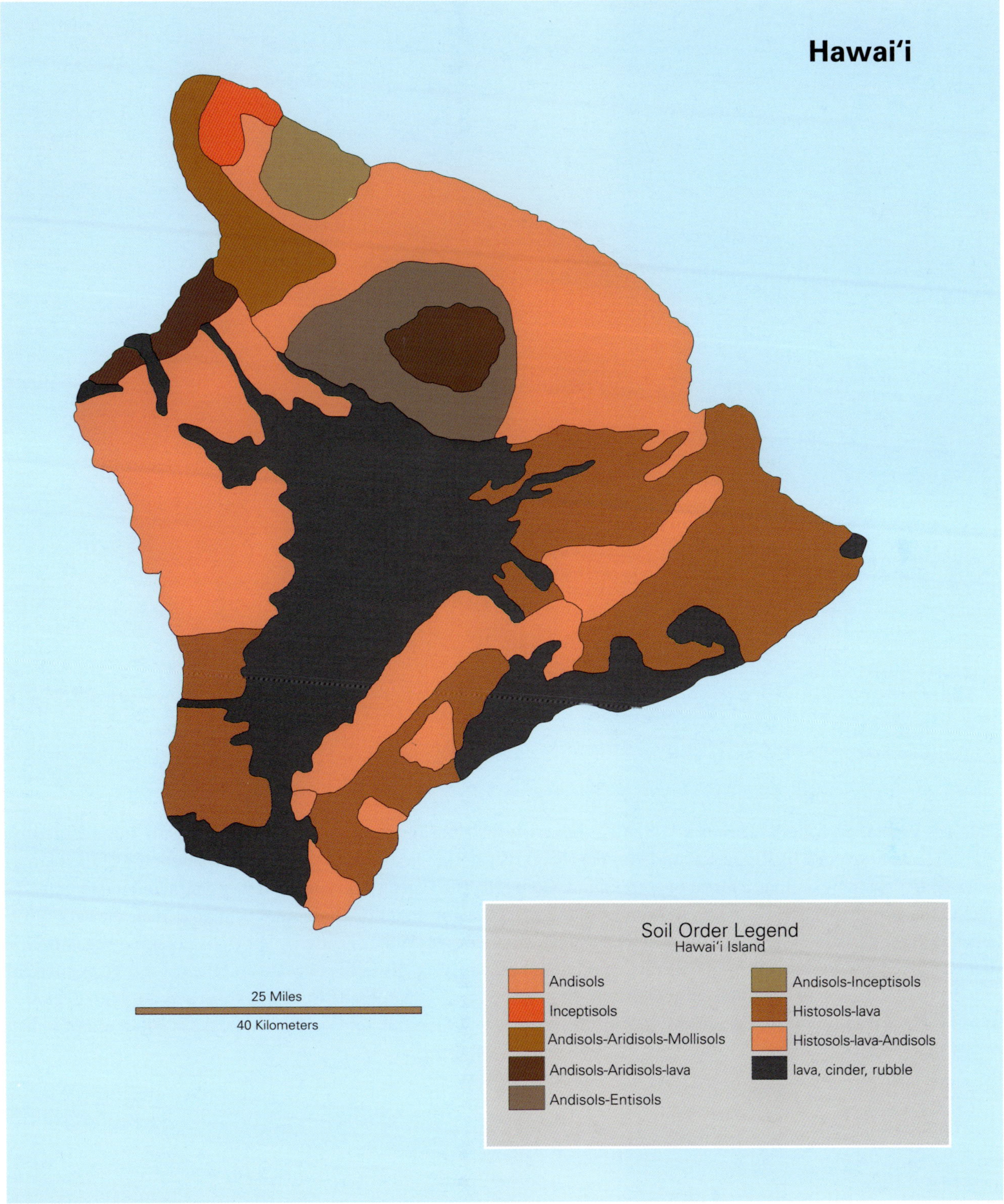

Hawai'i
25 Miles
40 Kilometers
Soil Order Legend
Hawai'i Island
Andisols
Inceptisols
Andisols-Aridisols-Mollisols
Andisols-Aridisols-lava
Andisols-Entisols
Andisols-Inceptisols
Histosols-lava
Histosols-lava-Andisols
lava, cinder, rubble

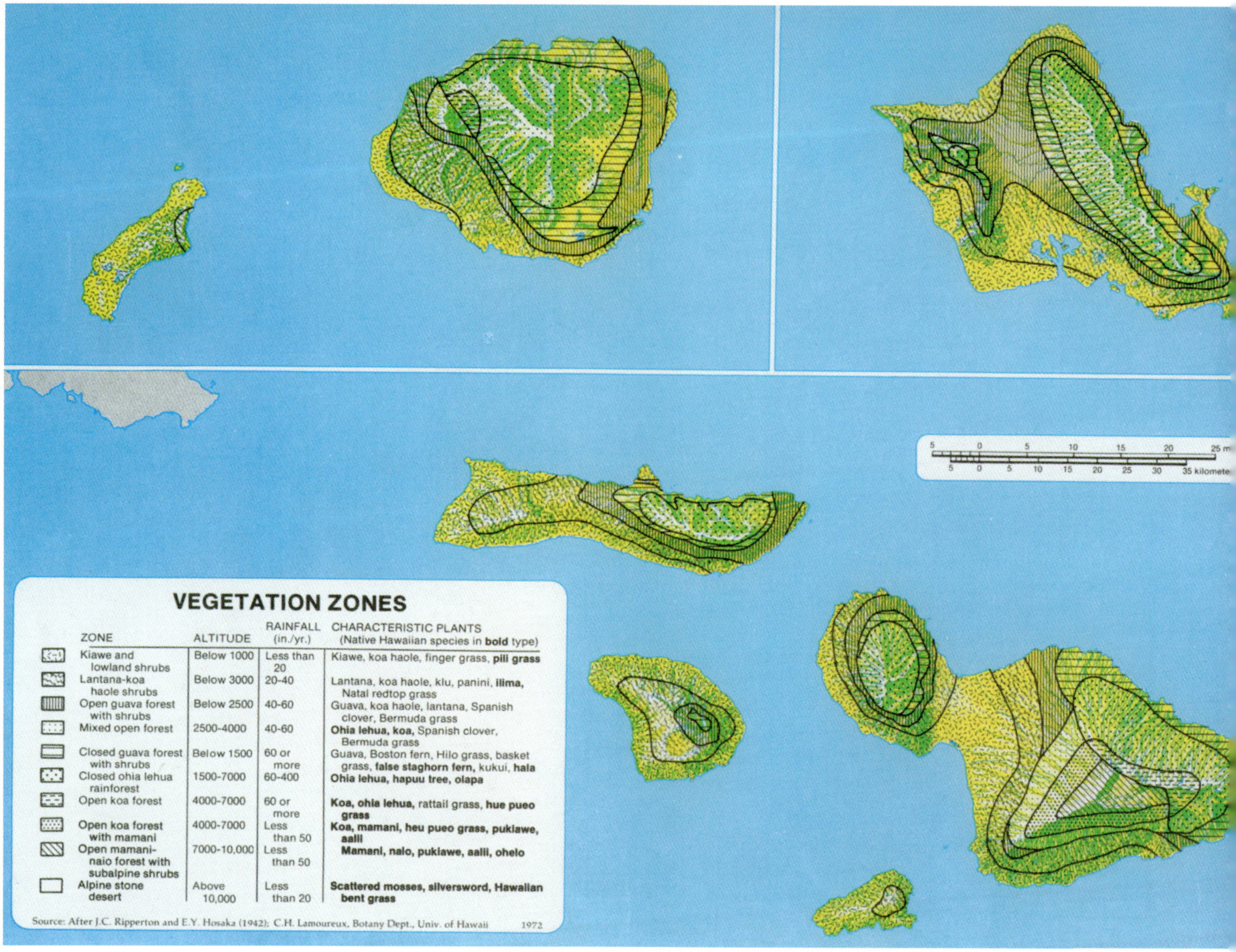


Plate 5.2. Vegetation zones of Hawaiʻi. "The map shows the distribution of vegetation zones in Hawaii. It is adapted from the one developed by Ripperton and Hosaka (1942). Vegetation is a term used for the physiognomy of the plant cover of an area, for example, a forest or a grassland. Generally the vegetation type is further characterized by reference to the dominant species. As can be seen from the map, the dominant or characteristic plants in all the vegetation zones at lower elevations are species introduced to Hawaii since 1778. What is actually shown is potential vegetation, that is, the vegetation that one might predict could develop in an area, given the climate and the plant species currently present in that area. It does not take into account the vegetation, or lack thereof, in urban areas and does not depict modifications resulting from plantation agriculture, pasture development, and similar factors. Also, it does not reflect

differences in vegetation as related to different ages of lava flows or different stages of soil development. Although such factors are of considerable importance in governing the distribution of vegetation types, they cannot be depicted accurately on maps at this scale. The zonation of plants is closely related to climatic factors, as a comparison of the vegetation map with climate maps will reveal. The most important climatic factors governing plant distribution in Hawaii seem to be average annual rainfall at elevations below about 5,000 feet, and temperature and rainfall at higher elevations." (Reprinted from Lamoureux 1973 with permission from University of Hawai'i Press)

Plate 7.1. Quality-regulated marine water bodies in Hawai'i. (Adapted from Parks 1998)

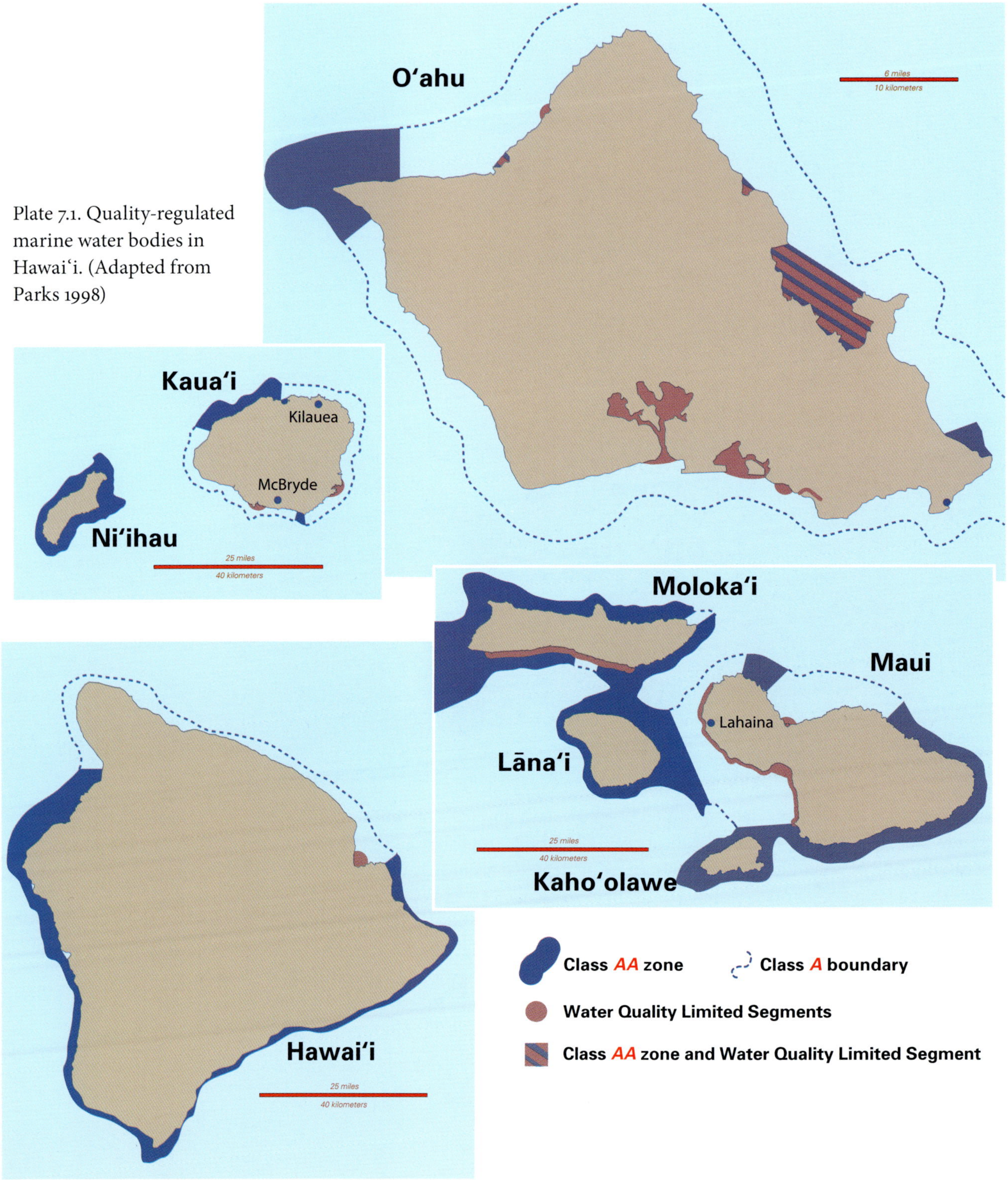

Plate 7.2. *(Left)* Typical headwaters environment in Hawai'i streams. (From Hawai'i Commission on Water Resource Management 1993; courtesy of Dan A. Polhemus)

Plate 7.3. (*Bottom right*) Hydraulic simulation of usable stream surface area for *Lentipes concolor* in upper reach of Nānue Stream, Hawai'i Island. (Adapted from Kinzie et al. 1986); (*bottom left*) *Lentipes concolor.* (From Hawai'i Commission on Water Resource Management 1993; courtesy of John Ford)

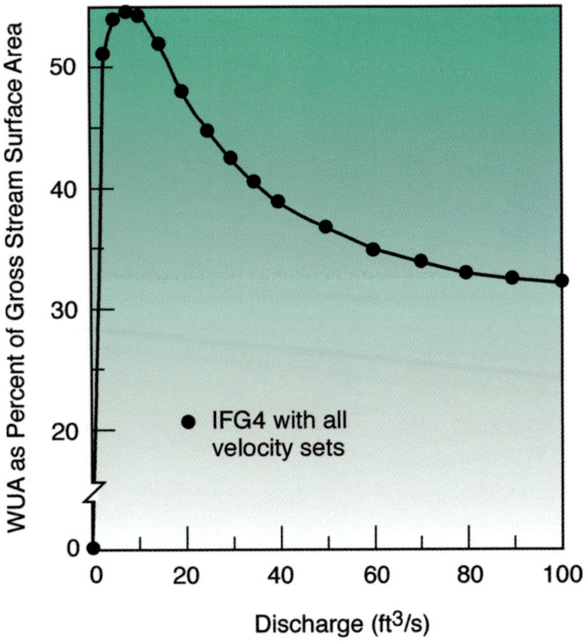

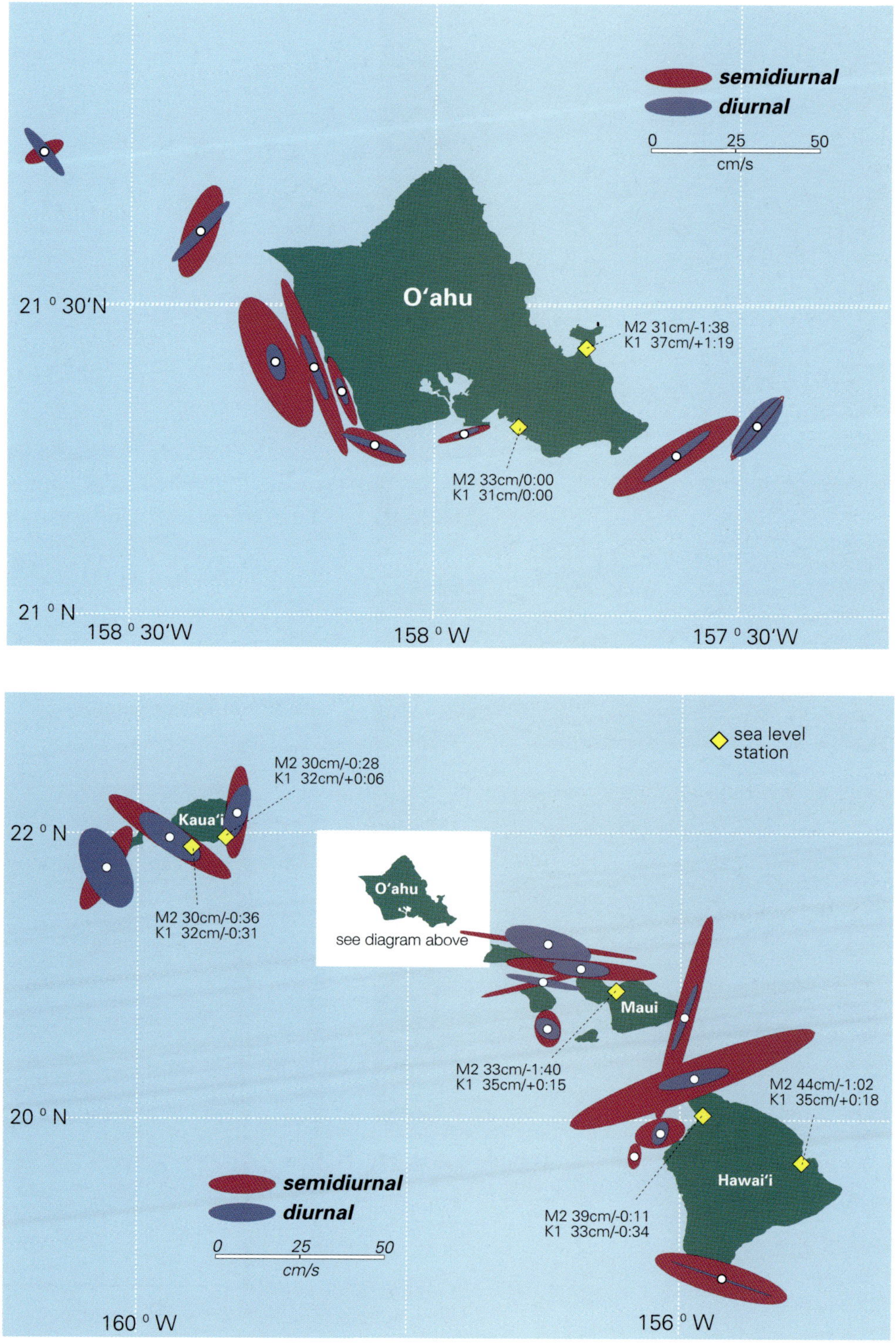


Plate 8.1. Hawai'i observed tidal currents: Ranges and phases. 1 ft/s = 30.5 cm/s. (Reprinted from Flament et al. 1998 with permission from J. Juvik and University of Hawai'i Press)

Plate 8.2. Some common coral species of Hawai'i. *Top, Pocillopora meandrina* (cauliflower coral); *middle, Porites lobata* (lobe coral); *bottom, Porites compressa* (finger coral). (Courtesy of Anthony R. Russo)

Plate 8.3. *(Right)* Coral and coralline algal growth on ocean sewer outfall, Wai'anae, O'ahu, May 1994. (The photo was provided by the City and County of Honolulu, Department of Environmental Services, and is being published with permission from Russo 1994)

Plate 8.4. *(Below)* Beach and coral reef inside the breached volcanic crater at Hanauma Bay, O'ahu. (Reprinted from Macdonald et al. 1983 with permission from E. Abbott and University of Hawai'i Press)

Table 6.4. Applicable Drinking Water Standards, Possible Health Effects, and Potential Sources of Groundwater Contamination

| Common Name | Applicable Drinking Water Standards[a] (in μg/l) | | Possible Noncarcinogenic Effects (Based on Ingestion Effects on Animals)[b] | EPA Carcinogen Rating[c] | Potential Contamination Sources |
|---|---|---|---|---|---|
| Alachlor | 2 | MCL | Damaged red blood cells, causing kidney and spleen damage | Probable | Herbicide |
| Ametryn | 60 | LHA | Liver damage | Unclassified | Herbicide |
| Atrazine, Desethyl Atrazine, Despropyl Atrazine, Diamino Atrazine | 3 | MCL | Heart and liver damage; fetal/child development retarded | Possible | Herbicide |
| Carbon Tetrachloride | 5 | MCL | Liver, kidney, and lung damage | Probable | Solvent, dry-cleaning agent |
| Chlordane | 2 | MCL | Central nervous system, liver, and kidney damage | Unclassified | Pesticide (termiticide) |
| 1,2-Dibromo-3-chloropropane (DBCP) | 0.04 | SMCL | Male reproductive system, liver, and kidney damage | Probable | Pesticide (soil fumigant) |
| 1,1-Dichloroethylene (DCE) | 7 | MCL | Central nervous system depression; a heart effect; liver and kidney damage | Possible | Solvent |
| 1,2-Dichloropropane (DCP) | 5 | MCL | Gastrointestinal irritation, liver and kidney damage | Probable | Pesticide, solvent |
| Dieldrin | 0.002 | $10^{-6}$ | Central nervous system, liver, and kidney damage | Probable | Pesticide |
| Diuron | 10 | LHA | Central nervous system depression; damaged red blood cells, causing spleen damage; altered fetal development | Unclassified | Herbicide |
| Ethylene dibromide (EDB) | 0.04 | SMCL | Male reproductive system, liver, gastrointestinal, and adrenal gland damage | Probable | Gasoline additive, soil fumigant, solvent |
| Hexazinone | 200 | LHA | Reduced body weight or possibly reduced growth | Unclassified | Herbicide |
| Lindane | 0.2 | MCL | Nerve damage and central nervous system seizures; liver and kidney damage; suppression of the immune system | Probable | Insecticide |

Table 6.4. *(continued)*

| Common Name | Applicable Drinking Water Standards[a] (in μg/l) | | Possible Noncarcinogenic Effects (Based on Ingestion Effects on Animals)[b] | EPA Carcinogen Rating[c] | Potential Contamination Sources |
|---|---|---|---|---|---|
| Simazine | 4 | pMCL | Liver, kidney, and brain damage | Possible | Herbicide |
| Tetrachloroethylene (PCE) | 5 | MCL | Central nervous system depression; liver and kidney damage | Probable | Solvent, dry-cleaning agent |
| Trichloroethylene (TCE) | 5 | MCL | Central nervous system depression; a heart effect; liver and kidney damage | Probable | Solvent |
| 1,2,3-Trichloropropane (TCP) | 0.8 | MCL | Decreased red blood cells; liver and kidney damage | Unclassified | Solvent, trace contaminant in certain pesticides |

SOURCE: Hawai'i Department of Health 1996.

[a]MCL, a maximum contaminant level or the maximum permissible level of a contaminant in water that is delivered to any user of a public water system (MCLs are the only federally enforceable drinking water standards); pMCL, the U.S. Environmental Protection Agency (EPA) has proposed an MCL for that contaminant; LHA (Lifetime Health Advisory) , a nonregulatory concentration of a drinking-water contaminant at which adverse health effects would not be anticipated to occur over a lifetime exposure of 70 years duration (the advisories are based on data describing noncarcinogenic risk from such exposure); SMCL, the State's maximum contamination level (which may be more stringent than the MCL) of a contaminant in water (applies to community and nontransient noncommunity water systems, as defined in Hawai'i Administrative Rules, Title 11, Department of Health, Chapter 20—Rules Relating to Potable Water Systems); $10^{-6}$, those chemicals that EPA considers to be potential human carcinogens (EPA estimates a "cancer risk level" as the level at which an individual who consumes water over his or her lifetime [70 years] would have no more than a one-in-a-million chance of developing cancer as a direct result of drinking water containing the contaminant).

[b]Based on Health Advisories from the EPA's Office of Drinking Water.

[c]Based on estimates from the EPA's Health Hazard Assessment Group.

zone are essential (Lau et al. 1975). California grass and sugarcane are similarly effective, as was demonstrated in field tests (Lau et al. 1975, 1989). See Appendix 5.3 for additional discussion.

High temperature, desiccation, and ultraviolet radiation from sunlight are powerful environmental factors that inactivate viruses. Sewage-borne human enteric viruses were effectively inactivated within a 0.9-m (3-ft) vegetated soil mantle in controlled field investigations directed at reclamation and reuse of wastewater in Mililani in central O'ahu (Lau et al. 1980).

Aquifer types vary in their vulnerability to surface contamination. Confined aquifers, especially artesian aquifers, are protected from surface contaminants, but poorly confined and nonartesian confined aquifers can be vulnerable because downward leakage can cause contamination in areas where the hydraulic head in the confined aquifer is weakened. Unconfined aquifers are clearly vulnerable, as has been evidenced in the Pearl Harbor Aquifer Sector and the Wahiawā Aquifer System. In both cases, the water table is deep at the site of contamination (Mililani, 195 to 265 m [640 to 870 ft]; Schofield Barracks, 220 m [721 ft]). This indicates that great depth to the water table alone is not a guaranteed barrier to all contaminants; instead, it merely requires a longer transport time.

## Principal Aquifers: Saturated Zone

In the saturated zone of principal aquifers, natural remediation continues but with limited results. In Maui and O'ahu, the DBCP concentration was diminished by one order of magnitude in the saturated zone as compared with four orders of magnitude (from 1.0 mg/l to 0.1 μg/l) in the unsaturated zone (Green et al. 1986; Orr and Lau 1988). In the saturated zone, advection and dispersion are effective in reducing the level of contamination while hydrolysis continues to degrade organic chemicals, but only slowly. Long residence or transit time in the saturated zone enhances microbial die-off, but validating data in Hawai'i have been difficult to secure.

The advance and retreat of the front of the contaminated groundwater plume are manifestations of the overall transport process in the saturated zone. During a period of approximately 5 years (1987 to 1992), the detected front of organics advanced as much as about 1.7 m/d (5.5 ft/d) in the Waialua Aquifer System in the North Aquifer Sector and about 0.6 m/d (2.0 ft/d) in the Pearl Harbor Aquifer Sector. The values for these aquifers are within the range of estimated global average groundwater velocity. The shifts (see Figure 6.11) can be explained by the natural groundwater flow regimes as affected by pumping. The nonpoint-source contamination was widespread over about a 16,700-ha (64 mi$^2$) area.

The TCE plumes detected in the Wahiawā Aquifer System were well defined horizontally and vertically during the remediation investigation. The groundwater body in the system is unique in the sense that its water table level is 82 to 84 m (270 to 275 ft) above sea level. Two widely separated plumes of TCE at greater than 5 μg/l were discovered. The smaller one, 47 ha (116 ac) in area, is located in the vicinity of a former landfill, the suspected source. The larger one, of unknown source, covers 1,400 ha (5.4 mi$^2$) and is positioned near Wheeler Airfield.

Solute contamination plumes in O'ahu are on the order of 76 to 84 m (250 to 275 ft) thick. At the Punanani well field where return water from excess irrigation of sugarcane had occurred, the measured thickness of the layer having relatively high dissolved solids was 61 to 76 m (200 to 250 ft) (Orr and Lau 1987). At the old Schofield Barracks landfill, the TCE plume extended to 57 m (187 ft) below the water table and the TCE concentration was vertically uniform (9 to 11 μg/l). The other plume surrounded major pumping wells and was at least 75 m (245 ft) thick, but the vertical concentration distribution was not uniform.

The reasons for the considerable thickness and the vertical concentration distribution have not been sufficiently investigated. The influence of pumping and the daily fluctuation of the measured barometric pressure (4.6 cm [0.15 ft]) have been speculated as the causes (Harding Lawson Associates 1996). The structural features and discontinuous nature of the lava flows provide ample passages for vertical transport if these driving forces are of sufficient magnitude.

Small pockets of petroleum hydrocarbon have been reported in sedimentary sequences in the coastal plain. For instance, a plume of benzene, toluene, and ethylbenzene detected at the corner of Ward Avenue and Auahi Street in Honolulu was about 0.2 ha (0.5 ac) in size.

Transport by advection and dispersion in a basal lens was evaluated based on injection experiments of dissolved helium in Waipahu, O'ahu (Gupta et al. 1994a, 1994b). Helium is conservative and has no detectable natural background in the saturated zone on O'ahu. The experimental site consisted of pumped and sampled production wells, 28.4 mld (7.5 mgd) for one experiment and 18.9 mld (5.0 mgd) for another. The helium was introduced in boreholes and drawn by the converging flow toward the pumped well. The breakthrough and elution curves are well defined at long travel distances and for long durations: 26 m (85 ft) in 1-day duration for one experiment and 119 m (390 ft) in 40-day duration for the

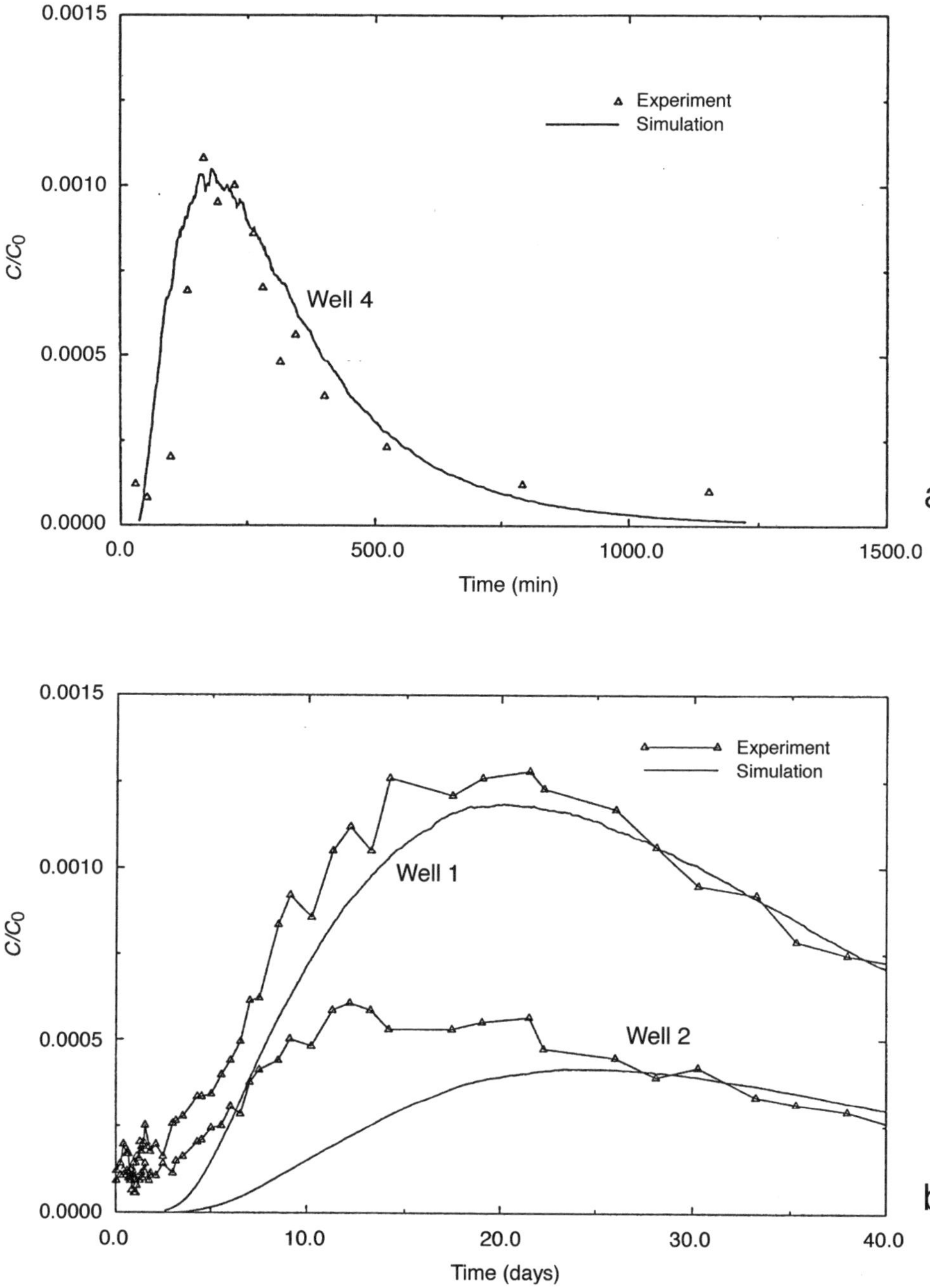


Figure 6.12. Measured and simulated helium breakthrough and elution in the Waipahu aquifer, O'ahu, Hawai'i: *a,* Experiment 3—injection duration 0.5 h, travel distance 26 m; *b,* Experiment 4—injection duration 38 h, travel distance 119 m. The poor match in Experiment 4 is attributed to insufficient information about the flow field. 1,500 min = 1.04 d. (Reprinted by permission from Gupta et al. 1990)

other (see Figure 6.12). The longitudinal dispersivity values apparently increase with travel distance: 28 m for the first experiment and 250 m for the other. The indicated dependence on travel distance supports the generalization by Gelhar et al. (1992).

## Protection Strategies

Strategies for the protection of groundwater from surface contamination transcend remediation by technology alone. Prevention is fundamental and involves regulatory actions and land management. The experience of non-point-source pesticide contamination of potable aquifers in O'ahu has provided the needed rationales for protective policies and measures. Also, research has resulted in science and technologies that are uniquely appropriate for the Hawaiian environment and may be useful by others to forecast or meet similar contingencies (Lau and Mink 1987, 1995b). Federal regulations for the protection of groundwater abound (Saner and Pontius 1991; Pontius 1992). They have been the ultimate driving force behind the remarkable advances made since the 1980s.

### *Remediation*

Remediation technologies aim to reduce the contaminant level of the polluted soil and groundwater to below maximum contaminant level set by regulatory agencies. Monitoring is an integral part of remediation. The cleanup of a contaminated aquifer by natural action takes a long time (i.e., years and even decades), but today's remediation demands quicker results. Innovative technologies have been developed at a fast pace. A reference guide lists forty-eight available remediation technologies—in situ and ex situ, biological, thermal, and physical/chemical processes (U.S. Environmental Protection Agency 1993). *Groundwater Monitoring and Remediation* is an often-referenced journal of the subject. The choice of remediation technology begins with characterization of the site and contaminants. Single technologies are usually sufficient, but a train of them may be necessary. Cost leads off a long list of selection criteria, which include time, residual produced, and community acceptability.

For the basal water in Pearl Harbor aquifers contaminated with organic chemicals, granular activated carbon technology has been used since 1986 to treat a total of 49.2 mld (13 mgd) of municipal well water at three plants—one each at Mililani, Waipahu, and Kunia. Aquifer-wide monitoring in 1991–1993 indicated the unlikelihood of quick aquifer cleanup, confirming model prediction (Lau and Mink 1995b). The appropriate and realistic remediation strategy should focus on "point-of-use." This means that the contaminated groundwater is pumped and treated in centralized facilities and distributed directly for drinking. Aquifer cleanup is not an achievable short-term goal.

In Hawai'i remediation of contaminated soils has been uncommon. In 1990, a total of 50,000 tons of contaminated soils was excavated from the Del Monte Kunia Superfund site for off-site disposal followed by natural attenuation. The first recorded soil remediation in Hawai'i involved chemical treatment (oxidation and reduction) of chromium-contaminated soil that had been excavated from the Pearl Harbor naval facilities (Dugan et al. 1984).

A flurry of regulatory actions was executed in Hawai'i to deal with the huge backlog of soil and groundwater contamination that mostly resulted from leaks of many underground storage tanks. Specific technologies include in-situ bioremediation-filtration-adsorption train and recovery of free (floating) products (Hawai'i Department of Health 1994). Some actions resulted in the economic demise of service stations and costly remediation for the Honolulu police headquarters. Subsequently, the Health Department adopted a method for quick screening based on SESOIL (see Models section) before requiring corrective actions and formal and costly risk assessment (Hawai'i Department of Health 1995a). Risk assessment concerns exposure of humans to contaminants by various pathways: ingestion, inhalation, and dermal adsorption. For example, carcinogenic risk, which is defined as the incremental probability of an individual human developing cancer over a lifetime as a result of exposure to a carcinogen, is kept very low: $10^{-6}$ by the Health Department and $10^{-4}$ to $10^{-6}$ by the U.S. Environmental Protection Agency. The Schofield Barracks remediation investigation reported an estimated carcinogenic risk of $4.9 \times 10^{-7}$ to $3.0 \times 10^{-5}$ due to TCE.

### *Prevention*

Prevention consists of both technical and nontechnical measures. They are aimed to deal with, first, the contaminants; second, the water, which is the premier carrier of the contaminants; and third, the land, on which some human activities may result in contamination of the groundwater.

Certain chemicals used for various purposes at different times in Hawai'i have been identified to be toxic to human health. Ordinarily, their presence in a drinking water source is limited to very low levels (see maximum contaminant levels [MCLs] in Table 6.4). The State of Hawai'i initially was forced to create its own water-quality standards for DBCP and EDB and chose to base them on the detection limit of laboratory instruments. In 1990, after the U.S. Environmental Protection Agency issued its maximum contaminant levels based on health risk assessments, Hawai'i relaxed its own standards but still made them more stringent than those of the Environmental Protection Agency (Lau and Mink 1995b).

Assessing the concentration and risk of chemicals after application on land and the time required for reaching a groundwater source has been attempted with models, such as at Mililani (Orr and Lau 1987, 1988), Schofield Barracks (Harding Lawson Associates 1996), and Waiawa (Loague et al. 1995). The assessments involve many assumptions and yield results with various uncertainties. Special difficulties arise where the water table lies at great depths because of the lack of parameter values for the thick unsaturated zone.

It has been stated that all aquifers are vulnerable to surface contamination (National Research Council 1993). However, certain types, such as unconfined aquifers, are more vulnerable than others. In Hawai'i, an aquifer classification for the six major islands has been available and in use by the State water commission and the State Department of Health since 1990. The relative vulnerability to contamination of the aquifers is included in the classification (Mink and Lau 1990a, 1990b, 1992a, 1992b, 1993a, 1993b).

The exclusion of potentially polluting land uses in groundwater recharge areas can provide the ultimate protection. Conservation zoning dedicates lands for the protection of watersheds and water sources. Conservation districts in Hawai'i were formally created in 1961 with the passage of the land-use law (Hawai'i Revised Statutes 1985), the first such comprehensive zoning legislation in the United States. The designated conservation lands preclude urban, agricultural, and rural uses. Not all of the conservation districts were designated to protect water resources, but on O'ahu in particular, where groundwater sources are relied upon for most of the drinking-water supply, much of the conservation zones envelopes the wet portions of the Ko'olau and Wai'anae Mountain Ranges, where hydrologic considerations are dominant. Of the total O'ahu landmass, 40 percent has been placed in conservation districts. Additional areas have been designated conservation for water protection in the 1992 five-year boundary review (Mink and Lau 1997).

In Hawai'i, an underground injection control line delineates groundwater sources of potential drinking quality. No injection of contaminants is permitted within the underground injection control line (Hawai'i Department of Health 1992). The underground injection control and conservation lines provide considerable groundwater protection. The inner line (conservation) excludes all pollution sources; the outer line (underground injection control) excludes point-source pollution.

Notwithstanding these land-management regulations, more is needed. The reason is that some landmass located between these two lines is vulnerable to nonpoint-source pollution. In Hawai'i, some of these lands have already been developed for urban and agricultural uses. The dilemma of balancing groundwater protection and land use may be alleviated by applying as a control the concept of wellhead protection areas. Wellhead protection area is defined as the surface and subsurface areas tributary to potable water wells. Within the wellhead protection areas potentially contaminating activities need to be controlled but not necessarily totally banned by regulations.

In Hawai'i, wellhead protection area guidelines have been promulgated as a voluntary program (Hawai'i Department of Health 1995b). The delineation of a wellhead protection area is, for the most part, hydrologically correct, but its proper application depends on accurate field measurements.

## Hawai'i Data and Models

### Data

Usual groundwater data consist of water level, water quality, and groundwater extraction rates. In Hawai'i data are located in several governmental agencies: Honolulu Board of Water Supply, Hawai'i State Commission on Water Resource Management, Hawai'i Department of Health, University of Hawai'i, and U.S. Geological Survey. The private sector, such as sugarcane and pineapple plantations as well as consultants (geologic, hydrologic, environmental), is also an important source. The rationale of acquisition and mode of information dissemination vary considerably among these sources.

Data stem from investigations, monitoring activities, programmatic surveys, and research. Investigations are conducted typically for the purpose of producing water supplies and identifying water pollution. Monitoring includes operations of groundwater extraction facilities, regulatory requirements of contaminant remediation, and deep wells probing the transition zones. Programmatic surveys are usually the network types (Anthony 1997).

Groundwater observation and sampling are conducted principally in wells and boreholes, and at springs, and in wetlands. Remote sensing may be performed on the ground surface or from the air by geophysical techniques.

Data repositories include the U.S. Geological Survey for water-level and water-quality parameters generated from programmatic surveys, the Hawai'i Department of Health for required regulatory monitoring conducted by permittees, and the Commission on Water Resource Management for inventory of water wells.

Data summaries are published annually by the U.S. Geological Survey. Investigation and survey reports are issued occasionally by the U.S. Geological Survey and University of Hawai'i (Water Resources Research Center and Hawai'i Institute of Geophysics). Monitoring data and reports are presumably open file in nature. Private-sector data and information may be restricted but are most often available. Scientific journal articles provide succinct information but rarely sufficient data for many purposes including modeling. Historical accounts regarding groundwater technology of Hawai'i are documented (R. H. Cox 1981).

### Models

Groundwater systems usually behave in complex ways that are difficult to describe effectively in a classical mathematical framework unless great simplifications are made. These simplifications, such as assigning well-posed boundary conditions and linearizing the equations of groundwater flow, are useful in explaining particular aspects of behavior but generally fail if employed in describing the dynamics of the whole system. The analytical approach falters on the unsolvability of the second-order nonlinear partial differential equations defining flow in porous media, and on the heterogeneity of aquifer properties that frequently defies averaging. However, analytical solutions facilitate understanding of the physical phenomena and can be of practical use.

To overcome the limitations of exact methods of solving the groundwater equations, physical and numerical mathematical models are commonly used. Physical models are highly constrained in simulating real aquifer behavior, whereas numerical models can be made as flexible as computer efficiency allows. Numerical modeling is the preferred technique for explaining and predicting aquifer behavior because it accommodates heterogeneities and nonisotropism to whatever degree is economically acceptable by dividing the aquifer into an array of cells and nodes, each of which may display different properties. Numerical models, however, require a reliable data record of substantial length for calibration, which is not often available. The process of composing a model and testing it against the historical record requires a clear description of the groundwater system within a rational framework, an exercise that sharpens knowledge of and provides insights to aquifer behavior. The final objective of modeling is to promote optimal development of an aquifer.

Many kinds of groundwater models have been used in Hawai'i. In terms of solving the governing mathematical equations, they range from analytical, physical, and analog to numerical. Most of them are deterministic. At least two efforts in Hawai'i attempted to address the stochastic nature of the system. The modeling work cited is intended to be representative rather than complete. Original models or any models (by kind or type) that were applied for the first time in Hawai'i are presented only briefly because numerous models have been employed, especially in the last decade. In a review of Hawai'i groundwater modeling from a historical perspective, the current and future roles of models are critically addressed (Lau and Mink 1995a).

In Hawai'i, groundwater modeling has been used for at least four purposes: (1) simulating responses to known stresses and predicting the impact of proposed stresses, (2) understanding the natural processes and evaluating parameter values, (3) implementing public policies, and (4) monitoring remediation. In each of these applications, models are used as a tool to integrate theories and site-specific data.

## Simulation and Prediction

Models have been used for simulation and prediction of aquifer responses to stresses in Hawai'i. Many models deal with aquifer behavior and sustainable yield, some address injection of wastewater, and others consider surface contamination by chemicals and viruses.

### *Sustainable yield*

Sustainable yield is the average rate of extraction of groundwater of specified water quality that can be sustained indefinitely without creating undesirable results. Unacceptable depletion of storage in basal lenses and seawater encroachment have been the chief reasons for assigning sustainable yield.

The Robust Analytical Model (RAM) is an original analytical model formulated and calibrated for the Pearl Harbor Aquifer Sector (Mink 1980). It has also been applied for planning purposes to all of the aquifers in the six major Hawaiian Islands by the Commission on Water Resource Management.

RAM is a transient flow model for a vertical section of a thick lens with a sharp interface, assuming horizontal flow, that is, the Ghyben-Herzberg relationship. The transient solution to the flow continuity equation is obtained for recharge and draft, both of which may vary with time. The simulated head values match satisfactorily the measured maximum head values in southern O'ahu aquifers over a 100-year period (1880–1980). The derivation of RAM is given in Appendix 6.3.

For the determination of aquifer sustainable yield (steady state, $t \to \infty$) aquifer recharge, $I$, must be greater than aquifer draft, $D$. For $I > D$, the general solution for transient state is

$$h_{i+1} = h_0 \left[\frac{I-D}{I}\right]^{1/2} \times \left\{ \frac{\left[(I-D)^{1/2} + \frac{h_i}{h_0}(I)^{1/2}\right] \exp\left[\frac{2T([I-D]I)^{1/2}}{V_0}\right] - (I-D)^{1/2} + \frac{h_i}{h_0}(I)}{\left[(I-D)^{1/2} + \frac{h_i}{h_0}(I)^{1/2}\right] \exp\left[\frac{2T([I-D]I)^{1/2}}{V_0}\right] + (I-D)^{1/2} - \frac{h_i}{h_0}(I)} \right. \tag{6.6}$$

where $T = t_{i+1} - t_i$.

The steady-state form for constant $D$ and $I$ is (see Figure 6.13)

$$\frac{h_e}{h_0} = \left(1 - \frac{D}{I}\right)^{1/2} \tag{6.7}$$

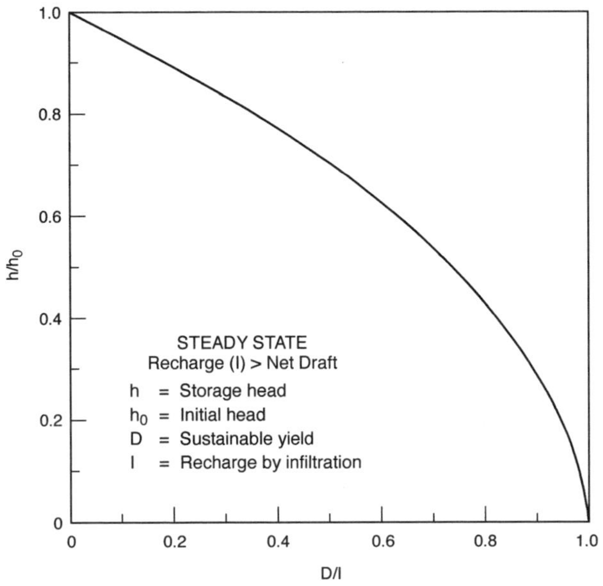


Figure 6.13. Aquifer sustainable yield by Robust Analytical Model (RAM). (Adapted from Mink 1980)

where $h_e$ is steady-state aquifer storage head for which the produced water is of acceptable quality for $D$, and $h_0$ is the aquifer initial head. Leakage from the lens is recharge less draft.

The value of storage head is a management decision that takes into consideration other factors such as deep wells that are already in place. Given aquifer recharge and initial head, aquifer sustainable yield may be determined by the use of the equation or the graph. The determination treats the aquifer as a whole and does not address the specifics of individual wells.

Should a greater draft be desired, management must be content to accept a thinner lens, a smaller storage, a lowered head, and possibly greater salinity. The increased draft is obtained at the expense of reduced leakage, but the freshwater lens would cease to exist if the draft is allowed to equal recharge.

Several computer programs (codes) have been applied to various Hawaiian aquifers to estimate the aquifer responses to stresses, including SUTRA, SHARP, AQUIFEM-SALT, and some without an acronym. Among them, SUTRA is the only code that predicts salinity distribution as isochlors in the transition zone.

SUTRA is a two-dimensional, finite-element code coupling flow and transport phenomena of a varying density fluid (Voss 1984). It was applied to predict salinity changes scenarios (Souza and Voss 1989) in the vicinity of pumping centers in the Pearl Harbor Aquifer Sector resulting from pumping and recharge. SUTRA was calibrated for the Pearl Harbor Aquifer Sector. For this application, it was refined and discretized to 2 m (6 ft) vertically.

SUTRA results indicate that the net system discharge (aquifer leakage) controls seawater intrusion. The computed salinity distribution, for example, the 2 percent isochlor (about 380 mg/l chlorides), stabilizes when the sustainable yield is kept at a level not exceeding 70 to 80 percent of the recharge. During a long (10-year) drought, the computed aquifer salinity rises rapidly, but it drops back to the previous long-term concentration after the drought. SUTRA was also applied to a single scenario in the Nu'uanu Aquifer System (Liu et al. 1991*a*).

AQUIFEM-SALT is an areal two-dimensional, finite-element code that assumes a sharp interface and no saltwater motion (Voss 1984). In Hawai'i, it was first applied to predict head changes resulting from increased pumping in the Wai'alae Aquifer System, southeastern O'ahu (Eyre 1985). It has also been applied elsewhere, including the 'Ewa-Kunia Aquifer System on O'ahu, to test for the effects of increased pumping (Souza and Meyer 1995); Hāwī, Hawai'i, for an initial groundwater extraction (Underwood et al. 1995); and Kualapu'u, Moloka'i, for siting a monitoring well (Oki 2000).

SHARP is a quasi-three-dimensional, finite-difference code that assumes a sharp interface but allows saltwater flow (Essaid 1990). SHARP was applied to simulate and predict steady-state effects of additional pumping in southern O'ahu (Oki 1998). For calibration, average heads during the decade of the 1950s were considered steady state. The computed average heads differ from corresponding measured average heads by more than a preset 0.3-m (1-ft) limit at seventeen of forty sites. Comparison of measured and computed depths to the interface was not possible because of lack of data. SHARP was also applied elsewhere, including Kona (Oki 1999).

The first computer model in Hawai'i was devised for the Pālolo Aquifer System in southern O'ahu by GE-Tempo in 1968 (Meyers et al. 1974). The long-term (40 years) record of head was successfully simulated with many fitted parameter values. The transition zone was treated as a sharp interface. Difficulties in simulation were encountered for the period of high demand during the World War II years.

Computer models are essential for handling complex problems because they can describe spatially varying aquifer parameters and complex spatial and temporal trends in hydrologic variables, that is, water levels and concentrations. On the other hand, numerical models should be used cautiously for groundwater systems having no abundant record of behavior with which to validate the model and having little or poor knowledge of aquifer parameters and aquifer boundaries. Most aquifer systems in Hawai'i fall in this category.

### *Injection of wastewater*

Brackish basal lenses in Hawai'i have received injectants of various composition, including treated wastewater, return brackish water after use as cooling water, and return ocean water after aquacultural uses. Two different models have been used to determine the distribution and fate of the injectants (Wheatcraft and Peterson 1979; Mercer et al. 1980). Mercer's numerical model was applied to coastal injection of Wailuku-Kahului wastewater effluent by wells into and below the transition zone near Kahului, Maui. Mercer's modeling is closely parallel to SHARP (i.e., it assumes a sharp interface and allows saltwater motion). Its code is finite difference for two-dimensional areal flow. Wheatcraft's work is a study of buoyancy and entrainment behavior.

### *Surface contamination*

Groundwater contamination by chemicals applied on the ground surface has been simulated and predicted with the MOC (method of characteristics) numerical model and an analytical model of multiple mixing cells (Orr and Lau 1987, 1988). The concern was DBCP transport in the deep aquifer from Mililani downgradient toward Pearl Harbor. MOC is a two-dimensional, finite-difference code for transient transport of solutes (Konikow and Bredehoeft 1978). The multiple mixing cell model is a lumped parameter model based on Mercado's (1984) work on mass balance of water and solute in each cell. It provides an analytical solution for the contaminant concentration in the groundwater—one for the pollution period and another for the recovery period. Simulations with the two models reproduced the lower bound of observed concentrations without calibration and checked each other well, despite vastly different assumptions. A comparison between the observed and predicted concentrations, however, suggests that a future use of a statistical framework for both modeling and monitoring of nonpoint-source pollution would be appropriate.

CANVAS is a composite analytical-numerical model developed for and used by the U.S. Environmental Protection Agency for virus and solute transport. It contains two coupled computational models: one-dimensional vertical water flow and solute transport in the unsaturated zone, and two-dimensional areal water flow and virus transport in the saturated zone. The main processes that control virus fate and transport in the subsurface (e.g., adsorption, inactivation advection-dispersion, and filtration) are included in the model. It is used as a screening tool to predict virus concentration at a wellhead.

CANVAS was applied to several sugarcane fields near Kunia that were irrigated in 1983–1986 with wastewater effluent (Orr and Li 1997). CANVAS was modified to include the temperature-dependent virus-inactivation rate and Monte Carlo simulation in the application. The simulated results indicate a very low probability that viruses will reach drinking-water wells. A long setback distance is necessary for the areas where the surface soils are not thick and the subsurface temperature is not high enough to become factors in the natural disinfection processes. A 3-year (1983–1986) groundwater monitoring program indicated that viruses and changes in bacterial quality were not detected (Fujioka and Lau 1987).

## Processes and Parameters

### *Saltwater phenomena*

Two different models were used to predict the flow pattern and chloride distribution in the thick lens in the Pearl Harbor aquifer: a physical sandbox model (Lau 1962; Lau and Mink 1967) and the SUTRA numerical model (Voss and Souza 1987). The sandbox is a large three-dimensional model constructed and operated in accordance with hydraulic similitude to reproduce water table, spring discharge, and salinity-depth data in a deep well in Kalauao springs, O'ahu. Both models revealed that the flow lines in a vertical section are generally parallel to the simulated isochlors in a thick lens under steady-state conditions. This phenomenon has yet to be validated by field data. An inference of this result is that the steady-state behavior of the flow is insensitive to longitudinal dispersivity but highly sensitive to transverse dispersivity (Voss and Souza 1987).

Sandbox experiments and numerical modeling were conducted in tandem in an interpretive study involving injection of used freshwater into and below the transition zone. The results support the concept of a buoyant plume, disprove the hypothesis of entrainment of ambient water, and suggest a displacement process (Wheatcraft et al. 1976).

### *Storage and leakage*

Two models, one analytical (RAM) and the other numerical (SUTRA), were applied to address the relationship of storage and leakage in a thick basal lens. Both models deduced the same conclusion that the lens shrinks as a result of increased aquifer draft and that leakage is reduced under the steady-state condition (Mink 1980; Souza and Voss 1989). This important relationship is the basis for estimation of sustainable yield in Hawai'i.

### *Effects of saltwater motion on freshwater head*

It has been commonly assumed that saltwater is static in aquifers, but it has not been resolved whether the effects of saltwater motion on freshwater head are negligible. These effects have been demonstrated to be negligibly small (0.03 m [0.1 ft]) in the course of calibration of SHARP (saltwater motion allowed) for the period from 1935 to 1975 for the Wai'alae Aquifer System (Essaid 1986). The evaluation was made possible by comparison with Eyre's (1985) work using AQUIFEM-SALT (no saltwater motion). This result is especially important because of the known high sensitivity of the aquifer responding to variations in input and stress.

The effects of saltwater motion on freshwater head in southern O'ahu aquifers were further examined with Essaid's reference case. The simulated heads for both steady and transient states were virtually identical for SHARP and RAM, suggesting that RAM is an accurate global surrogate for SHARP (Mink 1998). It should be noted that all of the reference cases tested used aquifer parameter values that are common for southern O'ahu.

### *Rising of immiscible interface*

This phenomenon has been studied in two dimensions with SHARP (Essaid 1986) and in one dimension with an analytical model (Ogata and Lau 1990). SHARP results are indicated by graphs of head variation over time for a reference case for each of the following four parameters: seaward boundary leakance (hydraulic conductivity divided by aquifer thickness), aquifer transmissivity, aquifer storativity, and aquifer anisotropy. The freshwater head appears little affected by the variations of the parameter values except the seaward boundary leakance. This result is hydrologically sensible.

Ogata and Lau derived a theoretical function that predicts the transient position of an immiscible interface set in vertical motion from an initially static equilibrium state. The function is proportional to the square root of time since commencement of motion. The proportionality constant is an explicit function of porosity and permeability of the aquifer, the difference in density of the two liquids, and the amount of ultimate head change. The theory is generally verified by laboratory sand-column experiments; however, precise experiments are needed to elucidate the constant. This phenomenon is relevant to issues resulting from sea-level rise as well as reduction in lens storage due to extraction by pumpage.

### *Regional flow*

Chemical and isotopic evidence that occurs over time is suited to complement the traditional physical approach to analysis of large-scale regional flows. Modeling of the regional flow in the Pearl Harbor aquifers (Voss and Wood 1993) and Wahiawā Aquifer System (Turnbull et al. 1994) successfully used this approach. The models took advantage of data collected and information gained from prior work, including temperature variations of the hydrologic transport cycle in leeward O'ahu (Mink 1964), chemical and tritium in the fast-circulating dike-impounded high-level water in O'ahu (Hufen et al. 1974), and radiocarbon as an additional parameter in the relatively slowly circulating basal water in Honolulu and Pearl Harbor aquifers (Hufen et al. 1980).

### *Dike-impounded high-level water flow*

The occurrence of dike-impounded water bodies was first evaluated as a result of field investigations initiated by the pioneering geologist W. O. Clark in the 1920s (D. C. Cox 1981). Flow in the rift zones has been limited to conceptual modeling. The analytical solution of a simplified flow model in tunnels that drain the dike aquifers has been calibrated with discharge data (Takasaki and Mink 1981).

### *Stochastic processes*

The stochastic nature of contaminant transport has been incorporated in Hawaiian models such as that for TCE migration from Schofield Barracks (Evans et al. 1995; Harding Lawson Associates 1996) and that for virus transport in Kunia, O'ahu (Orr and Li 1997).

In the Schofield Barracks work, the Monte Carlo method was used to examine the sensitivity in using the mean value of hydraulic conductivity ($\bar{K}$) for prediction, assuming $K$ follows a lognormal probability distribution. Comparison of the predicted pressure heads using $\bar{K}$ values with the average pressure heads obtained from the 300 Monte Carlo simulations indicates that the differences in pressure head exceeding the maximum allowable error occur only at 3.8 percent (224 of 5,964) of the finite-element nodes. This result suggests adequacy in the use of $\bar{K}$ for prediction and places the modeling in a probabilistic framework.

The Monte Carlo method was also used to modify CANVAS, a U.S. Environmental Protection Agency deterministic model for simulating virus transport (Orr and Li 1997). The work was an attempt to evaluate potential virus contamination of basal groundwater in a 3-year study in Kunia of sugarcane irrigated with effluent from a secondary treatment plant (Fujioka and Lau 1987). The model was partially validated with soil column experiments. The use of Monte Carlo simulations resulted in a range of predicted average concentration of viruses and accounted to some degree for the uncertainties and heterogeneity of the subsurface.

### *Evaluation of parameters*

The conventional approach to the evaluation of aquifer parameters is by the inverse method. Pioneering works on a regional scale started nearly 50 years ago when a labor strike in 1958 shut down virtually all extraction of groundwater from the Pearl Harbor aquifers, which at that time were overlain with approximately 4,800 ha (12,000 ac) of sugarcane. The recovery and drawdown data were put to use for evaluation of global aquifer parameters by Mink (1980) and Souza and Voss (1987). On a lesser scale, pumping test data from individual wells in O'ahu were analyzed with established formulas by Williams and Soroos (1973). This work provided a collection of basic aquifer parameter values.

Field experiments with injected helium yielded data for the determination of global longitudinal dispersivity of a basaltic aquifer in southern O'ahu (Gupta et al. 1990). The MOC numerical model was applied for data fitting.

The hydraulic conductivity of a highly permeable recent basalt flow (1801) from the Hualālai Volcano was evaluated with data from a field experiment conducted at Keāhole Point, island of Hawai'i (Lau and El-Kadi 1995). The experiment involved recharging the brackish basal lens with ocean water in an open trench at a rate of 46.3 mld (12.2 mgd) for 6 hours, raising the trench water level by 1.2 m (4 ft) within the first hour. The model used was SUTRA.

Barometric pressure variations are known to cause an inverse change in water table elevations in small permeable islands (Todd 1980). Oki (1997) determined the value of the parameter, vertical hydraulic diffusivity (ratio of vertical hydraulic conductivity to specific storage), by fitting well water level data in the Wai'anae Aquifer Sector with a model developed for assessing the response of barometric pressure in unconfined aquifers (Rojstaczer and Riley 1990). For the Pearl Harbor Aquifer Sector, the parameter value was evaluated using different phenomena, models, and data (Souza and Voss 1987).

Tidal effects on groundwater offer an independent way to determine hydraulic conductivity and transmissivity. This method can account for the parameter values averaged over a much larger portion of the aquifer than water well pumping tests can. The development of such a method was made by means of analytical, electrical analog, and hydraulic models for both one-dimensional and two-dimensional aquifers (Williams and Liu 1975). A parallel effort with mathematical models resulted in a catalog of type curves for one-dimensional aquifers of various boundary conditions and forcing functions (Dale 1974). Reviewing his and Dale's work, Williams concluded that although the tidal method is valid, the well test method is more reliable. Oki (1997) applied only $O_1$ and $M_2$ constituents of the tidal data to exclude barometric pressure effects in his tidal analysis of the Ko'olau aquifers (Wai'alua and Kawailoa Aquifer Systems) in northern O'ahu.

## Public Policies

Models have been used for implementing public policies concerning subsurface waters in Hawai'i, for example, the risk-based corrective action (Hawai'i Department of Health 1995a) and the wellhead protection program (Hawai'i Department of Health 1995b).

SESOIL is applied in the risk-based corrective action as a screening tool for assessment of the contaminated flux that may be leached on a monthly basis from contaminated soil into a shallow water table. SESOIL is scientifically rational, but the relevant natural processes—for example, mobility, adsorption, volatility, and degradation—have been highly simplified. The results are considered conservative. In the wellhead protection program, various simple flow models are stipulated for computing the wellhead protection area for drinking-water wells.

## Remediation

Monitoring remediation of contaminated groundwater entails the use of models. The first major case in progress in Hawai'i is at Schofield Barracks, which is located in the Wahiawā Aquifer System. The remediation has involved point-of-use pump and treatment since 1986. The monitoring is long term as mandated by the U.S. Environmental Protection Agency (Harding Lawson Associates 1997). The concern is the migration of the chemicals TCE and $CCl_4$, both used as solvents, from the two defined contaminated plumes in the high-level water in the Wahiawā Aquifer System to drinking-water wells that extract basal water in the Pearl Harbor aquifers.

The models used for simulation and prediction are FEMWATER for flow and FEMWASTE for solute transport. These are three-dimensional, finite-element codes for variably saturated flows. Historical water levels, groundwater age as determined by carbon 14 measurements, and drawdown data from pumping tests were used for model calibration. Another code applied was TOUGH for three-dimensional multiphase flow and transport of water vapor, air, heat, and liquid water in variably saturated fractured media.

Prediction of travel time was made with the solute transport model and also by particle tracking. Solute transport results indicate that TCE concentration would be observed in Mililani wells at a concentration slightly above the maximum contaminant level (5 ppb) after 100 years for the conservative case, assuming no decay and no retardation of the chemicals. On the other hand, no detectable impact would be experienced for the case of high decay and high retardation. Particle tracking resulted in a travel time longer by a factor of two than that for the solute transport model. Particle tracking was performed with GMS-TRACE code and groundwater velocity computed by FEMWATER.

A total of thirty-one wells, including ten deep wells drilled for the remediation investigation in the Wahiawā Aquifer System, has been sampled since 1997, some as frequently as quarterly. The results indicate no changes from the existing pattern. The monitoring will continue for at least 5 years; after that, a site review will be performed.

It is clear that monitoring data are needed to assess the effectiveness of remediation measures. In addition, they can be used as a basis for a postaudit to improve the models and revise prediction.

# Appendix 6.1. Storage Head in Ghyben-Herzberg System

In a basal lens, the pressure exerted by the column of basal water at the interface is balanced by the pressure of seawater at the interface. In a strict sense, the equality applies only to a static balance, but it is true also where the Dupuit conditions of horizontal flow and vertical equipotentials prevail. These conditions are accurate to within a few percent at a distance from the discharge front of the lens of about 1.5 to 2.0 times the depth of flow in the lens (Bear 1979).

The equality of pressures at the interface, $p_b = p_s$, allows the Ghyben-Herzberg ratio δ to be calculated for the basal lens–seawater system. Figure 6.14 is a reference diagram for such a system where points *b* and *s* are, respectively, above and below the interface. After substituting pressure with the product of thickness of water column, water density, and gravitational acceleration *g*, the equality becomes

$$(h\rho_f + m_f\rho_f + m_j\rho_j)g = (m_f + m_j)\rho_s g \tag{6.8}$$

where $h$ is the water table level above sea level (head), $m_f$ is the thickness of the freshwater core below sea level, $m_j$ is the thickness of the transition zone, $\rho_f$ is the freshwater density, $\rho_s$ is the saltwater density, and $\rho_j$ is the average density in the transition zone. Density in the transition zone is assumed to vary symmetrically at about the midpoint of the zone where salinity is equal to one-half that of freshwater and seawater.

Substituting, rearranging, and simplifying lead to

$$h = \frac{\rho_s - \rho_f}{\rho_f}\left(m_f + \frac{1}{2}m_j\right) \tag{6.9}$$

or

$$h = \frac{1}{\delta}\left(m_f + \frac{1}{2}m_j\right) \tag{6.10}$$

or

$$\text{Storage head} = \frac{\text{Depth to the 50\% isochlor}}{\text{Ghyben–Herzberg ratio}} \tag{6.11}$$

The head calculated in the preceding equation is the "storage head," which is the true balance head in contrast to the "operating head," which is the elevation of the water table above sea level that is influenced by pumping drawdown. A similar calculation may be made when the density distribution in the lens is not symmetric, or when the seawater head is not zero. The condition—prevailing Dupuit horizontal flow under dynamic equilibrium in a Ghyben-Herzberg system—is the basis for derivation of the statement that the Ghyben-Herzberg ratio holds at the middepth of the transition zone where the salinity distribution with depth is symmetrical (Lau 1962, 1967).

The ʻĪao Aquifer System in West Maui affords a good example of the use of this equation. As measured in Waiehu monitoring well, the thickness of the freshwater core below sea level, $m_f$, is about 183 m (600 ft) and the thickness of the transition zone $m_j$ is about 85 m (280 ft). For the standard density of seawater at 1.025 and of freshwater at 1.000, head calculates to be 5.6 m (18.5 ft). This value is the same as that obtained by applying the 40:1 ratio to the midpoint of the transition zone at 225 m (740 ft).

Although the respective standard densities of seawater (1.025) and freshwater (1.000) are commonly employed in estimating the depth of a basal lens, the given densities are for seawater at a temperature of 16°C (61°F) and freshwater at a temperature of 5°C (41°F). In thick basal lenses in Hawaiʻi, the temperature of the freshwater core is usually about 20°C (68°F), for which the density is 0.9982. The usual temperature of the underlying seawater is approximately 25°C (77°F), for which the density is 1.02261 (data from de Marsily 1986). Inserting these values in the equation yields a head of 5.5 m (18.10 ft), which is 0.1 m (0.4 ft) lower than the head computed with standard densities.

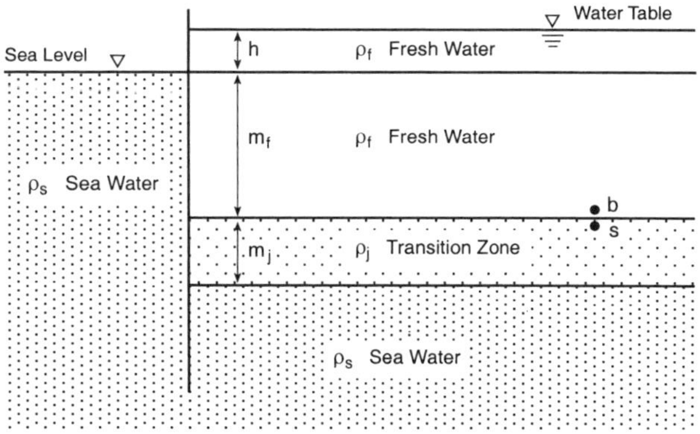


Figure 6.14. Storage head in Ghyben-Herzberg system.

# Appendix 6.2. Aquifer and Status Codes for O'ahu, Hawai'i

| Island | | Aquifer Sector | | Aquifer System | Aquifer Type | Aquifer Code | Status Code | Quadrangle Number |
|---|---|---|---|---|---|---|---|---|
| 3 | 01 | Honolulu | 01 | Pālolo | 116 | 30101116 | 23321 | 13 |
| | | | | | 121 | 30101121 | 11113 | |
| | | | | | 111 | 30101111 | 11111 | 13 |
| | | | | | 212 | 30101212 | 11111 | 13, 15 |
| | | | 02 | Nu'uanu | 116 | 30102116 | 13321 | 13 |
| | | | | | 121 | 30102121 | 11113 | |
| | | | | | 111 | 30102111 | 21111 | 13 |
| | | | | | 212 | 30102212 | 11111 | 13 |
| | | | 03 | Kalihi | 116 | 30103116 | 13321 | 13 |
| | | | | | 121 | 30103121 | 11113 | |
| | | | | | 111 | 30103111 | 11111 | 13 |
| | | | | | 215 | 30103215 | 11111 | 12, 13 |
| | | | 04 | Moanalua | 116 | 30104116 | 23321 | 10, 13 |
| | | | | | 121 | 30104121 | 11113 | |
| | | | | | 111 | 30104111 | 11111 | 10, 12, 13 |
| | | | | | 212 | 30104212 | 21111 | 12, 13 |
| | | | 05 | Wai'alae | 116 | 30105116 | 23421 | 13, 15 |
| | | | | | 121 | 30105121 | 21113 | |
| | | | | | 111 | 30105111 | 11111 | 13, 15 |
| | | | | | 212 | 30105212 | 21111 | 13, 15 |
| | 02 | Pearl Harbor | 01 | Waimalu | 116 | 30201116 | 12211 | 9, 10 |
| | | | | | 121 | 30201121 | 12212 | |
| | | | | | 111 | 30201111 | 11111 | 9, 10, 12 |
| | | | | | 212 | 30201212 | 21111 | 9, 12 |
| | | | 02 | Waiawa | 116 | 30202116 | 12211 | 9, 10 |
| | | | | | 121 | 30202121 | 12212 | |
| | | | | | 111 | 30202111 | 11111 | 8, 9 |
| | | | | | 212 | 30202212 | 21111 | 8, 9, 11, 12 |
| | | | 03 | Waipahu | 116 | 30203116 | 12211 | 5, 6, 9, 10 |
| | | | | | 121 | 30203121 | 12212 | |

| Island | Aquifer Sector | | Aquifer System | | Aquifer Type | Aquifer Code | Status Code | Quadrangle Number |
|---|---|---|---|---|---|---|---|---|
| | | | 04 | ʻEwa | 116 | 30204116 | 13321 | 6 |
| | | | | | 121 | 30204121 | 13213 | |
| | | | | | 111 | 30204111 | 11111 | 5, 6 |
| | | | | | 212 | 30204212 | 21111 | 5 |
| | | | 05 | Kunia | 111 | 30205111 | 21112 | 5 |
| | | | | | 212 | 30205212 | 21111 | 5 |
| | 03 | Waiʻanae | 01 | Nānākuli | 116 | 30301116 | 23421 | 2, 5 |
| | | | | | 122 | 30301122 | 23423 | |
| | | | | | 112 | 30301112 | 23321 | 5, 6 |
| | | | | | 212 | 30301212 | 21121 | 5 |
| | | | 02 | Lualualei | 116 | 30302116 | 13311 | 2,5 |
| | | | | | 122 | 30302122 | 23323 | |
| | | | | | 112 | 30302112 | 23321 | 2, 5 |
| | | | | | 212 | 30302212 | 11111 | 2, 5 |
| | | | 03 | Waiʻanae | 116 | 30303116 | 13311 | 2 |
| | | | | | 122 | 30303122 | 23223 | |
| | | | | | 112 | 30303112 | 11111 | 2 |
| | | | | | 232 | 30303232 | 11111 | 1, 2, 4, 5 |
| | | | 04 | Mākaha | 116 | 30304116 | 13321 | 2 |
| | | | | | 122 | 30304122 | 11113 | |
| | | | | | 112 | 30304112 | 11111 | 2 |
| | | | | | 232 | 30304232 | 11111 | 1, 2, 4 |
| | | | 05 | Keaʻau | 116 | 30305116 | 33421 | 1 |
| | | | | | 122 | 30305122 | 11212 | |
| | | | | | 112 | 30305112 | 21211 | 1, 2 |
| | | | | | 212 | 30305212 | 21111 | 1, 2 |
| | 04 | North | 01 | Mokulēʻia | 116 | 30401116 | 13221 | 1, 4 |
| | | | | | 121 | 30401121 | 11113 | |
| | | | | | 111 | 30401111 | 11111 | 1, 4 |
| | | | | | 212 | 30401212 | 21111 | 1, 4, 5 |

| Island | Aquifer Sector | | Aquifer System | | Aquifer Type | Aquifer Code | Status Code | Quadrangle Number |
|---|---|---|---|---|---|---|---|---|
| | | | 02 | Waialua | 116 | 30402116 | 12211 | 4 |
| | | | | | 121 | 30402121 | 11213 | |
| | | | | | 111 | 30402111 | 11111 | 4,8 |
| | | | 03 | Kawailoa | 116 | 30403116 | 12211 | 3, 4 |
| | | | | | 121 | 30403121 | 12313 | |
| | | | | | 111 | 30403111 | 11111 | 3, 4, 7, 8 |
| | | | | | 112 | 30403112 | 11111 | 3, 7 |
| | | | | | 116 | 30403116 | 12211 | 3, 7 |
| | | | | | 112 | 30403112 | 21112 | |
| | | | | | 212 | 30403212 | 21111 | 3, 7, 8 |
| | 05 | Central | 01 | Wahiawā | 212 | 30501212 | 11111 | 4, 5, 8, 9 |
| | | | 02 | Koʻolau | 212 | 30502212 | 11111 | 8 |
| | 06 | Windward | 01 | Koʻolauloa | 116 | 30601116 | 12211 | 7, 8, 11 |
| | | | | | 121 | 30601121 | 12213 | |
| | | | | | 111 | 30601111 | 11111 | 7, 8, 11 |
| | | | | | 212 | 30601212 | 21111 | 7, 8, 11 |
| | | | | | 112 | 30601112 | 11111 | 7 |
| | | | | | 116 | 30601116 | 22221 | 7 |
| | | | | | 122 | 30601122 | 21122 | |
| | | | 02 | Kahana | 116 | 30602116 | 12211 | 11 |
| | | | | | 122 | 30602122 | 11113 | |
| | | | | | 112 | 30602112 | 11111 | 11 |
| | | | | | 212 | 30602212 | 11111 | 8, 11 |
| | | | 03 | Koʻolaupoko | 116 | 30603116 | 12211 | 12 |
| | | | | | 122 | 30603122 | 11122 | |
| | | | | | 212 | 30603212 | 11111 | 11, 12, 13 |
| | | | 04 | Waimānalo | 116 | 30604116 | 12211 | 14, 15 |
| | | | | | 122 | 30604122 | 11113 | |
| | | | | | 212 | 30604212 | 11111 | 12, 13, 14, 15 |

SOURCE: Mink and Lau (1990a).

NOTE: Island numbers are 1 (Niʻihau), 2 (Kauaʻi), 3 (Oʻahu), 4 (Molokaʻi), 5 (Lānaʻi), 6 (Maui), 7 (Kahoʻolawe), and 8 (Hawaiʻi).

# Appendix 6.3. Modeling Storage in Ghyben-Herzberg System

Groundwater movement is defined by hydraulic flow equations, the solution of which requires information obtained by composing a hydrologic balance. The simplest balance consists of equating input and output for the natural, undeveloped condition of the aquifer. Infiltration to the saturated zone is computed as the difference between rainfall, which is the only external source of water, and losses due to evapotranspiration and direct runoff to the sea.

Expressed as a balance equation, the relationship is

$$I = P - \mathrm{DRO} - \mathrm{AET} \tag{6.12}$$

where $I$ is infiltration to the aquifer, $P$ is rainfall, DRO is direct surface runoff lost to the sea, and AET is actual evaporation and transpiration lost to the atmosphere. Under development practices the equation is modified chiefly by incorporating pumping as an output and irrigation return as an input. The equation applies to long-term averages, but it can be discretized into time intervals. It also may be transformed to a transient equation by including gain or loss of storage over a time interval, for which the balance equation is written as

$$L = P - \mathrm{DRO} - \mathrm{AET} \pm \Delta U \tag{6.13}$$

where $L$ is leakage, which is identical to $I$ before development, and $\Delta U$ is the change in storage.

A hydrologic balance is important in mathematical modeling because it provides a value for infiltration (recharge). It also defines the magnitude of system fluxes and suggests limits to groundwater development. It is usually the first computation made and quite often is the only model of reasonable validity.

Mathematical models are hydraulic flow models that combine Darcy's Law of flow in porous media with a continuity equation. Steady-state solutions are the easiest to obtain because time is eliminated as a variable. The reliability or reasonableness of the assumptions necessary to complete a model controls the validity of output.

Analytical mathematical models provide direct solutions of aquifer behavior but are limited unless modified because for complex systems the nonlinear equations cannot be solved by exact methods. Numerical modeling avoids this handicap by decomposing the governing partial differential equations into an array of algebraic equations whose solutions give an approximate description of flow. Finite difference and finite element are the numerical techniques most commonly employed. But because the solution methods are approximations, the results of numerical modeling have to be tested against a historical record to ascertain closeness of fit between actual behavior and model behavior before the model can be accepted as a valid representation of reality. If the simulation is within acceptable limits, the model is presumed to be capable of predicting future behavior for different aquifer scenarios.

Although sophisticated numerical models are more comprehensive, a simpler analytical model called RAM (Robust Analytical Model) is useful for providing a global perspective on aquifer behavior. The model describes a two-dimensional homogeneous and isotropic section with one-dimensional horizontal flow. It relies on an analytical solution of the equation of groundwater flow in a Ghyben-Herzberg lens but is discretized by time intervals. Spatial discretization may also be employed, but the databases do not warrant this complication. It assumes changes that are averaged throughout the aquifer. It does not take into consideration solute transport; instead, it assumes a sharp interface between the freshwater and seawater. Although subject to limiting assumptions, it is straightforward and gives an exact accounting of water balances. Even though it refers to global rather than to local behavior, it is useful for defining the limit of aquifer yield and for predicting the consequences of different levels of extraction.

Derivation of the RAM equations is given by Mink (1980). Even with an extensive database a freshwater-saltwater system with two moving boundaries eludes practical modeling. All of the basal lenses in Hawai'i have "soft bottoms" although some are confined from above by the caprock and other lithologies, so that the free upper surface and the lower surface are mobile. These physical boundary conditions place serious limitations on numerical models that simulate contraction and expansion of a lens.

The continuity equation for the groundwater balance may be written as

$$Q\,dt = dU = An\,dh \tag{6.14}$$

in which $Q$ is rate of flow, $dt$ is change in time $t$, $dU$ is change in water volume $U$, $A$ is horizontal area of the flow domain, $n$ is effective porosity, and $dh$ is change in head $h$. For a Ghyben-Herzberg lens equation (6.14) is written as

$$Q\,dt = An(1 + \delta)\,dh \tag{6.15}$$

in which $\delta = \rho_f/(\rho_s - \rho_f)$ where $\rho_s$ and $\rho_f$ are densities of saltwater and freshwater and $A$ is the horizontal area of the flow domain.

The flow rate term $Q$ can be decomposed into its balance elements to give

$$I - L - D = An(1 + \delta)\, dh/dt \tag{6.16}$$

in which $I$ is steady recharge to the aquifer, $L$ is leakage from the aquifer, and $D$ is draft (pumpage). For standard freshwater and seawater with densities of 1.000 and 1.025, respectively, $\delta$ is 40, and $(1 + \delta)$ is 41. Hereafter, the value of 41 will be used in the derivations. Combining Darcy's Law with equation (6.16) produces a hydraulic flow model.

Darcy's Law for a basal lens with unidimensional flow is written as

$$Q = 41Kh\, dh/dx \tag{6.17}$$

where $Q$ is discharge of flow through the depth of the lens in the aquifer of unit width, $K$ is hydraulic conductivity, and $x$ is horizontal distance along the gradient. When integrated between the limits $h_1$, and $h_2$, and $x_1$, and $x_2$, equation (6.17) yields

$$Q = \frac{41K}{2}\frac{h_2^2 - h_1^2}{x_2 - x_1} \tag{6.18}$$

For simplicity let $h_1$ and $x_1$ be zero, which are conditions at the hypothetical discharge front, then

$$Q = 41Kh^2/2x \tag{6.19}$$

This equation is valid no matter where or what the discharge front is. The actual distance, $x$, is immaterial in the final equations because it is subsumed in a constant term along with other variables.

At initial conditions, leakage, $L_o$, is equal to total input, designated $I$, and equation (6.19) can be restated as

$$I = L_0 = 41Kh_0^2/2x \tag{6.20}$$

in which leakage, $L_o$, and head, $h_o$, are initial conditions.

In equation (6.20), $41K/2x$ is a constant for a selected location $x$ in the Ghyben-Herzberg lens such that $c = 41K/2x = I/h_o^2$. This conversion eliminates $K$ from the final equations but requires that $I$ be held constant, a condition that permits transient equations to be solved.

Restating the balance equation, equation (6.16), as amended by Darcy's Law gives

$$I - ch^2 - D = 41nA\, dh/dt \tag{6.21}$$

where $I$ includes both vertical infiltration and external recharge, $D$ is the net extraction, $ch^2$ is the leakage, $n$ is the effective porosity, and $A$ is the horizontal area of the flow domain. Designating the constant $41nA$ as $b$, the initial volume of water in the lens $U_o$ is calculated as the product of $b$ and a function of the spatial distribution of the initial head. For a relatively flat groundwater table, which is characteristic of aquifers of high permeability, the average initial head, $h_o$, can be used in place of this function, yielding $b = U_o/h_o$. This constant $b$ does not change over time because the relationship between $U$ and $h$ remains constant. Effective porosity is eliminated from the final equation when $b$ is used. Further discussion on the computation of water volume in basal lenses is given in Appendix 6.4.

Equation (6.21) is an ordinary differential equation readily solved by separation of variables. Integration over the limits $(h_i, h_{i+1})$ and $(t_i, t_{i+1})$, in which $h_i$ is head at the start and $h_{i+1}$ is head at the end of an interval, and $t_{i+1} - t_i = T$ is the fixed length of the interval, and in which the constants $c$ and $b$ are replaced by initial system values, yields the following equations:

For $I > D$

$$h_{i+1} = h_0\left[\frac{I-D}{I}\right]^{1/2} \times \left\{\frac{\left[(I-D)^{1/2} + \frac{h_i}{h_0}(I)^{1/2}\right]\exp\left[\frac{2T([I-D]I)^{1/2}}{V_0}\right] - (I-D)^{1/2} + \frac{h_i}{h_0}(I}{\left[(I-D)^{1/2} + \frac{h_i}{h_0}(I)^{1/2}\right]\exp\left[\frac{2T([I-D]I)^{1/2}}{V_0}\right] + (I-D)^{1/2} - \frac{h_i}{h_0}(I}\right. \tag{6.22}$$

where $T = t_{i+1} - t_i$ and

for $I < D$

$$h_{i+1} = h_0\left(\frac{D-I}{I}\right)^{1/2}\tan\left\{\arctan\frac{h_i}{h_0}\left(\frac{I}{D-I}\right)^{1/2} - \frac{T[(D-I)I]}{V_0}\right. \tag{6.23}$$

The steady-state form for $I > D$ with constant $D$ is

$$h_e = h_0\{1 - (D/I)\}^{1/2} \tag{6.24}$$

where $h_e$ is the equilibrium head that eventually becomes established. There can be no steady state for either $I < D$ or $I = D$.

Equations (6.22) and (6.24) are identical to equations (6.6) and (6.7), respectively.

In equations (6.22) through (6.24), $D$ may vary among intervals but $I$ is fixed. However, variations in transient recharge can be accommodated by adjusting the actual draft values to reflect increases or decreases in recharge without tampering with the initial value of $I$. The model defines water balances but cannot predict local water table depressions or mounds resulting from pumping and local recharge. A computer spreadsheet is employed to make the interval calculations.

# Appendix 6.4. Computation of Water Volume in the Basal Lens

The curve of the surface of an unconfined basal lens, as well as the curve of the contact of the lens with underlying seawater, is parabolic. Assuming static balance and referring to sea level as the datum, a point on the surface of the lens is matched on its base at a depth 40 times the height of the surface point above sea level. From this simple geometry combined with effective porosity of the aquifer, the volume of water in the lens can be calculated.

The geometry of the portion of the lens above sea level is illustrated in Figure 6.15. The portion below sea level is an exact match of the upper curve but at a depth 40 times greater. For generality, the origin of the coordinates is taken as $x = 0$, $h = h_1$. In Hawai'i this construction simulates an unconfined basal lens just inland of the caprock.

The parabolic surface is expressed as

$$h = h_1 + ax^{1/2} \tag{6.25}$$

For a unit strip the volume above sea level is equivalent to area under the curve $A$, which is obtained by integration as follows (Mink 1980):

$$A = \int_{x_1}^{x_2} h\,dx = h_0 x_2 + \frac{2}{3} a x_2^{3/2} \tag{6.26}$$

Substitute $a$ in equation (6.26) for the explicit expression of $a$ from equation (6.25) and simplify

$$A = \frac{1}{3}(2h_2 + h_1)x_2 \tag{6.27}$$

Define $x_m$ so that the truncated parabolic area A is equal to the area of the rectangle $h_m x_2$ (i.e., $A = h_m x_2$), resulting in

$$h_m = \frac{2}{3}h_2 + \frac{1}{3}h_1 \tag{6.28}$$

Substitute $h_m$ and $h_2$ in equation (6.28) for the form of $h$ in equation (6.25) and simplify

$$x_m = \left(\frac{2}{3}\right)^2 x_2 \quad \text{or} \quad x_m = 0.444x_2 \tag{6.29}$$

This results in the selection of $h_m$ at the distance $0.444\ x_2$ (Mink 1998).

In the 11.26-km (7-mi) reach between the inner margin of Pearl Harbor and the Wahiawā high-level terminus of the lens, the head at the distance 11.26 × 0.444 or 5.0 km (3.1 mi) inland of the Pearl Harbor margin is used to determine the volume of water in the lens. For the initial conditions of $h_1$ = 10.2 m (33.5 ft) and $h_2$ at 11.3 km (7 mi) = 12.2 m (40 ft), $h_m$ and $x_m$ are computed to be 11.5 m (37.8 ft) and 5.0 km (3.1 mi), respectively.

The volume, $U$, of water in storage in the lens is calculated by

$$U = h_m \times 41 \times n \times \text{surface area of the aquifer} \tag{6.30}$$

in which $n$ is effective porosity. For the Pearl Harbor systems (Waipahu-Waiawa and Waimalu), total surface area is about $1.86 \times 10^8$ m² (72 mi² or $2 \times 10^9$ ft²). Assuming effective porosity of 0.05 and the initial head conditions, the initial volume $U_o$ is computed to be

$$U_o = 37.8 \times 41 \times 0.05 \times 2 \times 10^9 = 1.55 \times 10^{11}\ \text{ft}^3 = 1.16 \times 10^{12}\ \text{gallons} = 4.4 \times 10^9\ \text{m}^3$$

The volume at any time after the lens has undergone development is calculated using head at the same point $x_m = 0.444x$.

Further derivations from the basic equations prove that the rise or fall in heads is symmetrical over the full length of the parabolic curves such that a change in head at one point on the curve is matched by a proportional change elsewhere on the curve. For example, reducing the head at the Wahiawā boundary from 12 m to 6 m (40 ft to 20 ft) would be accompanied by an identical 50 percent decrease in head at Waipahu, from 10 m to 5 m (33.5 ft to 16.75 ft).

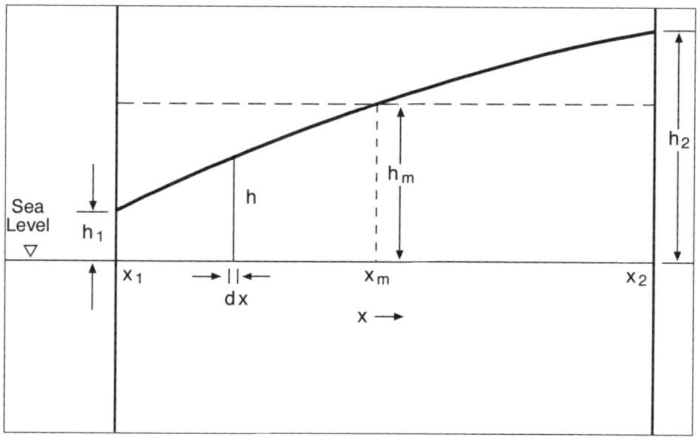


Figure 6.15. Volume of water in the basal lens above sea level.

CHAPTER SEVEN

# Surface Water

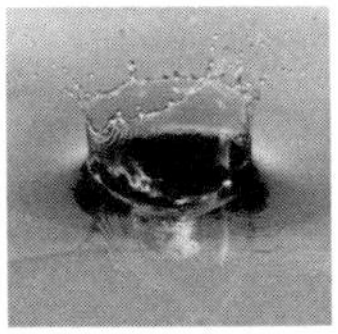

Surface water, like rainfall, is always welcome except when excess creates flooding. Human settlements are inevitably attracted to places of abundant surface waters. Historically, waterworks with open channels are believed to date back to 3200 B.C., when King Scorpion ceremonially cut the first sod for an irrigation canal in Egypt (Biswas 1972). Humans have traditionally valued surface waters for swimming and fishing. Yet, only in the last few decades have humans begun to protect the aquatic ecosystems from overdevelopment and contamination and judiciously strike a balance between development and conservation.

Atmospheric precipitation is the ultimate source of surface water. The fraction that does not infiltrate flows downhill over the land surface. Quickly the overland flow (also known as direct surface runoff) seeks out and enters the smallest rivulets and fills the shallowest depressions. Eventually, the overland and channel flows on a local scale accumulate and move as streamflow to lakes and the sea. Apart from direct surface runoff from rainfall and the melting of snow and ice, streamflow also originates from groundwater discharges.

Surface water phenomena are easy to comprehend and can be accurately measured because they are visible and accessible. During the seventeenth century, Perrault and Marriotte of France provided computational proof with field data that rainfall and snow are sufficient to account for the water flow in rivers and springs. The study of surface water is a multidisciplinary field. In the course of performing their duties, civil engineers and scientists pioneered, and have continued to contribute substantially to, the science and technology of surface waters.

## Nature of the Processes

### Drainage Basin

At a given location along a stream channel, a drainage basin is a tract of land that contributes to surface water flowing past that location. The location is called the outlet of the contributory basin. The size and shape of the basin are uniquely delineated on a topographic map with a water divide, which is a curved and closed line starting from and ending at the outlet. The line is drawn from the outlet point toward and perpendicular to the nearest contour and then on to successively higher contours. Within the drainage basin water flows downhill (see Figure 5.4). The choice of the outlet location from which to divert water from the stream is determined by the intended use of the water resource.

In hydrological terminology, a drainage basin is often considered synonymous with a watershed or catchment. Usage of the term "watershed" can be quite loose. For instance, in Hawai'i it denotes land that produces substantial groundwater consequences (Mink et al. 1993). The term "catchment" is favored in England.

The network of a stream channel and its tributaries appears orderly and somewhat like tree branches—with the smaller ones joining to form larger ones. Since Horton's pioneering work in the 1930s and 1940s, a whole science has been developed that ties together hydrology and geomorphology (Strahler 1964; Chow et al. 1988). The relationship between hydrology and geomorphology is intrinsic and appears intuitive. For example, water flows freer and faster in a deep channel than it does overland as a thin sheet that is more strongly resisted by surface roughness. As a result, a basin with a dense network of streams drains off surface water much faster than one with a sparse network.

The rational approach to hydrogeomorphology has advanced from statistical correlation to dynamics of water flow. The progress of the latter, which began in the late 1970s, is exemplified by the geomorphic instantaneous unit hydrograph, which is design oriented (Yen and Lee 1997), and by fluvial response to climate changes, which is more analysis oriented (Tucker and Slingerland 1997). The geomorphic instantaneous unit hydrograph has been adapted for Molokaʻi water basins (Diaz et al. 1995).

Along a stream course, different processes are at work: erosion in the highlands, transport in the midcourse, and deposition in the downstream course. These processes result in different landforms such as rills and gullies, waterfalls and pools, channels and terraces, and natural levees and alluvial fans (Leopold et al. 1964; Leopold 1997). Streamflow and geomorphology constitute an integral unit that governs riparian traits, including the biota.

## Rainfall-Runoff Events

The ultimate response to a rainfall event is visible by observing the level of the water surface in the stream channel, known as stage (measured above a datum), which rises, crests, and then recedes. At a given location along the stream, the runoff event is represented by a stage hydrograph or more commonly a discharge hydrograph (i.e., a plot of the volumetric rate of flow against time). The volume of water passing the channel cross section during a specified period is represented by the area under the discharge hydrograph. For a stable site in the channel, stage is measured and expressed as a rating curve that reflects the hydraulics of the site and converts the stage to discharge (see Appendix 7.1).

Surface water is the residual of rainwater remaining after hydrologic abstraction. The residual is also known as rainfall excess. During the initial period, the major abstractions are interception by vegetation, surface retention as puddles, water needed to create a thin (less than 1 cm [0.4 in.]) layer of overland flow, and infiltration. After the initial period, the only remaining abstraction is the continuing but ever-diminishing infiltration. Overland flow moves laterally and quite rapidly on the surface toward a channel. Interflow may also take place in the shallow subsurface if the infiltrated water is perched and moves laterally over relatively impermeable materials such as dense rocks, hardpan, or plow pan. Interflow has been regarded as lateral seepage, which is addressed in Chapter 5 (Green et al. 1993). Where the subsurface consists of deep permeable soils, deep percolation to recharge groundwater is favored over interflow. Interflow may end up in the channel but moves slower than overland flow. The full conceptual water disposal or accounting has been diagrammed to include minor processes such as rain falling directly in the channel and evapotranspiration (Linsley et al. 1982; Chow et al. 1988).

During dry weather, water that continues to flow in a stream is known as base flow. Its source is mainly regional groundwater discharge. Base flow is indicated as a steady baseline in the hydrograph, whereas direct surface runoff resulting from a rainfall event is manifested as bumps or peaks (Figure 7.1). Base flow persists throughout a rain event. Some channel water may seep into the banks during high stream stages, forming bank storage. As the stream stage subsides, bank storage drains back to the channel.

Streamflow discharge as measured in a channel is the sum of overland flow, interflow, and base flow (bank storage). Judicious analysis of a discharge hydrograph can reasonably separate these components using well-established algorithms (Chow et al. 1988).

A gaining stream is one in which streamflow increases because the regional groundwater table is higher than, and intersected by, the streambed. Conversely, a losing stream is one in which streamflow decreases because of leakage into a groundwater table that is lower than the streambed.

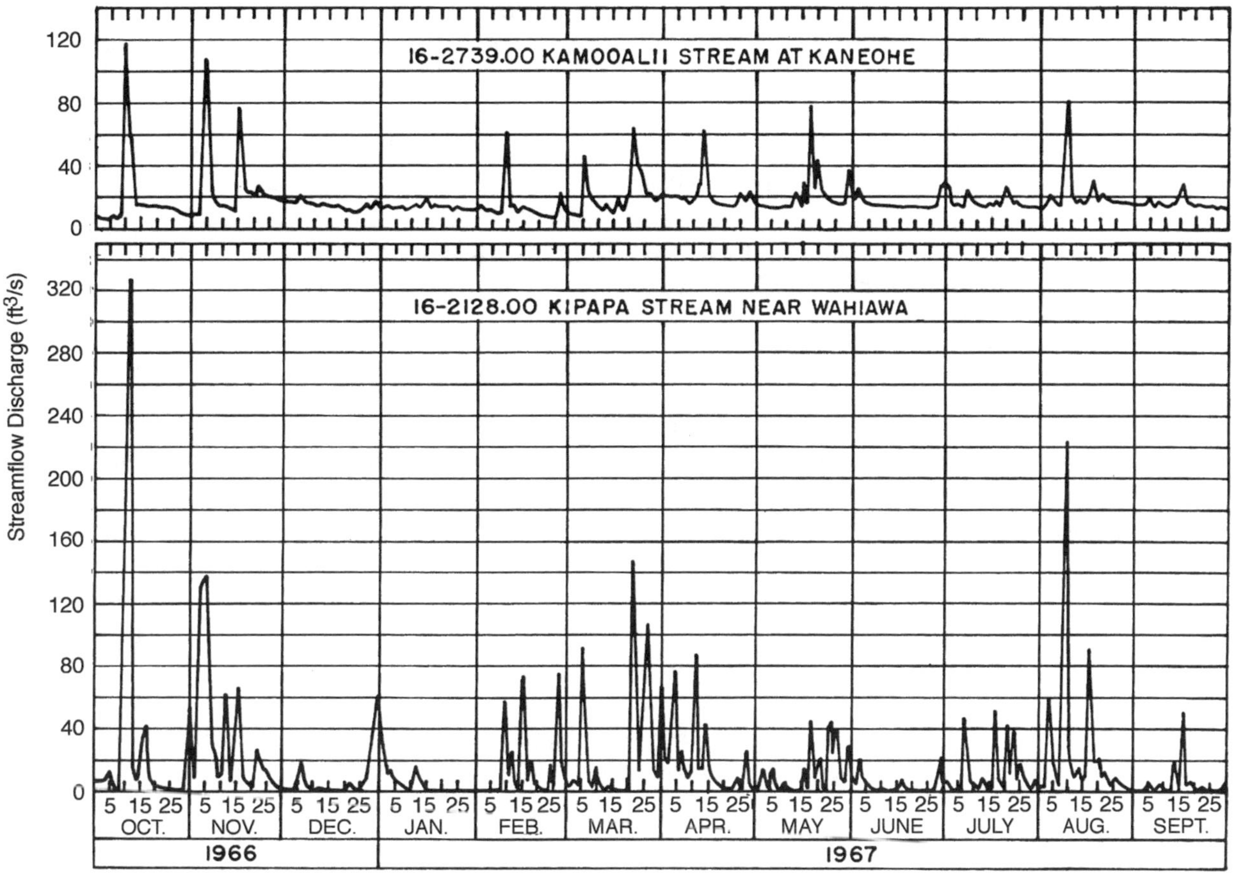


Figure 7.1. Streamflow hydrographs indicating prominent base flow (*upper*) and predominate direct surface runoff (*lower*). 1 $ft^3/s = 0.0283\ m^3/s$. (Adapted from Jones et al. 1971)

The behavior of overland flow has become better understood since the concept of saturation overland flow was introduced in the 1970s. According to the pioneering Horton's theory, overland flow is produced when rain intensity exceeds the infiltration capacity. Such a condition may occur on poorly pervious surfaces in urban areas and on natural surfaces where soil layers with sparse vegetal cover have low infiltration capacity, as often occurs in semiarid and arid lands. In the 1970s several studies reported that Hortonian overland flow rarely occurs on vegetal surfaces in humid regions. Under such a condition, the infiltration capacity of the soil may exceed rainfall intensities for all but very intense rainfalls. Typically, in hillslope areas and hollows, the subsurface becomes locally saturated, easily forcing overland flow to occur. The surface areas producing saturation overland flow vary in size during the rain event. Such variable source areas are of more academic interest than practical importance.

Hydraulics, which treats the dynamics of flowing water, can be dauntingly complicated if rigorously applied to a natural environment that is highly nonuniform. Water flow in a fixed-bed channel has been adequately considered as turbulent flow in open-channel hydraulics and is amenable to empirical analyses (Chow 1959). Overland flow hydraulics, whether laminar or turbulent, assumes kinematic wave (uniform flow) with the conditions that bed slope equals friction slope and Manning's equation applies (Chow et al. 1988).

## Sustainability of Streamflow

Inquiry into the variability of streamflow requires frequency analysis of the time series of mean daily flow. Mean daily flow is the average discharge of streamflow, usually expressed in cubic meters per second, over a 24-hour period at a specified location on the stream. As a random variable, it assumes a value or variate every day and forms a time series.

Streamflow is probabilistic, correlating with rainfall but also affected by basin characteristics. Frequency (probabilistic) analysis is employed to make some sense of the variability. In such analysis, the order of occurrence is ignored. If sequencing is of concern, stochastic analysis is appropriate. Frequency analysis in hydrology is well established and is typically documented in hydrology textbooks (e.g., Chow et al. 1988). The analysis transforms a time series into frequency (probability distribution) functions and yields statistical parameters. Flow duration curve and flood frequency curve are the two most common hydrologic applications of frequency analysis.

A commonly used probability function is exceedence probability, which is a complementary cumulative frequency. Exceedence probability is defined as the percentage of the time or probability that the random variable (such as mean daily flow) is equal to or exceeds an indicated magnitude (variate). For example, the 90 percentile exceedence flow means that the listed flow is equaled or exceeded 90 percent of the time, or an average of 329 days of the year. The 50 percentile exceedence flow is identical to the median flow. In the normal distribution, the median flow equals average flow.

The reliability of frequency studies in hydrology hinges on the length of the record. Twenty years is commonly regarded as the absolute minimum for flood frequency studies.

Stochastic analysis of a streamflow record addresses sequential flows and leads to synthesizing long-term streamflow time series. A branch of hydrology often called "synthetic hydrology" has been developed to analyze and model historical time series.

The origins of synthetic hydrology lay in the goal of establishing a relationship among sequential streamflow, storage in reservoirs, and permitted outflow. Hurst (1951) analyzed the flow record of the Nile River in studying long-term reservoir storage requirements, resulting in Hurst's empirical model:

$$R_n = s_n \left( \frac{N}{2} \right)^h \tag{7.1}$$

where $R_n$ is the maximum range of the reservoir storage required to produce a steady outflow equal to the mean over a period of $N$ years in a stream, and $s_n$ is the standard deviation of the inflow. The exponent $h$ is the Hurst coefficient. For the Markov process $h = 0.5$. An especially clear exposition of synthetic hydrology is given by Fiering (1967).

Other early inquiries include studies of wet and dry years (Yevjevich 1963) and augmentation of hydrologic data (Matalas and Jacobs 1964). It has since expanded to analysis of rainfall and water quality. Operational publications include monthly streamflow synthesis known as HEC-4 (U.S. Army Corps of Engineers 1971) and applied stochastic techniques (Lane and Frevert 1989). A full summary is provided by Salas (1993).

## Floods

Floods are surface waters that inundate lands not commonly underwater. They are feared for many of their traits: stage, velocity, volume, and duration. Flood cresting can last for days in mammoth drainage basins, and water stage rises rapidly and torrent sweeps forcibly during flash floods in small basins. Generally speaking, small basins are susceptible to flash floods because of three factors: prior wetting, rain intensity, and urban land use (Leopold 1997).

Flood peaks receive special attention in the form of flood frequency analysis. The commonly used random variable is the highest momentary peak discharge in 1 year: the series thus formulated from data is known as the annual flood series. A well-developed related theory is known as extreme value distributions (Chow et al. 1988). The resulting probability, *P*, is commonly plotted as the recurrence interval or return period, *T*, which is the reciprocal of exceedence probability. *T* is the average of the intervals (in years) between flood peaks when a specified magnitude is equaled or exceeded.

Flood frequency analysis is well accepted and is complemented by advances introduced since the 1960s. Regional flood approaches are now commonly used for ungaged sites where flood peak information is lacking but needed for planning purposes. The premise is based on the existence of geomorphic and climate homogeneities in a region that includes both gaged and ungaged sites. The statistical correlation approach must be moderated by exercising hydrologic common sense rather than by blindly following a statistical procedure.

The development of human-made flood-control structures always involves some risk because historical records are not long enough to include all rare floods, and economic reality must be balanced with absolute safety. Methodologies are available to assess uncertainty and risk in hydro-economic analysis (Chow et al. 1988). A recent advance is that of L-moments proposed by Hosking (1990) for summarizing statistical properties of hydrologic data. The theory of L-moments (sum) parallels the theory of conventional moments (product). But it suffers less from sample variability and is thus more robust to outliers in the data and allows for better inference from small samples of data. It has begun to gain acceptance (Vogel and Wilson 1996).

A 100-year rain may not necessarily produce a 100-year flood. For instance, if a basin has already become thoroughly wet by antecedent rains, the runoff can become disproportionally large. As a result, a rain event can be outranked by the resulting runoff event in respective frequency analysis. The flood may be rarer than the 100-year flood.

## Stream Water Quality

Water in rivers, lakes, and estuaries is used for drinking and domestic purposes, irrigation, livestock water, hydroelectric power generation, navigation, fisheries, and innumerable other purposes. Human uses of water and land modify the natural water quality, which is endowed by the geological and biological environments. Although all uses are beneficial from a human perspective, the altered quality may render the water unsuitable for the same or other uses. Drinking water after human use is returned as sewage, which has acquired bacteria and other objectional parameters (see Table 7.1).

Water-quality criteria are used to evaluate the acceptibility of a water source for various purposes. Specifically, the criteria stipulate pertinent water-quality parameters and their concentrations. A principal document on water-quality criteria was first issued in 1968 and was known as the green book. The current version is often known as the gold book, *Quality Criteria for Water*, issued by the U.S. Environmental Protection Agency (1987).

In 1849, John Snow was able to associate the disastrously high incidences of human death from cholera with fecal contamination of drinking water that was drawn from the polluted reaches of the Thames River. Today, many typical water-related diseases are known (see Table 7.2) (Tchobanoglous and Schroeder 1985). Also, a multitude of contaminants—including organic chemicals from industries; heat from power generation; heavy metals after use; nutrients and pesticide residues after agricultural uses; and oil, radioactivity, and acidity from urban cities—manage to find their way to surface and coastal waters. These contaminants can potentially cause acute and chronic toxicity of aquatic life (Laws 1993).

Water-quality criteria address specifically human health: diseases caused by microorganisms and toxic chemicals. *Escherichia coli* is generally accepted as an indicator of fecal contamination by human wastes because it is abundant in the intestinal tract of humans and other warm-blooded animals. Fecal coliform, fecal streptococci, and *Clostridium perfringens* are similarly found and used as indicators. The latter is considered the most

Table 7.1. Generalized Water Quality of Various Waters

| Characteristic | Water Source | | | Selected Drinking Water Quality Objectives | |
|---|---|---|---|---|---|
| | Typical Surface Water | Typical Groundwater | Domestic Wastewater (U.S.) | Raw Water Source | Drinking Water |
| Physical | | | | | |
| Turbidity (NTU) | — | — | — | — | < 1 |
| Solids, total (g/m³) | — | — | 700 | — | 500 |
| Suspended (g/m³) | > 50 | — | 200 | — | — |
| Settleable (ml/l) | — | — | 10 | — | — |
| Volatile (g/m³) | — | — | 300 | — | — |
| Filtrable (dissolved) (g/m³) | < 100 | > 100 | 500 | — | — |
| Color (units) | — | — | — | < 150 | < 15 |
| Odor (number) | — | — | Stale | — | < 3 |
| Temperature (°C) | 0.5–30 | 2.7–25 | 10–25 | < 20 | — |
| Chemical: Inorganic matter | | | | | |
| Alkalinity (eq/m³) | < 2 | > 2 | > 2 | — | — |
| Hardness (eq/m³) | < 2 | > 2 | — | — | — |
| Chlorides (g/m³) | 50 | 200 | > 100 | 250 | 250 |
| Calcium (g/m³) | 20 | 150 | — | — | — |
| Heavy metals (g/m³) | — | 0.5 | — | — | — |
| Nitrogen (g/m³) | < 10 | < 10 | 40 | — | — |
| Organic (g/m³) | 5 | — | 15 | — | — |
| Ammonia (g/m³) | — | — | 25 | — | — |
| Nitrate (g/m³) | < 5 | 5 | 0 | — | 10 |
| Phosphorus, total (g/m³) | — | — | 12 | — | — |
| Sulfate (g/m³) | — | — | — | — | 250 |
| pH | — | 6.5–8 | 6.5–8.5 | — | 6.0–8.5 |
| Chemical: Organic matter | | | | | |
| Total organic carbon (TOC) (g/m³) | < 5 | — | 150 | — | — |
| Fats, oils, greases (g/m³) | — | — | 100 | — | — |
| Pesticides (g/m³) | < 0.1 | — | — | — | 0.2 |
| Phenols (g/m³) | < 0.001 | — | — | < 0.005 | 0.001 |
| Surfactants (g/m³) | < 0.5 | < 0.5 | — | — | 0.5 |
| Chemical: Gases | | | | | |
| Oxygen (g/m³) | 7.5 | ≈ 7.5 | < 1.0 | > 4.0 | > 4.0 |
| Biological | | | | | |
| Bacteria (MPN[a]/100 ml) | < 2,000 | < 100 | $10^8$–$10^9$ | < 5,000 | < 1.0 |
| Viruses (PFU[b]/100 ml) | < 10 | < 1 | $10^2$–$10^4$ | — | — |

SOURCE: Tchobanoglous and Schroeder 1985.
[a]MPN, most probable number.
[b]PFU, plaque-forming units.

Table 7.2. Typical Water-Related Diseases

| Category and Method of Contraction | Disease | Causative Agent | Symptoms |
|---|---|---|---|
| Waterborne: ingesting contaminated water | Amebiasis (amoebic dysentery) | Protozoan (*Entamoeba histolytica*) | Prolonged diarrhea with bleeding, abscesses of the liver and small intestine |
| | Shigellosis (dysentery) | Bacteria (*Shigella*, 4 spp.) | Severe diarrhea |
| | Cholera | Bacteria (*Vibrio cholerae*) | Extremely heavy diarrhea, dehydration, high death rate |
| | Gastroenteritis | Virus (enteroviruses, parvovirus, rotovirus) | Mild to severe diarrhea |
| | Giardiasis | Protozoan (*Giardia lamblia*) | Mild to severe diarrhea, nausea, indigestion, flatulence |
| | Infective hepatitis | Virus (hepatitis A virus) | Jaundice, fever |
| | Leptospirosis (Weil's disease) | Bacteria (*Leptospira*) | Jaundice, fever |
| | Salmonellosis | Bacteria (*Salmonella*, ~1,700 spp.) | Fever, nausea, diarrhea |
| | Typhoid fever | Bacteria (*Salmonella typhosa*) | High fever, diarrhea, ulceration of small intestine |
| Water-washed: washing with contaminated water | Shigellosis (dysentery) | Bacteria (*Shigella*) | Mild to severe diarrhea |
| | Scabies | Mite | Skin ulcers |
| | Trachoma | Virus | Eye inflammation, partial or complete blindness |
| Water-based: worm infections involving water as one stage in cycle | Filariasis | Worm | Blocking of lymph nodes, permanent damage to tissue |
| | Guinea worm | Worm | Arthritis of joints |
| | Schistosomiasis | Worm (schistosomes) | Tissue damage and blood loss in bladder and intestinal venous drainage |

SOURCE: Tchobanoglous and Schroeder 1985.

reliable indicator of sewage contamination (Fujioka and Shizumura 1985). Toxic chemicals are usually evaluated in tests using animals such as mice and rats. Levels of intake are determined for dosage and duration of exposure. The results are expressed in terms of risks. Similar but more routine tests are performed using aquatic animals, usually fishes in water, for specific contaminants and in whole effluent samples.

Water-quality criteria are principles and measures based on scientific research. Water-quality standards are enforceable by law; they are based on water-quality criteria but may have been tempered by considerations of practicalities such as technologies and economics (McCutcheon et al. 1993).

Wastewaters are waters that in the process of being used for beneficial purposes acquire either unacceptable parameters or a concentration of extant parameters that render the water unfit for its specified use. Wastewaters come from domestic, agricultural, and industrial return flows. A wastewater can be upgraded by treatment (reclamation) to become suitable for beneficial uses again. For example, sewage after treatment is used for irrigation of many crops in Hawai'i (see Appendix 5.3). Likewise, treatment of natural surface (raw) water has been the approach to meet drinking-water standards. Although technologies can upgrade a water to any high degree in quality, the cost of treatment is a premier factor in determining the feasibility of water and wastewater management.

U.S. drinking-water standards have been established since 1914. Currently, maximum contaminant levels (MCLs) have been promulgated for many primary health-based (enforceable) parameters and a few secondary (nonenforceable) parameters involving aesthetic qualities such as taste, odor, and appearance. Another nonenforceable standard is the maximum contaminant level goal (MCLG), which is a best estimate of concentration that protects against adverse human health effects and allows an adequate margin of safety. Current U.S. Environmental Protection Agency maximum contaminant levels for drinking waters are stipulated for ninety-seven substances: sixty-four organic chemicals, twenty inorganic chemicals, four radionuclides, and nine microorganisms (Pontius 2003). Many of these organic chemicals are identified as contaminants in drinking-water sources in Hawai'i. Their possible health effects and applicable drinking-water standards are listed in Table 6.4.

Where wastewater is discharged into natural water bodies, ambient standards and/or effluent standards are applied to limit pollution. The use of ambient, also known as receiving water, standards takes into account the intended use of the water body and its ability to assimilate wastes. On the other hand, effluent standards specify the maximum contaminant level of pertinent parameters in the discharge water and disregard the size and use of the receiving water body. In addition, methods of treatment of the effluent may sometimes be specified. In some states, including Hawai'i, a mixing zone of finite size and shape in the receiving water may be allowed in cases where the discharge would violate the receiving water standards. Monitoring of the receiving water is required.

Surface waters are recreational waters for swimming and fishing and are an integral part of the ecosystem. All pollutant discharges into these waters are regulated by the U.S. Environmental Protection Agency and a state health and/or environmental agency. The National Environmental Protection Act of 1969 grants such authority. The Environmental Protection Agency is empowered with a National Pollutant Discharge Elimination System that establishes that all pollutant discharges into U.S. water are illegal unless authorized by a permit. Permitting requires the use of best-available technology to protect receiving water quality.

The 1972 Clean Water Act was enacted for the improvement of water quality for swimming, fishing, and protecting aquatic and riparian habitats. In 1994, about 60 percent of the nation's rivers, lakes, and estuaries were considered safe for fishing and swimming, as compared with only 36 percent in 1972, according to the Environmental Protection Agency. The results are attributed primarily to regulations of point discharges of wastewaters. To improve the remaining 40 percent, cleanup strategies must address prevention of nonpoint-source pollution and management of watersheds. Nonpoint-source pollution results from the discharge of return waters and storm runoff from large land areas—typically agricultural and urban lands.

Surface water is wide open to contamination. The water column can be swiftly polluted but can also recover quickly by dilution. Stream sediments, on which certain contaminants such as heavy metals and pesticides are absorbed, remain contaminated for extended time in terms of years. The aquatic community is a true indicator of environmental stresses.

Surface water differs considerably from groundwater

in quality. Generally, surface waters have a low concentration of total dissolved solids, chlorides, and other major ions, especially during high flow. However, surface waters have a high concentration of suspended solids, turbidity, microorganisms, and organic forms of nutrients. Surface waters contain dissolved gases, carbon dioxide, and dissolved oxygen and are easily oxygenated by aeration at the water surface. Surface water temperature is subjected to considerable seasonal and depth variations. The pH is slightly higher than 7.0. The biochemical oxygen demand level is extremely low, not higher than 1 to 2 mg/l. Odor, taste, and color can be noticeable in impounded surface water. Algae grow in slowly moving or stagnant waters, especially during the summer months.

Although groundwater is generally preferred as a drinking-water source, many headwaters offer high-quality water sources when protected from land development. Table 7.1 contains a generalized numerical coding of various waters (Tchobanoglous and Schroeder 1985).

## Traits of Hawai'i Surface Waters

### Drainage Basins

Hawai'i's drainage basins range from very young to moderately mature. The spectacular landscapes on the older islands and the absence of drainage patterns in the very young terrains on the island of Hawai'i present opportunities for illuminating observations and creative theories on surface water runoff (see Figure 1.5). A set of maps indicating geographic features and streams on the major Hawaiian Islands is provided in Figure 5.10.

Continuous erosion was proposed as the creator of Hawai'i's drainage basins by Wentworth (1928). In his pioneering studies, he reasoned that even precipitous *pali* (cliffs) are the work of predominantly chemical erosion rather than of faulting (see Figure 1.5). More recently, the creation of *pali* has been attributed to massive landslides (Moore et al. 1994). Sonar detection of enormous piles of debris off the coasts appears to substantiate Moore's theory.

Hawai'i's drainage basins are small, on the order of a few square miles upstream of gaging stations. Exceptionally large basins include the Wailuku River in Hilo, Hawai'i (USGS Station No. 7130), 663 $km^2$ (256 $mi^2$); Waimea River near Waimea, Kaua'i (USGS Station No. 0310), 150 $km^2$ (57.8 $mi^2$); and Waikele Stream in Waipahu, O'ahu (USGS Station No. 2130), 118 $km^2$ (45.7 $mi^2$).

As early as 1928, drainage nets were mapped and qualitatively described by Wentworth for Honomū, Hawai'i (a young pattern) and Waimea-Makaweli, Kaua'i (a mature pattern) (Figure 7.2). Many parameters of Hawai'i's drainage basins and streams, such as basin area and stream slope, have been used in correlations between flow behavior and geomorphology since Wu's (1967) initial work. Parameters of streams and drainage basins have been used in correlations between Hawai'i's fish habitats and geomorphology (Parham 2002).

### Natural Streamflow in Hawai'i

Streamflow may be defined as all waters that accumulate and travel in a stream channel. It includes direct surface runoff, groundwater seepage, and bank storage. Direct surface runoff is the component of rainfall that moves overland on the surface (overland flow) and through a shallow layer of soil and debris before joining a stream (interflow). Groundwater is the infiltrated water that accumulates in a saturated aquifer after passing through the unsaturated (vadose) zone. Bank storage is the channel water that seeps into the banks during high stream stages, remains in the banks, and drains back into the stream during low stream stages.

The volume of direct surface runoff depends on the intensity and persistence of the rain, as well as the size, geology, and morphology of the drainage basin. In Hawai'i direct surface runoff lasts for a short time—no more than a few days even in large drainage basins. Groundwater seepage originates as overflow and underflow from dike and perched aquifers and as outflow from basal aquifers at the inland margins of coastal plains. Bank storage drains slowly and requires frequent rainfall for replenishment because storage volume is small.

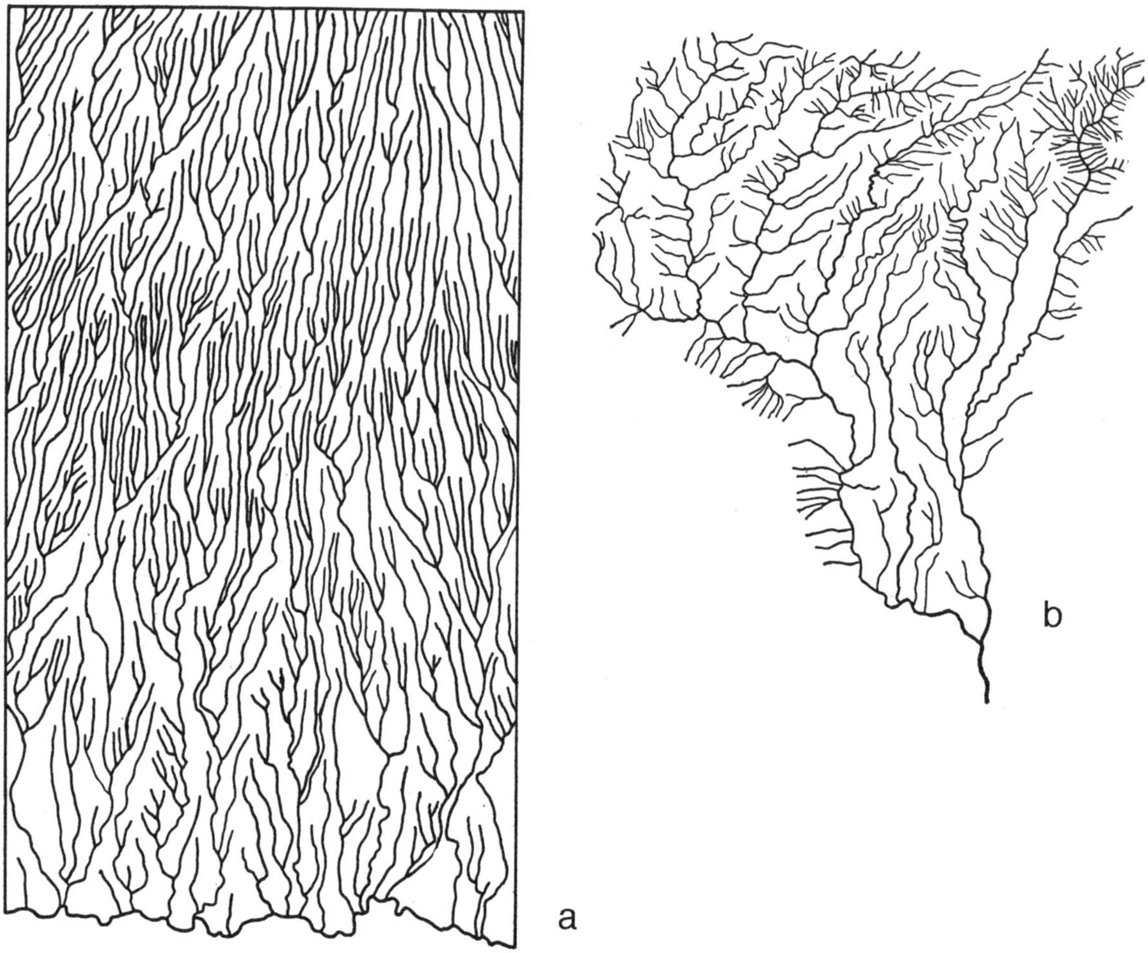


Figure 7.2. Two drainage nets in Hawai'i: *a,* Honomū quadrangle (portion), northeastern coast of Hawai'i Island, a young drainage pattern; *b,* Waimea-Makaweli basin, island of Kaua'i, a mature drainage pattern. (Reprinted from Wentworth 1928 with permission from University of Chicago Press)

In Hawai'i volcanic rocks and the accompanying overburden are an efficient infiltration medium that extracts a large fraction of rainfall to percolate to deep groundwater bodies. Once in the zone of saturation, groundwater moves seaward unless it is interrupted by a stream channel that acts as a drain. In rift zones, streams incise dike compartments to allow drainage, whereas in regions where poorly permeable andesitic-trachytic flank lavas cover the highly permeable primary basalts, perched aquifers are a source of seepage. The great basal aquifers contribute water to streams only at low elevations — usually below about 7.6 m (25 ft) above sea level.

The volume of groundwater lost to streams is far less than the volume that remains in the ground to eventually discharge into the sea, but the groundwater component of streamflow sustains the aquatic ecosystem and in many regions is a vital irrigation supply source. Most wetlands survive because of groundwater seepage.

Streamflow in Hawai'i is highly variable, and the statistics of flow are dominated by direct runoff from rainfall. The mean flow of a stream, which is perennial because of groundwater contribution, is about two to three times the base flow (dry-weather flow). In most instances base flow is taken as the flow at the 90 percentile exceedence. Flow-duration curves of natural streams generated solely by rainfall show a stronger resemblance to a lognormal distribution than to the standard normal distribution.

Only those streams that carry water at all times, no matter how little, are considered perennial. Streams that traverse rift zones containing high-level aquifers from headwaters to the sea are perennial throughout their length. In many cases, reaches of streams are perennial in high-level groundwater zones but nonperennial where channels pass over permeable flank lavas. In very high rainfall areas streams are perennial because of nearly constant overland flow and seepage from bank storage where high-level aquifers do not exist. Streams sustained by basal groundwater outflow occur only where the basal lens is thick and the head is high as a result of caprock impeding escape of groundwater.

## Stream Classification

Various classification schemes that address particulars of stream behavior, especially in terms of ecology in perennial streams, have been proposed but none has become standard. A simple classification should be based on the physical attributes of the stream and its drainage basin. Once the physical framework is established, ecological and other environmental considerations can be woven into the classification.

The simplest classification should refer to general stream behavior, geology of the drainage basin, and provenance of perennial flow. Once these features are established, ancillary characteristics follow naturally. The scheme suggested below accounts for basic stream attributes (Yuen and Associates 1992):

General Behavior
1. Perennial
2. Nonperennial
3. Ditch (human-made diversion)

Basin Geology (upstream of a stated point)
1. Rift zone
2. Flank lavas
3. Rift and flank
4. Sediments
5. Rift and sediments
6. Flank and sediments
7. Rift and flank and sediments

Perennial Flow Provenance
0. Not applicable (nonperennial)
1. Dike aquifers
2. Perched aquifers
3. Dike and perched aquifers
4. Basal aquifer
5. Bank storage

Each stream can be coded using a single number from each of the three categories. For example, a perennial stream that traverses the rift zone and whose flow derives from dike aquifers would be coded 111, where the first 1 refers to perennial under General Behavior, the second 1 to rift zone under Basin Geology, and the third 1 to dike aquifers under Perennial Flow Provenance.

To these, a status code can be added. An elementary status code describes the condition and use of the stream, in particular whether a stream is diverted, receives flow, or has been modified. The simple status code for this differentiation is as follows:

Diversion
1. No
2. Yes

Receive Flow
1. No
2. Yes

Modified
1. No
2. Yes

A stream that is diverted but does not receive flow and has not been modified would have the status code 211. The basic classification code for the stream would then be 111.211, in which 111 refers to its physical attributes and 211 to its usage and condition.

Employment of the classification scheme outlined here will permit the addition of more complex stream parameters.

## Analysis of Runoff Events

The volume and peak discharge of runoff resulting from rainfall events are the most important streamflow behavior characteristics of interest for engineering applications. Volume relates to water supply and peak discharge to floods.

### Runoff Volume from Individual Rainfall Events

Forested basins in the leeward Koʻolau Mountains on Oʻahu yield a surface runoff volume equal to about 35 percent of the rainfall of a moderate to heavy rainstorm. Mink (1962b) conducted a 3-year study of the forested Kīpapa basin, where he determined a runoff:rainfall ratio of 36 percent for several isolated cyclonic rainfall events. In a 13-year study of the nearby and more highly instrumented forested Moanalua basin, the average yield of surface runoff was about 34 percent for thirty rainfall events, all of which exceeded 50.8 mm (2 in.) of rainfall (Shade 1984). That antecedent rainfall enhances surface runoff is manifested in the Moanalua data. Preceded by a rainy period, each of two 152-mm (6-in.) rain events yielded almost 102 mm (4 in.) of runoff. Neither study is intended to provide an accurate assessment for all drainage basins, but the assessments appear to be reasonably applicable to the leeward mountains of south-central Oʻahu.

St. Louis Heights, an urbanized domestic community near the University of Hawaiʻi at Mānoa, was instrumented for a rainfall-runoff study (Fok et al. 1977). As expected, the surface runoff yield was high. Subsequent studies at Mililani Town supported the results obtained at St. Louis Heights (Murabayashi and Fok 1979).

Table 7.3. Generalized Runoff Curve Numbers for Sugarcane and Pineapple Covers

| | Hydrologic Soil Group | | | |
|---|---|---|---|---|
| Cover Condition | A | B | C | D |
| Sugarcane cover | | | | |
| Bare | 54 | 80 | 89 | 92 |
| Limited cover[a] | 50 | 71 | 81 | 86 |
| Partial cover[b] | 45 | 61 | 72 | 78 |
| Complete cover | 39 | 49 | 58 | 64 |
| Pineapple cover | | | | |
| Bare | 67 | 80 | 87 | 90 |
| Limited cover[a] | 49 | 61 | 71 | 76 |
| Partial cover[b] | 39 | 50 | 60 | 66 |
| Complete cover | 28 | 38 | 48 | 55 |

SOURCE: Cooley and Lane 1982.

[a]Less than 50%.

[b]Greater than 50%.

Surface runoff yields from agricultural land (sugarcane and pineapple) were measured in plot sizes of 0.8 to 2.8 ha (2 to 7 ac) by Cooley and Lane (1982). The sugarcane plots were located on the island of Hawaiʻi (Laupāhoehoe and Honokaʻa) and on Oʻahu (Waialua). The pineapple plots were located in central Oʻahu (Mililani and Kunia). The results are generalized and expressed in curve number for hydrologic soil groups A, B, C, and D and for various cover conditions (Table 7.3). The curve number approach to runoff is presented in Chapter 5. Generally, soil group A denotes low runoff and soil group D, high runoff. Group A-1 has lower runoff than group A. A curve number value of 100 represents total runoff and a value of 20, almost no runoff. Hawaiʻi soil classification by hydrologic soil groups is listed in Table 5.3 (U.S. Natural Resources Conservation Service 1993).

### Event Peak

Maximum discharge in a surface water runoff event in Hawai'i streams happens abruptly and lasts only for an instant (Wu 1967). These event peaks traditionally are important in planning because drainage facilities such as channels and culverts are designed to convey peak discharges. Hawai'i data indicate that peak discharges have not exceeded 2,548 $m^3/s$ (90,000 $ft^3/s$) (maximum recorded is 2,472 $m^3/s$ [87,300 $ft^3/s$] in the south fork of Wailua River, Kaua'i, on April 15, 1963). Few events have exceeded 283 to 566 $m^3/s$ (10,000 to 20,000 $ft^3/s$) (see Table 7.4). But severe flood damages have been inflicted by far lesser magnitudes—on the order of only several thousand cubic feet per second, especially in urban areas.

### Unit-Hydrograph Theory

Introduced by Sherman in 1932, a unit hydrograph provides the basin response as a discharge hydrograph of direct surface runoff resulting from 2.54 cm (1 in.) of excess rainfall generated uniformly over the basin at a constant rate for an effective duration. Excess rainfall is that rainfall that is neither infiltrated into the soil nor retained on the land surface. The basin is treated as a black box with all of its traits (cover, soils, terrain) embodied in its unit hydrograph. It has gained wide acceptance for application even though the assumptions cannot be fully satisfied under natural conditions. Theoretically, it is a simple linear system (Chow et al. 1988).

Wang and Wu (1972) inquired into the applicability of unit-hydrograph theory in Hawai'i's small basins by way of an instantaneous unit hydrograph using data from twenty-nine drainage basins on O'ahu. Their work demonstrated linearity between peak discharge and surface runoff volume, thus proving partial applicability. Chong and Fok (1973) approached the problem by testing linear and nonlinear watershed models for the Kalihi basin on O'ahu. The test results were inconclusive. Although a linear model gives the best estimation of time to peak, a nonlinear model best simulates peak discharge. Geomorpho-climatic instantaneous unit-hydrograph and related theories have been tested in the windward basins of East Moloka'i, the wet and precipitous Pelekunu, Wailau, and Waikolu (Diaz et al. 1995).

For common practical use, the unit-hydrograph theory for Hawai'i basins has not been disproved. An alternative approach to the rainfall-runoff relationship would be the use of physically based models that probe the major processes and parameters that transform rainfall to runoff.

## Sustainability

Hawai'i's perennial streams, whether at headwaters or lower reaches, are generally sustainable. However, they are susceptible to long, dry spells, such as a few months, of low rainfalls (see Chapter 3).

### Diversions

Diversion of streamflow for agricultural use was an ancient Hawaiian practice and engineering achievement (Kirch 1985). Excavations and radiocarbon dates for Kaua'i, O'ahu, Moloka'i, and windward Hawai'i all suggest that there was major construction of irrigated pondfield systems beginning about the fifteenth to sixteenth century. Typically, these systems dam perennial streams with boulder and cobbles, divert the water into stone-lined ditches, and distribute the water to terraced pondfields grown with taro and other crops. Sugarcane enterprises expanded on the ancient Hawaiian practice by diverting streamflow all from headwaters and transporting it by ditches and tunnels outside the basin to distant areas, especially to the dry lowlands (Wilcox 1996). In 1856, the first ditch diversion for sugarcane irrigation was used on Kaua'i. Since then, complex networks of diversion and transport structures have been built and vast quantities of water have been moved from one drainage basin to another.

An average flow of 27.4 $m^3/s$ (970 $ft^3/s$) is diverted from streams on five major islands, principally for irrigation and to a very small extent for drinking. Estimates of the major diversion by islands follow: Kaua'i, 10.9 $m^3/s$ (388 $ft^3/s$); O'ahu, 1.3 $m^3/s$ (47 $ft^3/s$); Moloka'i, 0.2 $m^3/s$ (8 $ft^3/s$); Maui, 11.6 $m^3/s$ (411 $ft^3/s$); and Hawai'i, 3.3 $m^3/s$ (116 $ft^3/s$). Flow diagrams of the major diversions are presented by

Table 7.4. Selected Instant Peak Discharge in Hawaiian Streams

| Local USGS Station No. | Name | Peak Discharge ($ft^3/s$) | Date |
|---|---|---|---|
| Kaua'i exceeding 20,000 $ft^3/s$ | | | |
| 0310 | Waimea | 37,100 | 02/07/49 |
| 0360 | Makaweli | 26,000 | 01/31/75 |
| 0490 | Hanapēpē | 39,000 | 04/15/63 |
| 0600 | S. Fork Wailua | 87,300 | 04/15/63 |
| 0710 | N. Fork Wailua | 53,200 | 11/12/55 |
| 1030 | Hanalei | 26,600 | 09/30/95 |
| 1080 | Wainiha | 40,000 | 02/17/56 |
| 0550 | Hulē'ia near Līhu'e | 26,800 | 11/28/70 |
| 0845 | Kapa'a | 30,330 | 10/03/94 |
| O'ahu exceeding 10,000 $ft^3/s$ | | | |
| 2130 | Waikele | 13,600 | 11/28/54 |
| 2160 | Waiawa | 27,900 | 10/28/81 |
| 2290 | Kalihi | 12,400 | 11/18/30 |
| 3300 | Kamananui | 16,800 | 11/20/90 |
| 2105 | Kaukonahua at Waialua | 15,600 | 12/26/92 |
| | | | 03/24/94 |
| 2471 | Mānoa-Pālolo | 10,100 | 12/18/67 |
| 3400 | Anahulu near Hale'iwa | 15,900 | 04/19/74 |
| Moloka'i exceeding 10,000 $ft^3/s$ | | | |
| 4000 | Hālawa | 26,900 | 02/04/65 |
| 4080 | Waikolu | 13,210 | 04/08/89 |
| Maui exceeding 10,000 $ft^3/s$ | | | |
| 5012 | 'Ohe'o Gulch near Kīpahulu | 14,700 | 09/18/94 |
| 5090 | Hanawī near Wahiku | 11,100 | 03/21/37 |
| 5003 | Hāwelewele Gulch near Kaupō | 13,600 | 01/08/80 |
| 5029 | Kawaipapa Gulch near Hāna | 16,880 | 08/01/82 |
| Hawai'i exceeding 10,000 $ft^3/s$ | | | |
| 7130 | Wailuku | 79,800 | 12/13/87 |
| 7170 | Honoli'i near Pāpa'ikou | 22,600 | 05/23/78 |

SOURCE: Fontaine et al. 1997.

NOTE: 1 $ft^3/s$ = 0.02832 $m^3/s$.

Yuen and Associates (1992). But in the 1990s, except for a few on Kaua'i and Maui, all sugarcane plantations were phased out for lack of economic profits.

## Volumes

Sustainability is relative and is measured or estimated in terms of a specified time period. Traits of stream and ditch flows on a mean daily basis are revealed by an analysis of flow-duration curves (Yuen and Associates 1992). Generally speaking, the flow-duration curves for Hawaiian natural streams resemble a lognormal distribution more than a normal distribution. Indeed, because of the skewness of the distribution, the average flow is much larger than the median flow ($Q_{50}$) even though it occurs less frequently. The average for many streams is approximated by the 25 percentile flow. This means that the daily flow equal to or exceeding the magnitude indicated occurs only 25 percent of the time or an average of 91 days per year. The median is a more reliable measure of sustainability than the average because half the time the flow is greater and half the time less than the indicated flow. Ditch flows that are captured from streamflow tend to follow the normal distribution because very high flows are truncated by ditch capacity.

Many productive Hawaiian streams have a median discharge on the order of 1 $m^3/s$ (35.3 $ft^3/s$). Rare cases exceed 3 $m^3/s$ (106 $ft^3/s$) (e.g., at Hanalei, Kaua'i [USGS Station No. 1030] with 3.68 $m^3/s$ [130 $ft^3/s$)] and Wailuku, Hawai'i [USGS Station No. 7130] with 4.53 $m^3/s$ [160 $ft^3/s$]). The relatively low flow values in consideration of the high rainfall in the upper reaches of the drainage basins are due to the high degree of infiltration in the permeable surface and the small size of the drainage basins. The median discharge may be expressed as flow per unit area of the drainage basin. Many productive basins yield 0.011 to 0.022 $m^3/s/km^2$ (1 to 2 $ft^3/s/mi^2$), and rare ones can yield close to 0.11 $m^3/s/km^2$ (10 $ft^3/s/mi^2$), such as at Hanalei even after water has been diverted to ditch flow. The great yields are derived from high rainfall, base flow of groundwater origin, and geological characteristics of the surface.

The shape of the flow-duration curve reflects the provenance of the different waters that supply the streamflow. Streamflow generated solely by rainfall shows a steep flow-duration curve, whereas streams draining groundwater as well as rainfall have a flatter trace. Hawai'i streamflow traits are illustrated in Figure 7.3. Kīpapa Stream near Wahiawā (USGS Station No. 2128) displays a steep slope throughout the whole range of discharges, suggesting the near absence of natural perennial sources in the drainage basin. Indeed, there are no perennial springs in the headwaters of streams in the Pearl Harbor region leeward of the Ko'olau Mountain crest, and these streams stop flowing a few days after rain stops (Hirashima 1971). Waikele Stream (USGS Station No. 2130), of which Kīpapa is a tributary, has a flow-duration curve with a flat lower end because the stream is perennial as a result of groundwater discharge from the basal water body (Visher and Mink 1964).

Low ditch flows sustained for durations longer than several weeks are a signal of drought conditions. Droughts during the summers of 1971 to 1975 resulted in heavy sugarcane crop losses in East Maui. Wailoa Ditch (USGS Station No. 5880), the largest of the ditches that collect water from the rain forests of East Maui, was analyzed with low-flow curves for six different periods: 1, 7, 14, 30, 60, and 90 days (Fok and Miyasato 1976).

One would expect that in Hawai'i's two-season year (summer: May to September; winter: October to April), the runoff volume would be greater during the wet season. This assumption was found to be valid by regression analysis in field experiments on O'ahu by Anderson et al. (1966). Their study was made in two adjacent small basins of about equal size (12 ha [30 ac]) from 1951 to 1955 near Helemano, leeward Ko'olau Mountains. One basin is tree covered and the other fern covered and had been burned.

The curve number approach was modified from its original purpose for individual events to a month period for a Pearl Harbor–Honolulu basin study that needed monthly surface runoff values for ungaged locations (Giambelluca 1983). The modified procedure required additional assumptions, manipulation of local data of soils and covers, generalization, and personal judgment. The

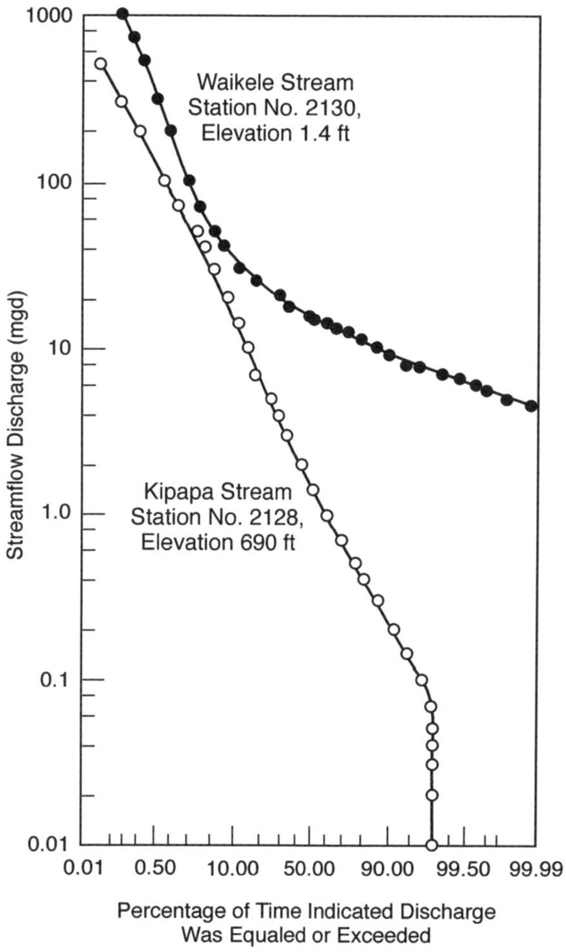


Figure 7.3. Flow-duration curve for Waikele Stream near Waipahu (USGS Station No. 2130) and its tributary Kīpapa Stream near Wahiawā (USGS Station No. 2128), Oʻahu. 1 mgd = 0.043 $m^3/s$ = 1.55 $ft^3/s$. (Adapted from Hirashima 1971)

resulting curve number values in Table 7.5 were derived for the Pearl Harbor–Honolulu basin and were used for water-balance estimates. Care should be exercised when applying the derived values to other basins. Each should be evaluated with field data if and when available.

The statistics of total surface runoff volume have been projected for an average year for Oʻahu. A study indicated that the ratio of surface runoff to rainfall is only 16 percent (i.e., 1,162 mld [307 mgd] out of 7,482 mld [1,977 mgd] [Shade and Nichols 1996]). This value differs appreciably from the 24 percent (1,627 mld [430 mgd] out of 6,813 mld [1,800 mgd]) reported in a previous study (Takasaki 1978). The 1996 study also avers that the region producing the lowest runoff ratio, 8 percent, is Waiʻanae, which is dry, and that producing the highest runoff ratio, 18 percent, is windward Oʻahu, which is wet. All of these projections involve many assumptions.

The runoff-rainfall ratio assumes smaller values as the time frame for the correlation increases from an event basis to a month basis, and on to an average year basis. This is expected because light rains do not produce runoff and because more light rains are included in extended time periods.

## Stochastic Analysis of Streamflow

The sequential nature of streamflow cannot be accounted for by flow-duration and low-flow curves, and yet it is the essence of drought. Sequences, but not their placement in time, can be generated from a record of stable statistics (stationary process) by techniques of stochastic hydrology. A popular sequence generator is the Lag-1 Markov model, which is based on correlation between a flow and its succeeding flow (Fiering and Jackson 1971). The stochastic nature of monthly ditch flow was examined for Wailoa Ditch on Maui (Fok and Miyasato 1976) and Kohala Ditch on the island of Hawaiʻi (Akinaka et al. 1975). Both works used the Lag-1 Markov model and the normal 100-year generated series based on 50 years of record. The Maui study concluded that the closely spaced 1971 and 1973 summer low flows were a rare combination of sequential low flows for a long time (100 years). None of the generated data displayed the historical combinations as in the 1971–1973 record. Subsequent analyses of Wailoa Ditch by Mink (1997) determined that the longest low-flow record occurred between July and November of 1984. The Hawaiʻi work, which examined a generated 100-year series on the basis of the 33-year historical record, concluded that the record already included the worst expectable 100-year low-flow traits. The Hawaiʻi work also discussed the ditch flows in terms of droughts variously defined by length, severity, and distribution.

Table 7.5. Generalized Curve Numbers for Pearl Harbor–Honolulu Basin

| Land Use Classification | Hydrologic Soil Group A | B | C | D |
|---|---|---|---|---|
| Sugarcane | 38 | 53 | 64 | 70 |
| Pineapple | 25 | 59 | 72 | 79 |
| Low-density urban | 51 | 68 | 79 | 84 |
| Medium-density urban | 68 | 79 | 86 | 89 |
| High-density urban | 89 | 92 | 94 | 95 |
| Mixed (medium-density urban/vacant) | 54 | 70 | 80 | 85 |
| Mixed (medium high–density urban) | 76 | 85 | 90 | 92 |
| Mixed (high-density urban/vacant) | 64 | 77 | 84 | 88 |
| Golf course/park | 39 | 61 | 74 | 80 |
| Forest/grazing/vacant | 36 | 60 | 73 | 79 |

SOURCE: Giambelluca 1983.

Without storage, three consecutive months of low flow is commonly considered as drought by any definition.

## Rainstorm Floods

Hawai'i's concern for flood problems is associated primarily with recent urban land use and development. Statistics for a period of about 100 years (1862 to 1965) show that only eight floods were reported by Honolulu newspapers during the first half of the period as compared with twenty-seven floods during the latter half of the period (G. Parvaras-Carayannis, Hawai'i Institute of Geophysics, 1967, in an unpublished survey of historical floods on O'ahu). Also, most of the later floods occurred in Honolulu and windward O'ahu. A substantial shift in land use from underdeveloped and rural to urban has taken place in both regions. There is no evidence of a significant change in rainfall from the early to the later period. Also, floods that occurred in uninhabited areas, especially in the early days, may not have been reported.

A loss of sixty-three human lives from rainstorm floods on O'ahu has occurred since 1867 (Fok 1967). The latest casualty statewide was in 2003 when two people lost their lives on Maui. Property damages from these floods have totaled many millions of dollars, including $34 million for 1988 alone (Dracup et al. 1991).

## Hawai'i's Flash Floods

Hawai'i floods result from intense rainstorms striking quickly in localized areas. The stream stage rises quickly, peaks sharply, and recedes only moderately slower than it rises. The quick response is due primarily to high rain intensity and the steep terrain in small headwaters areas. In the wet Hawai'i mountains, soil and vegetal cover abstract a large fraction of the rainfall (see Chapter 5). Less intense rainfall without prior wetting does not produce a large volume of overland flow. Peaking of Hawai'i flood flows reflects the traits of stream channels: their small sizes and steep channels combine to produce high-velocity surging floodwater. The rapid recession and short interval of inundation are consequences of the small natural storage capacities of the basins and their proximity to the ocean, which efficiently absorbs floodwaters except when peak flows coincide with high tides.

Typically, flood flows crest in about an hour and recede in a few hours. The time to peak for twenty-nine basins on O'ahu ranged between 0.55 and 2.58 hours, based on an analysis of nearly 200 hydrographs (Wu 1969). The most extreme recorded storm occurred on January 24–25, 1956, when over 965 mm (38 in.) of rain fell at Kīlauea, Kaua'i, within a 24-hour period. During the same storm, 152 mm (6 in.) of rain fell in a single 30-minute period and about 305 mm (12 in.) fell in 1 hour. Rainfall intensities and totals as high as these values can occur anywhere in Hawai'i without reference to orographic effects (Blumenstock and Price 1967). The most intense flood of record in Hawai'i occurred on November 18, 1930, in Kalihi Stream ($6.7 \times 10^6$ $m^2$ [2.6 $mi^2$]). The peak is among the highest recorded in the United States on a per-unit-area basis ($5.21 \times 10^{-5}$ $m^3/s/m^2$ [4,770 $ft^3/s/mi^2$]) (Linsley et al. 1982).

Flood situations are worsened by debris flow and sand plugs. Debris expands the volume of the flood flow and clogs the culverts and bridge openings. Debris basins are routinely required to reduce clogging, but the adequacy of the basin size has been questioned, as in the case of the 1987 New Year's Eve flood on O'ahu (Cheng 1992). Sand plugs occur in many ephemeral streams at the coastline as a result of littoral drift. The plugs often back up initial flash-flood water to inundate coastal riparian lands. A solution is diligent maintenance: breaching the plugs mechanically when severe storm rain is forecasted. Another solution is to use an automatic on-site hydraulic structure (without moving parts), which was demonstrated to be effective in breaching plugs (Nishimura and Lau 1978).

## Mitigating Flood Damages

### *Forecast and alert*

Alerting the public to imminent flood events saves lives and protects property. The U.S. National Weather Service has been actively pursuing advances in forecasting with satellite remote sensing, data automation, computer upgrades, graphic displays, and coupled hydrologic-hydrometeorologic modeling (Fread 1998). Some of these technologies were introduced to Hawai'i in the last decade to provide the data needed to make more reliable and timely forecasts of severe storms and flash floods. Included are weather satellites, Doppler radars, and a system of automatic rain gages.

### *Structural approaches to flood mitigation*

In the late 1950s rapid urbanization on O'ahu induced a great increase in flood problems, and the city government's response was to choose the structural approach using storm-drainage facilities for mitigation. The city retained Ven Te Chow in 1966 to provide the hydrologic basis for storm drainage standards. Because of the short record of data, he recommended envelope curves and the Rational method, and he proposed a framework for hydrologic studies (Chow 1966). His work was embodied in the 1969 standards and left an indelible imprint on subsequent standards (Lau 1982).

Storage of floodwater in surface reservoirs reduces peak discharge, but this approach by itself and in conjunction with drainage is generally neither feasible nor popular in Hawai'i because of the shortage of suitable reservoir sites and the high cost of land. A rare exception is the Kailua-Kāne'ohe reservoir located in the Ho'omaluhia Botanical Garden in Kāne'ohe, O'ahu (U.S. Army Corps of Engineers 1972).

Urban drainage facilities are intended only for flood mitigation. Hydraulically efficient concrete channels are not compatible with stream characteristics and riparian ecosystems. Attention was called to the ecological impact of channelization based on a statewide survey (Maciolek 1978). More than seventy streams have been channelized and most of them are on O'ahu (Timbol and Maciolek 1978). The Hawai'i stream assessment report included channelization as a priority item for investigation (Hawai'i Department of Land and Natural Resources 1990).

### *Floodplain management*

By nature, a floodplain is part of a natural channel (Leopold 1997). In contrast to flood control by structures, floodplain management is a nonstructural approach. It restricts or stipulates land use, development, and management practices in flood-prone lands. The coastal part of Hilo, Hawai'i Island, is a fine example of floodplain management even though the devastation responsible for establishing a management system was caused by tsunamis rather than rainstorms. In Hawai'i and elsewhere in the United States, the national flood insurance program requires mapping the zone of the 100-year expectation of flood inundation. Mapping of this hazard zone created flood insurance rate maps issued by the Federal Emergency Management Agency. Land within this zone is restricted to flood-compatible uses such as parks, car parking, and flood-proofed structures. These maps are continually revised. In the aftermath of the devastating New Year's Eve (December 31, 1987) flood in East O'ahu, maps as revised in 1987 were considered inadequate (Dracup et al. 1991). For example, Haha'ione, Kuli'ou'ou, and Niu Valleys sustained major flood damages, yet they were identified as D zones in the 1987 maps. In D zones, flood hazards are undetermined and the purchase of flood insurance is not required (Federal Emergency Management Agency 1988). The O'ahu maps were revised in 1995.

Other flood investigative approaches have been considered but are not generally applicable in Hawai'i. For instance, river forecasting, a common service provided by the U.S. National Weather Service for conterminous states, is not applicable in Hawai'i because of the very short transit time of flood flow to the ocean. Also, the small storage capacity of stream channels in Hawai'i has virtually no attenuating effects on the magnitude of flood flows.

### *Current status*

In summary, mitigating of damage caused by floods is effected through warning and drainage constructions. Containing and channeling the instantaneous peak discharge of a flood flow are principal design criteria. The drainage approach is acceptable because of the nature of flash floods, lack of reservoir sites, and proximity of the ocean. However, changing social values demand that environmental and other concerns be properly considered.

## Design Floods

Given the drainage approach currently practiced in Hawai'i, data on peak discharges and frequencies are needed for the design of flood-control facilities and floodplain management. The whole hydrograph is rarely required. However, for new urban development in Honolulu, data on flood volume are necessary for the assessment of water pollution. Nearly all basic computational methodology for flow and volume has been adapted from standard methods to reflect Hawai'i's geomorphic and climatic conditions.

### *Peak discharge formulas*

The envelope curve is an empirical summary of historical peak discharges in a group of basins having similar geomorphic and climatic features. As introduced by Chow, it is drawn to cover maximum instant discharges in a plot of discharge versus basin area (see Figure 7.4). The only explicit variable is the drainage basin area.

First introduced to the United States by Kuichling in 1889, the Rational formula is the most widely used method in urban drainage design for small basins. First used by the City and County of Honolulu in 1957, it was reaffirmed and strengthened by Chow in 1966. Basically stated, the increasing rate of surface runoff at the basin outlet reaches a maximum at a certain time, known as the time of concentration, when the whole surface of the drainage basin contributes to the flow. Rainfall is assumed uniform over the basin and lasts through the time of concentration. The resulting formula expressed in English units is $Q_p = C\,i\,A$ where $Q$ is the peak discharge (in cubic feet per second), $C$ is the runoff coefficient (dimensionless) that ranges from zero for a perfectly pervious surface to one for a perfectly impervious surface, $A$ is the drainage area (in acres), and $i$ is the average rain intensity during the time of concentration (in inches per hour). The rainfall intensity is derived from rain intensity-frequency analysis for various rain durations. Chow purposely limited the formula's applicability to small areas, 40.5 ha (100 ac) or less, so that the assumptions have a chance to appear reasonable. Criticism of the parameter values abounds (Lau 1982). The value of $C$ should be evaluated using Hawai'i field data accumulated since 1966 (see Chapter 5), and the formula should be assessed by exploring the probabilistic approach (Schaake et al. 1967; Pilgrim and Cordery 1993). For the time of con-

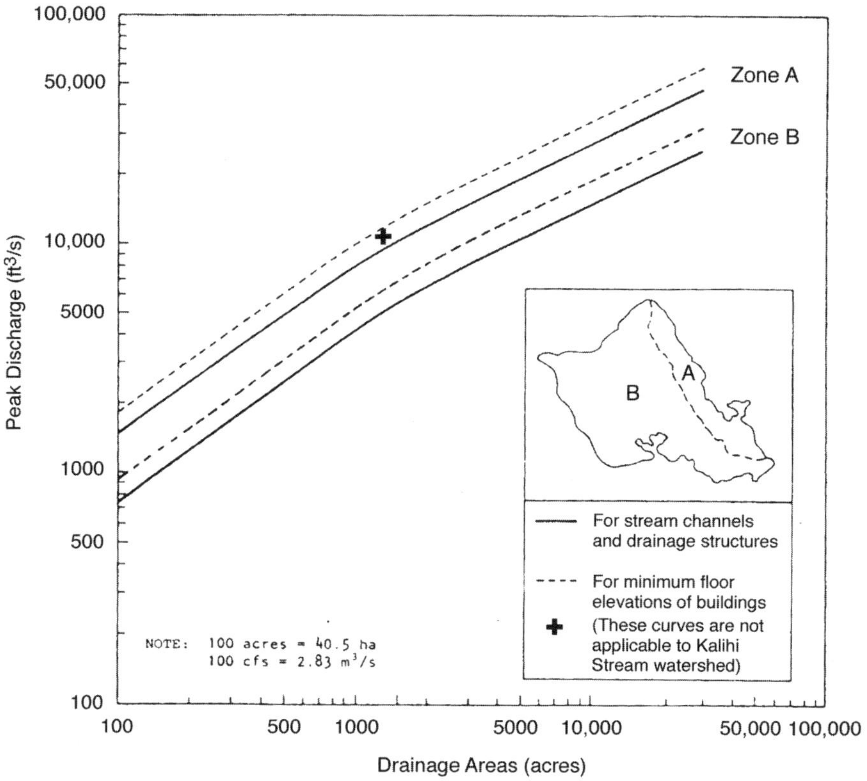


Figure 7.4. Envelope curves of maximum experience, O'ahu, Hawai'i. (Adapted from Chow 1966)

centration, the Kirpich method has been adapted and Hawai'i data incorporated. Rainfall intensity values are updated (Giambelluca et al. 1984).

The envelope curves do not signify return period of events. However, Wu (1967) estimated that the 1966 curves suggest 50-year to 100-year floods. Those in the 1988 standards were surmised to approximate 100-year floods (City and County of Honolulu 1988; Wong 1994).

The County of Maui (1995) encourages the use of the Rational method and the Natural Resources Conservation Service hydrograph analysis and stipulates different return periods by basin area and type of drainage structures. Further, as another instance of institutionalization, the minimum design flow cannot be less than the Federal Emergency Management Agency storm flows as determined in the 1995 flood insurance study.

### *Flood frequency*

Caveats are many in using frequency curves for analyses of floods. Chow (1966) did not encourage the use of short records for frequency analyses in preparing the storm drainage standards for O'ahu. Typically, the desired recurrence intervals are longer than the record. Even with a well-defined curve (Figure 7.5) care must be exercised to extend the curve to predict the 100-year flood (i.e., which theoretical frequency distribution may best fit the data). Wu (1969) chose the Gumbel distribution over the log-Gumbel and Pearson Type III distributions in his study of O'ahu 100-year floods and computed regional frequencies (100-year return period) using multiple regression with four parameters: drainage area, watershed length, watershed elevation, and rainfall frequency (100 years, 24 hours). Frequency curves were also computed using

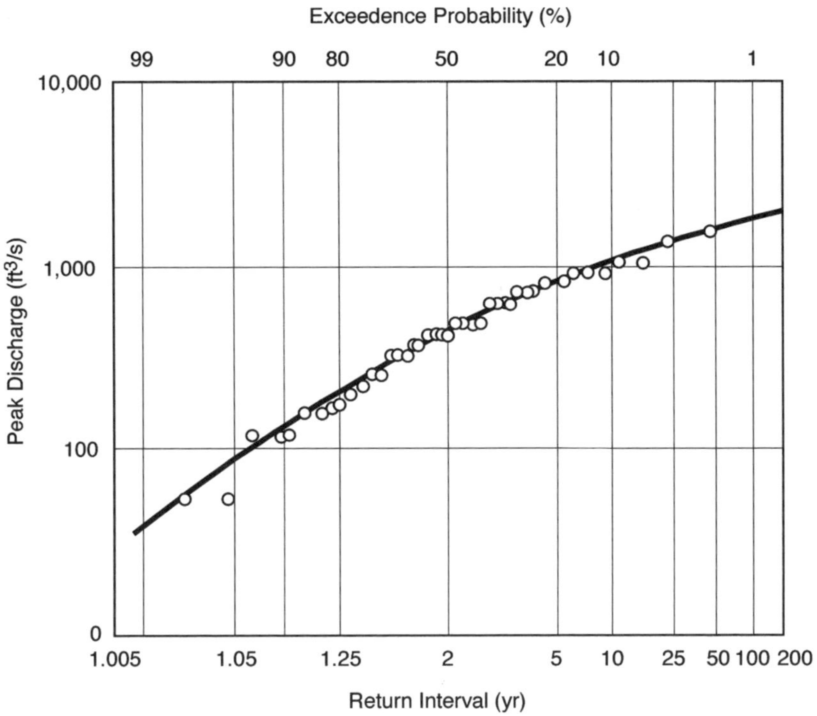


Figure 7.5. Flood frequency curve for Wai'ōma'o Stream near Honolulu. USGS Station No. 2460, years 1926 – 1968, drainage area 2.69 $km^2$ (1.04 $mi^2$), $n = 43$, standard deviation 0.357, skew factor −0.645. (Adapted from Lau et al. 1971)

log-Pearson Type III distribution for selected individual streams in Hawai'i (Hawai'i Division of Water and Land Development 1970).

The National Flood Insurance Act (1968) and the Flood Disaster Protection Act (1973) mandate using the approach of flood frequency and floodplain management. The latter legislation requires local governments to establish flood-control ordinances and to enforce land-control measures. Federal funds were committed to delineate flood-hazard areas by 1983 to estimate the 100-year flood elevations and boundaries from existing records or by analogy.

For O'ahu, the requirement for flood frequency prompted the first formalized analysis using methodology standardized in Bulletin 17 of the U.S. Water Resources Council (1976). The regional frequencies are computed with log-Pearson Type III distribution and records of seventy-four stations, with records ranging from 10 to 61 years. No more than two geomorphic and climatic parameters were selected from seven parameters on the basis of highest statistical correlation: drainage area, channel slope, channel length, mean annual precipitation, vegetal cover, mean basin elevation, and rainfall intensity (24-hour, 2-year frequency). The resulting parameters were drainage area for the windward region, drainage area and rain intensity for the North O'ahu–Wai'anae region, and drainage area and vegetal cover for the central O'ahu–Honolulu region. Application of these regional regression equations was limited to streams that are unaffected by water diversion and urbanization (Nakahara 1980). This work was recognized in the City and County of Honolulu's 1988 drainage design standard as the basis for constructing the design curves for the 100-year peak discharge for drainage areas larger than 40.5 ha (100 ac). Bulletin 17 sets the generalized skew coefficient at -0.05 for Hawai'i. This parameter has been independently computed to be -0.14 using the record for sixty-eight Hawai'i stream gaging stations. The difference was considered small and the value of -0.05 acceptable (Lee 1984). In 1994, the analysis was repeated with an additional 14 years of records and improved techniques, resulting in some modifications of the 1988 results (Wong 1994).

Individual frequency curves based on long record should always be preferable to regional frequency curves. For instance, Kalihi Stream (USGS Station No. 2290) experienced a record maximum peak of 351 $m^3$/s (12,400 $ft^3$/s). The computed 100-year flood is 317 $m^3$/s (11,200 $ft^3$/s), but it would be only 185 $m^3$/s (6,560 $ft^3$/s) if computed by regional regression (Nakahara 1980). Apparently, the regional regression equations are unable to capture the distinctive traits of the Kalihi basin.

### *Standard Project Flood*

The Standard Project Flood is estimated by applying the unit-hydrograph method to the standard project storm. The storm is the greatest storm that may be reasonably expected and can be derived by maximization and transposition of historical storms.

The U.S. Army Corps of Engineers' Standard Project Flood method was used for the design of the Kailua-Kāne'ohe flood-control reservoir (U.S. Army Corps of Engineers 1972). The design peak discharge is 424 $m^3$/s (15,000 $ft^3$/s), and the design flow volume is $4.25 \times 10^6$ $m^3$ (3,450 ac-ft) for a rainfall event of 706 mm (27.8 in.) in 24 hours for a drainage basin of 8.13 $km^2$ (3.14 $mi^2$). The design flood is 1.5 to 2.0 times larger than the greatest flood on record (February 1, 1969). The Standard Project Flood has an estimated recurrence interval exceeding 100 years. During the last year of construction, a storm occurred, dumping 246 mm (9.7 in.) of rainfall over a 24-hour period (March 18–19, 1980) at the project site. Although incomplete at the time, the facility apparently was able to function as designed — containing streamflow within the natural stream banks.

### *Physically based methods*

HEC-1 is a popular method for computing the direct runoff hydrograph in planning and design of urban storm runoff. The program is relatively easy to use and widely accepted in the United States. A computational example is given in a textbook by Chow et al. (1988). As a part of HEC-1, Synder's unit hydrograph continues to be used in Hawai'i. The National Resources Conservation Service hydrograph method is another popular method. In fact, the County of Maui has institutionalized its use. These methods shed little light on the nature of Hawai'i flood flows because they do not simulate historical events. Their applicability to conditions in Hawai'i is difficult to verify or disprove. At a minimum, the important parameters involved in the method should be calibrated. For example, the published curve number values should be calibrated against records of streamflow gaging stations.

The best that can be done with a physically based model of a drainage basin is to instrument the basin and subject the model to calibration and verification with the observed data. Once this is performed to satisfaction, the model may then be used for the subject basin and similar basins, assuming the model parameters can be secured for the other basins. The shortcoming of this approach is its high cost and time requirements. No ordinary development project can afford such an undertaking.

Only two basins have been physically modeled in Hawai'i: Moanalua basin with the Dawdy, Schaake, and Alley model (Shade 1984) and St. Louis Heights drainage basin with a model of the same name (Phamwon and Fok 1977). The two models are similar, differing only in details. They use the kinematic wave equation to describe both overland and channel flow routing and solve the equations with numerical methods. The model by Dawdy et al. is more detailed in the partitioning of rainfall by using soil moisture accounting and Philip's infiltration equation. In the application of the methods, the St. Louis Heights basin (0.39 $km^2$ [0.15 $mi^2$]) was more finely segmented than the larger Moanalua basin (8.65 $km^2$ [3.34 $mi^2$]). The calibrated models simulated the observed hydrographs reasonably well (Figure 7.6). The simulations revealed high sensitivity of runoff to infiltration.

## Hawai'i Surface Water Quality and Biota

The quality of Hawai'i surface water is prescribed by two flow regimes: base flow derived from groundwater and

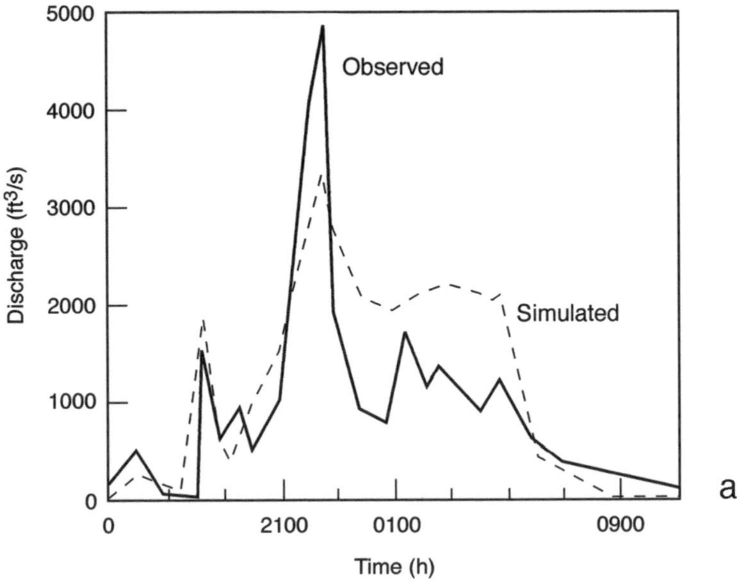


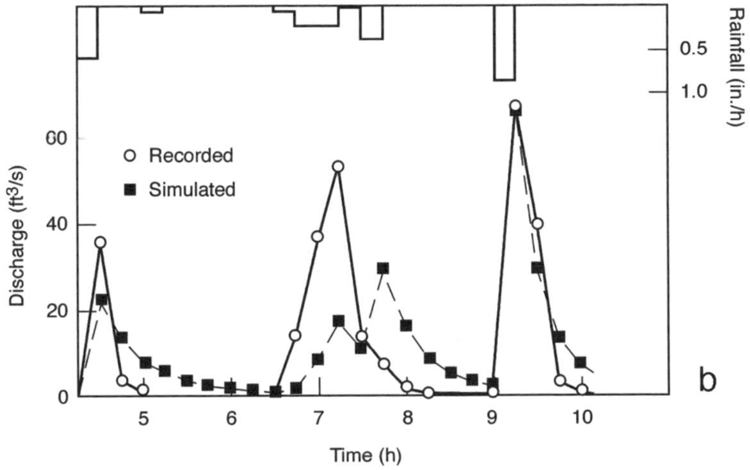


Figure 7.6. Simulated and observed storm runoff hydrographs, O'ahu: *a,* Moanalua basin, O'ahu, March 18–19, 1980, USGS Station No. 2282, observed rainfall volume 582 mm (22.9 in.), 10-day antecedent rainfall 163 mm (6.4 in.) (Adapted from Shade 1984); *b,* St. Louis Heights basin, Kānewai Park, O'ahu, October 19, 1974. (Adapted from Phamwon and Fok 1977).

direct runoff from rainfall. The low flow of perennial streams in the mountains depends on high-level aquifer discharge combined with overland flow resulting from trade-wind showers. Rainstorm runoff quality is temporary and highly varying, like the storm runoff itself.

The two most important issues concerning the quality of Hawai'i surface waters are their use as drinking-water sources, principally on the islands of Maui and Hawai'i, and their effects on coastal water quality. The second issue tends to draw public attention simply as turbid storm water entering the clear sea, but a broader perspective must take into account the relationship between land use and management practice and the receiving coastal water quality and the coral ecosystem (see Chapter 8).

## Sources of Dissolved and Suspended Solids

Most of the materials present in surface waters are derived from land and a fraction from the atmosphere. The land products include rock, soil, vegetation, and materials generated by animal and human activities. Additional sources include rainwater, sea aerosol, and groundwater.

The natural sources of dissolved solids are primarily the minerals in basaltic rocks and soils that have undergone chemical denudation, that is, dissolution and leaching by water. The abundance of the chemical elements in basaltic rocks is in the following order: $SiO_2$ (50 percent), $Al_2O_3$ (15 percent), $Fe_2O_3$ (3 percent) and $FeO_2$ (8 percent), CaO (9 percent), MgO (8 percent), $Na_2O$ (3 percent), $K_2O$ (<1 percent), and other elements including phosphorus, all <1 percent. Hawaiian basalts contain an insignificant amount of chloride. As indicated by the value of partition ratio of these elements, the rocks quickly lose the mobile calcium, magnesium, sodium, potassium, and much of the silica by leaching in regions of annual rainfall exceeding 1,524 mm (60 in.). As a result, basaltic rocks and soils contribute virtually all the dissolved but nonionic silica as $SiO_2$ as well as phosphate and fluoride, and substantial quantities of calcium, magnesium, sodium, and potassium present in the surface waters. Typical values of the major dissolved chemicals in the uncontaminated natural stream waters in the Ko'olau Range in southern O'ahu are shown in Table 7.6.

The concentrations of sodium and chloride that cannot be attributed to basaltic rocks originate from sea salts carried in by rainwater and sea aerosol. Rainwater also brings sulfate, pH, nitrogen species, and isotopes of hydrogen, carbon, and oxygen (see Chapter 3). Relative to other surface waters, seawater is dominated by concentrations of chloride, sodium, and total dissolved solids, but also contains small amounts of silica, nitrate, phosphate, and fluoride (see Table 3.2).

Bicarbonate is the most abundant ion in all freshwaters in Hawai'i. It is the end product of microbial biodegradation of organic matter and respiration of vegetative roots in the soils. High rainfall and temperature promote plant growth and, in turn, bicarbonate production at the pH values prevailing in natural surface waters.

Table 7.6. Chemical Composition of Uncontaminated Stream Water in the Ko'olau Range in Honolulu, O'ahu

| Water-Quality Parameters | Concentration (mg/l) |
|---|---|
| $SiO_2$ | 10 |
| $Ca^{+2}$ | 4.0 |
| $Mg^{+2}$ | 4.0 |
| $Na^{+1}$ | 12 |
| $K^{+1}$ | 1.0 |
| $HCO_3^{-1}$ | 38.5 |
| $SO_4^{-2}$ | 3.0 |
| $Cl^{-1}$ | 17 |
| $PO_4^{-3}$ | 0.11 |
| $F^{-1}$ | 0.06 |
| $NO_3^{-1}$ | 0.30 |
| Dissolved solids calculated | 90 |
| Specific conductance | 125 µmhos at 25°C |
| pH | 7.5 |

SOURCE: Visher and Mink 1964.

Chemical denudation is a method for determining the geological age of the Hawaiian Islands (Moberly 1963). Moberly's theory is based on the rate of removal of dissolved weathering products by surface water and groundwater discharges into Kāne'ohe Bay. His estimate of the denudation rate is 34 mg/cm²/year (1.5 ton/acre/year) or 1 meter in 7,700 years for the drainage basins surrounding Kāne'ohe Bay, island of O'ahu. Extending previous studies, Li (1988) estimated the rate of total (chemical and physical) denudation to range from 11 to 50 mg/cm²/year (0.5 to 2.2 tons/acre/year) for several basins on the islands of Kauai, O'ahu, and Hawai'i. As a result of the suggested continuous input of carbonic acid in Hawai'i surface waters, Li concluded that carbonic acid is the principal chemical weathering agent of basaltic rocks in Hawai'i.

Suspended solids in natural surface waters consist mainly of mineral particles, organic matters from soils and decaying vegetation, and microorganisms (bacteria and algae). In regard to sizes, they are usually larger than 1 μm ($4 \times 10^{-5}$ in.). Colloidal solids are smaller (between $1 \times 10^{-3}$ μm and 1 μm) and are often included as suspended solids in practice. Viruses and clays are of colloidal size. Dissolved solids are particles whose size is smaller than $1 \times 10^{-3}$ μm or $4 \times 10^{-8}$ in.

The amount of total organic carbon of natural origin in typical surface waters is small and usually less than 5 mg/l. The organic compounds of natural origin such as humic acid must be distinguished from those of anthropogenic origin. The residues of synthetic organic compounds used as pesticides and cleaning solvents occur only in trace amounts in surface waters. On the other hand, the total organic carbon in domestic wastewaters is on the order of 150 mg/l.

Soil particles are a major source of suspended solids in surface water. They are first detached from the soils by the splash of raindrops and then transported by overland flow in the rills. Rills are small channels created by concentrated overland flow on the soil surface. Rills are removable by normal tillage, but large channels and even gullies can evolve from rills over time by the work of water.

The total amount of the eroded materials from both sheet and channel erosion that reach the outlet of the drainage basin is called sediment yield. A considerable fraction of these materials is, however, redeposited in the flat portions of the basin. The ratio of sediment yield to the eroded amount is called the sediment delivery ratio. The value of the ratio ranges from 1 to 20 percent and tends to be high for small basins.

Tillage operation in agricultural land can increase the renewal rate of topsoils up to 112 mg/cm²/year (5 tons/acre/year) (Bennett 1939). This value is recognized by the U.S. Natural Resources Conservation Service as the upper limit of permissible soil loss from agricultural lands. The amounts of soil losses in six small (0.4 to 2.8 ha or 1 to 7 acres) drainage basins used as sugarcane or pineapple fields on the islands of O'ahu and Hawai'i were measured over a 5-year (1972 to 1977) period (El-Swaify and Cooley 1980). The soil losses ranged from 11 to 101 mg/cm²/year (0.5 to 4.5 tons/acre/year), all within the tolerable limit. By observation, much of the sediment eroded from the steep upper portion was redeposited in the flat lower portion of the field, thus not reaching the delivery and measurement points.

Soil loss by sheet (interrill) and rill erosion in plots may be predicted by statistical regression models and physically based models. The Universal Soil Loss Equation, a regression model extensively used in the United States, is intended for predicting long-term annual soil loss from agricultural lands (U.S. Agricultural Research Service 1961). The soil loss per unit area is correlated to several factors, namely, rainfall intensity, soil type, length and slope of the field, crop management, and soil conservation practice. The soil erodibility factor in the equation reflects the soil type. For the island of Hawai'i, for example, the value of the factor ranges from 0.02 to 0.55 depending on the soil series (U.S. Natural Resources Conservation Service 2001). Wishmeier, the principal author of the equation, clarifies its use only for long-term soil erosion that does not account for redeposition (Wishmeier and Smith 1978).

The physical model initiated by the U.S. Department of Agriculture, Water Erosion Prediction Project provides continuous simulation of sheet and rill erosion (Lane and Nearing 1989). The mathematical equations of this model are formulated to express mass conservation and dynamics of water flow and soil mechanics in the rills and interrill areas. The values of the soil parameters of the model are determined in field plots subject to simulated rainfall. The Griffith University Erosion and Sedimentation System is a lumped physical model, with an analytical solution of the mathematical formulation of detachment by rain, entrainment by runoff, and deposition of soil particles (Rose et al. 1983). Lo et al. (1988) adapted the model and calibrated it with data from erosion plots in Moloka'i and Kūka'iau soil series subject to simulated rainfall. The values of model parameters including stream power and entrainment efficiency were determined by curve fitting. A comparison of these two physical models is provided by Rose (1993).

Other sources of dissolved solids and suspended solids include massive amounts of materials dislodged by mass wasting and minute amounts of anthropogenic heavy metals and organic chemicals. Mass wasting is the downslope movement of large masses of rock, soil, and vegetation. The movement is caused by the weight of the mass and initiated by small displacements of the mass. In wet regions, mass wasting is induced by high rainfall and groundwater seepage. An example is the debris flow in southern O'ahu associated with the New Year's Eve storm of 1987–1988 (Ellen et al. 1993). The sliding masses of mostly weathered rock began on steep hillslopes as landslides, changed into slurries of mud and debris, then flowed along Kūpaua Valley for a distance of 3.2 km (2 mi), and finally rested on the coastal plain in Niu Valley. The total volume of the debris flow was estimated to be 34,400 $m^3$ (45,000 $yd^3$). Heavy metals principally originate from basaltic rock and soils in Hawai'i, but lead, copper, and zinc are enriched by urban activities (Lau et al. 1972; DeCarlo and Anthony 2002). Lead was an additive in gasoline and paint, copper is a paint additive, and copper and zinc are used in brake pads and tires. Organic chemicals are used as pesticides and cleaning solvents. They are persistent. For example, DDT was first introduced in the 1940s and banned in 1972 but is still detected in trace amounts (Lau et al. 1972; Brasher and Anthony 2000). Although these anthropogenic substances occur in minute concentrations and do not usually present direct risks to human health, they may affect aquatic biota through biomagnification and bioaccumulation.

The transport of dissolved and suspended solids in surface waters in Hawai'i can be reasonably assumed to follow advection and diffusion theories. These solids should remain little altered by other natural processes during the usually short transit time (less than a few hours even during low flow) to the coastal waters. The database of dissolved and suspended solids is neither continuous nor extensive, unlike that of streamflow discharge. Occasionally, efforts were sustained for several years in research projects at the University of Hawai'i and programmatic investigations of government agencies such as the U.S. Geological Survey's National Water Quality Assessment Program. Based on the results of stream monitoring of the H-3 Highway construction project during 1983 to 1991, the concentration of suspended solids frequently exceeded the State Water Quality Standards during construction, but the Standards were also exceeded occasionally before construction (Hill 1996). Similar exceedence was experienced in other forested drainage basins of the Ko'olau Range on O'ahu in a study of the 5-year (1973 to 1978) stream gage station data (Doty et al. 1981).

A sediment-rating curve is a correlation between suspended solids discharge and streamflow discharge. It is commonly used to estimate sediment load on days when no measurements are made. Instantaneous values are often used because daily values are laborious to obtain. A daily curve for Wailuku River at Hilo, island of Hawai'i (Li 1988), and an instantaneous curve for Kamo'oali'i Stream at Kāne'ohe, island of O'ahu (Jones et al. 1971), are illustrated in Figure 7.7.

A word of caution is noted about using field values of turbidity in Hawai'i surface waters as an index of suspended solids when the turbidity is caused by iron-coated aggregates of red clays (Ekern 1976). These measurements do not accurately reflect the concentration of suspended solids for values exceeding 100 mg/l. Apparently, the departure from the linear correlation between these two variables occurs as a result of high back scatter and low transmittance of light in the solution.

Total sediment load (that is, the rate of transport of all sediments) is either the sum of bed load and suspended load or the sum of bed-material load and wash load (Shen and Julien 1993). The need for this distinction stems from the observations that some of the bed material becomes suspended by turbulence at high streamflow. The rate of movement of this material is called suspended bed-material load. Suspended load equals suspended bed-material load plus wash load. The material in wash load is composed of fine sediment particles that cannot easily be seen individually on the streambed surface. These materials are supplied from upstream sediment, that is, sediment yield of the drainage basin. Wash load is not directly related either to the flow condition at the stream reach or to the sediment composition of the streambed. Einstein (1950) selected $d_{10}$, the diameter of sediment particle on the streambed surface of which 10 percent is finer

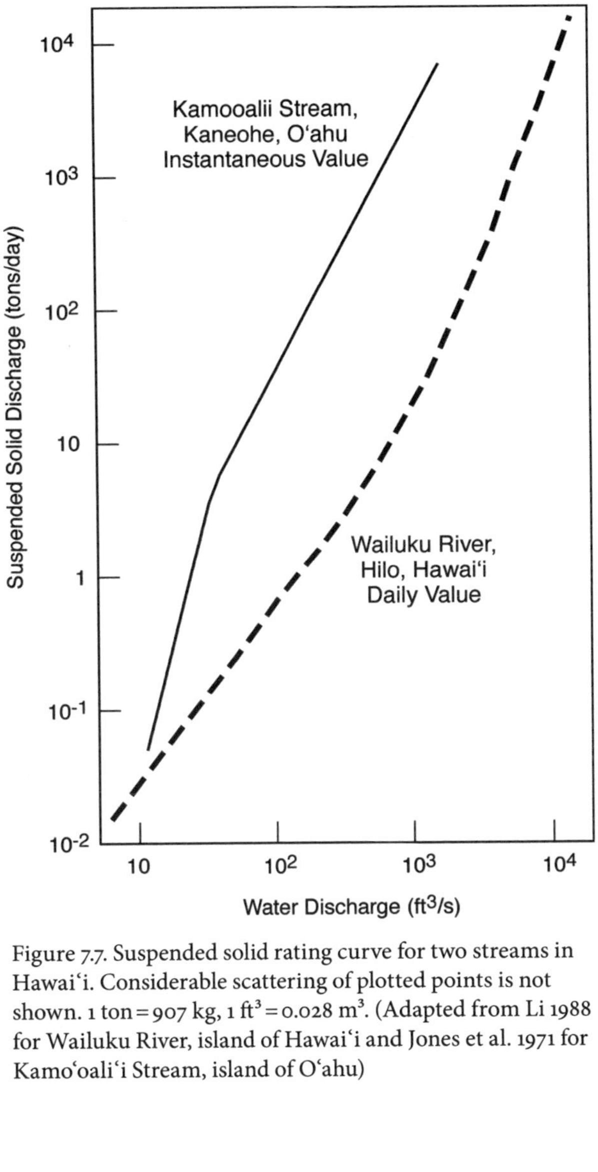


Figure 7.7. Suspended solid rating curve for two streams in Hawai'i. Considerable scattering of plotted points is not shown. 1 ton = 907 kg, 1 $ft^3$ = 0.028 $m^3$. (Adapted from Li 1988 for Wailuku River, island of Hawai'i and Jones et al. 1971 for Kamo'oali'i Stream, island of O'ahu)

than the indicated size, to define wash load. The value of $d_5$ has been suggested as a better parameter than $d_{10}$.

Transport of bed load or bed-material load in streams is reasonably understood, mathematically formulated, and tested based on the works of Einstein and others (Shen and Julien 1993). Sediment transport was investigated in Moanalua Valley, island of O'ahu, in connection with the construction of the H-3 Highway (Jones and Ewart 1973; Shade 1984). Moanalua Stream is a typical stream in the Pearl Harbor area, that is, perennial only in the upper and lower reaches. The streambed is strewn with moderately to poor sorted fine to very coarse particles of wide-ranging sizes from 1 to 1,000 mm (0.04 to 39 in.). Basic measurements were made including size distribution and specific gravity of the particles in suspended sediment and bed material for establishing some phenomenological concepts including the occurrence of wash load and suspended bed load, linkage between apparent channel meanders and massive landslides, and change in channel slope at dam-and-pool locations. The basic data were also used for the computation of bed load using various equations including that of Schoklitsch (1934). The solution to practical issues is essential, but it has been difficult to sustain continued investigations of sediment transport in Hawai'i-type streams.

## Water-Quality Standards

The current Hawai'i water-quality standards are based mainly on local experiences, although they did not start out that way. The principal regulation governing surface water-quality standards is Chapter 54, Hawai'i Administrative Rules, which is applicable to both inland and marine waters. Chapter 54 classifies state inland and marine waters and marine bottoms, assigns beneficial uses, and stipulates both narrative and numerical standards (Hawai'i Department of Health 2000). Chapter 54 is a stringent regulation. In the beginning, it was brief and uncomplicated, resulting from the work of an appointed master in 1955. It underwent substantial modifications in 1974 when regulations concerning the zone of mixing were amended, as well as in 1979 when the ecosystem approach was introduced. Two other Department of Health regulations are also relevant: potable water systems (Chapter 20) (1991a) and wastewater systems (Chapter 62) (1991b).

Inland waters can be fresh or brackish. Inland freshwaters are those in streams, springs and seeps, lakes and reservoirs, and elevated and low wetlands. Inland brackish waters are those in coastal wetlands, estuaries, and anchialine ponds (see Glossary). Marine waters are those in embayments, open coasts, and the ocean. Marine bottoms are classified as sand beaches, lava rock shorelines and solution benches, marine pools and protected coves, artificial basins, reef flats and reef communities, and soft-bottom communities.

Beneficial uses of waters are accommodated, including preservation and waste assimilation. Some of them appear incompatible at face value. Assignment of uses to various water bodies reflects social values, conflict resolutions, and, to some extent, scientific criteria. Inland water uses are classified as Class 1 and Class 2. Class 1 inland waters are reserved for preservation and as drinking-water sources. No waste discharges are allowed. Class 2 inland waters are protected for recreation and many other uses; waste discharges are permitted but only after a high degree of treatment. Inland brackish estuaries are off-limits for new industrial or sewage discharges.

Marine waters are either Class A or Class AA. Especially prohibited is new waste discharge in embayments. The prohibition reflects the lessons learned in Kāneʻohe Bay (see Chapter 8). Similar use classes for marine bottoms are Class I and II. For Class II (non-Northwestern Hawaiian Islands) marine bottom, wastewater effluent discharged from outfall structures may be allowed. This allowance reflects the experience gained from long-term monitoring in Māmala Bay, southern Oʻahu, which receives Honolulu's wastewater effluent (see Chapter 8).

For each water class, numerical standards are set for relevant quality parameters. The essential parameters are nutrients (nitrogen and phosphorus), water clarity (total suspended solids and turbidity), and the common physical and chemical traits vital to aquatic organisms (pH, temperature, salinity, and dissolved oxygen). Standards differ among the various classes of water, but within each class are two sets of standards, one for the nominal dry season (the 6 months from May 1 to October 31) and the other for the nominal wet season (the 6 months from November 1 to April 30). Three statistical parameters have been adopted: geometric mean, 10 percent nonexceedence, and 2 percent nonexceedence, as defined by lognormal distribution. Frequency analysis indicates that lognormal distribution best fits the data of many Hawaiʻi water-quality parameters, including nutrients, chlorophyll, suspended solids, and light absorption (Hawaiʻi Department of Health 1977).

Standards are also stipulated for stream bottoms. They limit newly deposited sediment from heavy rainstorms by thickness, grain size, and degree of biodegradation in the sediment. For instance, the thicknesses of new deposits are limited to less than 5 mm (0.2 in.) over hard bottoms and 10 mm (0.4 in.) over soft bottoms 24 hours after a rainstorm. The oxidation-reduction potential must not be less than +100 Mv in the top 10 cm (4 in.). In soft-bottom material in the pool section of streams, no more than 50 percent of the grain-size distribution of sediment shall be smaller than 0.125 mm (0.005 in.) in diameter, which is essentially fine-sand size. Obviously, the difficulty and practicality of monitoring and enforcing of these bottom criteria are largely beyond reasonable control.

The zone of mixing is an instrument to allow assimilation of wastewater if the discharge has received best practical treatment and if it does not interfere unreasonably with actual uses for which the water is classified. No zone of mixing is permitted in Class 1 freshwater and in Class AA marine water. Discharging into the zone of mixing is limited to a 5-year period, and permit renewal for discharging is contingent on monitoring by the dischargers and review and approval of the regulatory agency.

Estuaries including Pearl Harbor are considered brackish inland waters. Pearl Harbor is subject to a special set of standards. The water-quality standards for estuaries consist of parameters like those for streams, with minor variations such as inclusion of ammonia nitrogen and chlorophyll *a*. The latter is an indication of primary biological productivity.

Marine waters and marine bottoms are subject to different sets of water-quality standards. The three bodies of marine waters — embayments, open coastal waters, and ocean waters — are classified as either AA or A (with AA being higher quality), as delineated in Plate 7.1. Examples of Class AA waters are Kahana Bay, Kāneʻohe Bay, and Hanauma Bay.

Embayments are marine waters with restricted access to open coastal water. Numerically, embayments have a ratio of bay volume to cross-sectional entrance area of

700 or greater. Open coastal waters extend to the 183-m (600-ft) depth contour. Water-quality standards for open coastal waters are more stringent than those for embayments because open coastal waters are presumably more pristine.

Finally, marine bottom types consist of sand beaches, lava rock shorelines and solution benches, marine pools and coves, artificial basins, reef flats and communities, and soft-bottom communities. Each of these types is further classified by use as Class I or Class II. In East O'ahu, for example, Hawai'i Kai marina, an artificial basin, and Maunalua Bay, a reef flat to which the marina is connected, are Class II; whereas Hanauma Bay, also a reef flat, is Class I because it is a marine preserve. In Class II, harbors are permitted.

Where two classes of water meet, an academic problem arises. The regulated levels abruptly shift across the line, whereas the natural changes are far more gradual. The nitrate-nitrite limits for streams emptying into an embayment are 0.07 mg/l as N (dry season May 1–October 31) and 0.03 mg/l as N (wet season November 1–April 30). For embayments, the limits are 0.08 and 0.05, respectively. Further complicating the matter is the fact that the "wet" and "dry" definitions for embayments are different from those for streams. For embayments, "wet" criteria are defined not by seasons but by when the average freshwater inflow from the land equals or exceeds 1 percent of the embayment volume per day.

Besides the specific standards applicable to different waters, there are basic water-quality standards applicable to all waters. Toxicity is elaborately regulated. In total 97 chemicals are regulated by their concentrations in natural waters and in fish for human consumption. For example, the chronic toxicity standard for DDT is 0.001 μg/l (1 part per trillion) for water and 0.000008 μg/l for fish. Other chemicals commonly used or found in Hawai'i waters are also regulated. Toxicity standards are stipulated not only for receiving waters but also for effluents, such as those discharged from ocean outfalls. Heavy metals are included as toxics, but because a fraction of their concentration originates from common volcanic soils and rocks achieving the standards is often impossible.

Another important set of standards concerns microorganisms in marine recreational waters. The State of Hawai'i originally adopted fecal coliform as the standard for both inland and marine waters, assigning 200 colony-forming units (CFU)/100 ml for the geometric mean and 400 CFU/100 ml in not more than 10 percent of the samples collected in a 30-day period. A revelation came in 1983 as the U.S. Environmental Protection Agency concluded a 10-year epidemiological study of marine waters from beaches off New York City, Boston, and New Orleans (Cabelli 1983). Apparently, only enterococci in marine waters are reliably correlated with incidences of diarrheal diseases among swimmers. The State of Hawai'i not only embraced the results of the study, which excluded Hawai'i waters, but considered the Environmental Protection Agency recommendations of risks not stringent enough. The Hawai'i standards now stand at 7 CFU enterococci/100 ml, which corresponds to a risk level of 10 diseases per 1,000 swimmers (Fujioka 1996).

Soil erosion is an issue generally addressed by state regulations. Acceptable soil conservation practices are required, and any discharges must receive the best degree of treatment or control. The county governments have recently amended their soil erosion control standards (City and County of Honolulu 1996; County of Maui 1998).

In the early 1970s, certain inland and marine waters were discovered to be in violation of some of the then-applicable water-quality standards. Some of these water bodies are located in essentially undeveloped environments. A case in point is Kahana Bay and Kahana Stream, O'ahu (Lau et al. 1972; Lau et al. 1973), where a few parts per trillion of DDT was consistently detected (thirty-five out of thirty-six samples) even though the chemical was not known to have been used in Kahana. These waters would have violated the current toxicity standard for DDT in water, which is 1 part per trillion. In fact, DDT was ubiquitous in all of the study areas, including undeveloped, agricultural, urban recreational, and urban domestic lands on O'ahu and Kaua'i. Its transport is probably by air.

Other water-quality parameters also exceeded the class limits. Concentrations of both total nitrogen and total phosphorus in Kahana Bay waters were generally higher than the standards set for Class AA waters by the state. The open ocean likewise did not consistently meet the Class AA standards. It was noted in the early investigations that in all the study areas on O'ahu the maximum values of nitrogen and phosphorus that occur because of seasonal effects can exceed the standards by as much as sixfold.

Coliform organisms observed in samples taken by the State Department of Health from a small beach area entrance to Kahana Bay also indicated that the bay meets Class A standards rather than the more exacting Class AA requirements. Subsequently, these water bodies have been recognized by the state as water-quality-limited segments. They are defined as those coastal water areas that do not meet water-quality standards and will not meet applicable water-quality standards even after effluent limitation requirements are applied to point-source discharges of wastewater.

Standards must be subject to continual review as more applicable data become available. For example, a recent amendment was adopted pertaining to the specific standards of the west (Kona) coast of the island of Hawai'i only after an additional 10 years of data were secured and evaluated (Hawai'i Department of Health 2000). Water pollution may not be easily controlled or even identified unless the nature of the pollution is truly understood.

## Headwater Quality

Headwater areas in Hawai'i are typically located in high-rainfall forested mountains (Plate 7.2) and are usually in conservation zones. The two principal sources of surface water are overland runoff resulting directly from rain and groundwater discharge. The former is usually episodic and varies widely in intensity, depending on the type of rain. The latter is typically perennial with provenance from high-level dike water or perched water. The water quality of each source is substantially different and theoretically influences stream water quality according to the proportion of each. In terms of drinking water importance, the headwaters may not meet the microorganism and turbidity test of potability but most likely would pass the tests for chlorides and dissolved solids. The apparently undisturbed environment does not consistently ensure high stream water quality.

The only extant observed data of the overland flow quality in forest areas were obtained for a small area (0.73 ha [1.8 ac]) in the drainage basin of 'Aihualama Stream, a small tributary of Mānoa Stream in the leeward Ko'olau Mountains (Yim and Dugan 1975). Rainfall is very high (mean annual 3,556 mm [140 in.]). The results from seventeen rainfall episodes between September 1974 and March 1975 indicated high organic loads that were barely biodegraded in the fast-moving water, as was evident by the very low concentrations of nitrates and orthophosphates. The very high total solids were largely suspended solids (81 percent). Chlorides were typically low (13 mg/l), as were the dissolved solids. The direct overland flow in this virtually pristine area would fail to meet the current 1992 state stream water-quality standards for the most part, including the toxicity standards for heavy metals copper, nickel, and zinc (Table 7.7).

Three other studies produced water-quality data for the headwaters of four streams: three in the windward and one in the leeward Ko'olau Range (Lau et al. 1973; Dugan 1977; DeVito et al. 1995). The windward streams are Waihe'e, Luluku, and Kahana, and the leeward stream is 'Ōpae'ula-Anahulu in northern O'ahu. All basins are small, on the order of a few square kilometers, and have similar mean annual rainfall of about 2,540 mm (100 in.). In all cases, the measurements were made throughout the year. The median annual values of total nitrogen vary by one order of magnitude among the four basins (Table 7.8). The highest, which occurred in upper 'Ōpae'ula in 1992–1993, not only exceeded the geometric mean but also approached closely the 2 percent exceedence of the state water standards, which suggests the necessity for reexamination of the standards. Further, the organic nitrogen concentration is much greater than that of nitrate nitrogen. This inequality confirms the observations in Mānoa headwaters. These data show the natural variability of stream water quality in various forest reserves in the Ko'olau Mountains and likely in other reserves in Hawai'i.

Table 7.7. Overland Water Quality in a Forested Area, 'Aihualama Stream, O'ahu

| Water-Quality Parameters | | | Median Concentration (mg/l) |
|---|---|---|---|
| Physical-Chemical | | | |
| Total nitrogen | TN | as N | 2.21 |
| Total Kjeldahl nitrogen | TKN | as N | 2.10 |
| Nitrite and nitrate nitrogen | $NO_2^{-1} + NO_3^{-1}$ –N | as N | 0.03 |
| Total phosphorus | TP | as P | 0.86 |
| Orthophosphate | | as P | ND |
| Total solids | TS | | 546 |
| Total suspended solids | TSS | | 444 |
| Total dissolved solids | TDS | | 94 |
| Chlorides | | | 13 |
| Chemical oxygen demand | COD | | 235 |
| Heavy metals (dissolved) | | | (μg/l) |
| Copper | Cu | | 14 |
| Zinc | Zn | | 19 |
| Nickel | Ni | | 29 |

SOURCE: Yim and Dugan 1975.

NOTE: Median of seventeen rainfall episodes, October 1974–March 1975, drainage surface area 0.73 ha (1.8 ac), mean annual rainfall 3,556 mm (140 in.), elevation ~183 m (600 ft), terrain slope 10% to 45%, Tantalus silt loam. ND, not determined.

The Kahana basin study afforded an examination of the chemical composition (total dissolved solids, silica, major ions) of dike streams. A dike stream is one that cuts through dikes and releases dike-impounded groundwater. Dikes are prevalent along the windward Ko'olau Range, but only few exposed dikes are reported in Pearl Harbor and north of it in the leeward Ko'olau Range (Hirashima 1971; Takasaki and Mink 1981). Dike water has higher dissolved minerals, such as silica, than rain-derived direct runoff. During low streamflows, the silica concentration in windward streams between Punalu'u and Luluku ranges between 18 and 26 mg/l, whereas the concentration in leeward Ko'olau streams lacking dike water contribution is much lower, averaging only 4.3 mg/l. The chemical (mineral) quality of some dike water streams during minimum flows is indicated in Table 7.9. Typical chemical composition of waters of various origins is summarized in Visher and Mink (1964).

Rainwater may temporarily influence stream water quality, but the effects are likely slight and eventually overwhelmed by the thorough rinsing of the forest mantle. Rainwater in Hawai'i has a composition in which the concentration ratios approximate those of seawater, reflecting the sea salt nuclei around which the moisture collects to become raindrops (see Chapter 3).

Several heavy metals, lead, copper, chromium, and nickel, occur in considerable amounts in parent basalt and its weathered products and are released by solution. According to Patterson (1971), chromium concentrates in saprolite, whereas much of the lead, copper, and nickel are leached away. However, lead is complexed in the amorphous gel, which probably accounts for its sometimes unusually high concentration in suspended sediments. The heavy metals in Kahana Stream sediment and Kahana Stream water are summarized in Table 7.10. Analyses show that the solubility of heavy metals under prevailing

Table 7.8. Stream Water Quality in Forest Reserves, Koʻolau Mountains, Oʻahu

| Water-Quality Parameter | Upper Waiheʻe: Median of Monthly Medians Feb. 1974–Feb. 1975 | Luluku: Median of Monthly Medians Feb. 1974–Feb. 1975 | Kahana Median Mar. 1972–June 1973 | Upper ʻŌpaeʻula Median Apr. 1992–Apr. 1993 |
|---|---|---|---|---|
| Total suspended solids | — | — | 3.0 | 8 |
| Turbidity (NTU) | — | — | 1.7 | 4.7 |
| Total dissolved solids | — | — | — | 34 |
| Conductivity (μmho/cm) | — | — | 194 | 52 |
| $Cl^{-1}$ | — | — | 27 | — |
| pH | — | — | 7.3 | 5.8 |
| Temperature (°C) | — | — | — | 23.0 |
| Total nitrogen | 0.236 | 0.401 | 0.07 | 0.73 |
| Organic nitrogen | 0.154 | 0.226 | — | 0.54 |
| $NH_4$–N | 0.01 | 0.01 | — | < 0.05 |
| $NO_2$–N | | | — | 0.001 |
| $NO_3$–N | 0.072 | 0.165 | — | 0.10 |
| Total phosphorus | 0.033 | 0.035 | 0.01 | 0.039 |
| Orthophosphorus | 0.028 | 0.027 | — | — |
| Organic phosphorus | 0.005 | 0.008 | — | — |
| $SiO_2$ | 24.1 | 19.1 | — | — |
| Area (km²) | 2.41 | 1.14 | 9.69 | 7.72 |
| Elevation (m) | 43 | 67 | 9 | 341 |
| Median flow discharge (ft³/s) | 7.4 | 1.6 | 23 | 4.3 |
| USGS Station No. | 2840 | 2709 | 2965 | 3450 |

SOURCES: Dugan 1977 for Upper Waiheʻe and Luluku medians, Lau et al. 1973 for Kahana median, DeVito et al. 1995 for upper ʻŌpaeʻula median.

environmental conditions of Kahana Stream is very low. Only minute traces of such metals appear in the freshwater. As indicated in the tabulation in Table 7.10, heavy metals—especially lead, copper, zinc, chromium, and nickel—appear consistently and in a comparable range of a few to a few hundred milligrams per kilogram in stream sediments, watershed soils, and bay sediments. Their ubiquitous presence comes from no other sources than the parent geologic formations from which the soils and sediments are derived (Lau et al. 1973).

In summary, low streamflow generated mostly from groundwater seepage contains moderate concentrations of standard constituents that, during rainstorms, become obscured by dilution. Total nitrogen varies among streams and sometimes exceeds standards, whereas total phosphorus is low and less spatially variable. Nitrogen and phosphorus are high in particulate form and low in dissolved form. At low flow, total suspended solids are low but quickly increase when rainfall is intense. Dissolved oxygen is at or exceeds saturation, and pH is slightly acidic. Bacteria occur in moderate concentrations. Heavy metals, chiefly derived from volcanic rock and soils, are present in trace concentrations.

Table 7.9. Chemical Water Quality of Dike Streams at Minimum Flow, Ko'olau Mountains, O'ahu

| | Concentration (mg/l) | | |
|---|---|---|---|
| Water-Quality Parameter | Kahana Stream (USGS Station No. 2965) | Waihe'e Stream (USGS Station No. 2840) | Luluku Stream (USGS Station No. 2709) |
| Total dissolved solids (less $SiO_2$) | 74 | 88 | 96 |
| $Na^{+1}$ | 10 | 11 | 12 |
| $K^{+1}$ | 0.5 | 0.8 | 0.9 |
| $Ca^{+2}$ | 5.5 | 6.6 | 8.0 |
| $Mg^{+2}$ | 4.0 | 4.0 | 5.3 |
| $Cl^{-1}$ | 14 | 14 | 17 |
| $NO_3^{-1}$–N | 0.03 | 0 | 0 |
| $HCO_3^{-1}$ | 39 | 46 | 50 |
| $SO_4^{-2}$ | 2.2 | 3.0 | 3.6 |
| $SiO_2$ | 18 | 25 | 24 |

SOURCE: U.S. Geological Survey 1970 in Lau et al. 1973.

Table 7.10. Heavy Metals in Kahana Stream and Kahana Stream Sediments

| Constituent | Kahana Stream Water (mg/l) | Chronic Toxicity for Freshwater[a] (mg/l) | Kahana Stream Sediments (mg/kg) |
|---|---|---|---|
| Pb | ND–0.010 | 0.029 | 34–94 |
| Cu | 0.001–0.011 | 0.006 | 29–190 |
| Zn | ND–0.024 | 0.022 | — |
| Cd | ND–0.003 | 0.003 | — |
| Hg | ND | 0.00055 | — |
| Cr | ND–0.014 | 0.011 | 24–112 |
| Ni | ND–0.019 | 0.005 | 178–449 |

SOURCE: Lau et al. 1973.
NOTE: ND, not determined.
[a]Hawai'i water-quality standards (1993).

## Land-Use Effects

Nonpoint-source pollution refers to alteration of water quality by contaminants that originate from developed land surfaces. Wet weather washes out the contaminants to cause storm runoff pollution. Because they are diffuse, these sources are difficult to eliminate.

In Hawai'i, developed land uses are zoned as urban, agricultural, or rural. Contaminant sources have been identified by examining street sweepings, sampling street manholes, and collecting overland flow. For large land areas conventional upstream-downstream investigations have been conducted. These investigations involve assessing the quality of stream water at a point immediately downstream from an existing land development and comparing it with upstream quality. For many Hawai'i studies, the upstream sampling locations are at the forest reserve border and the downstream locations at the coastal terminus of the stream.

### *Urban land use*

Street refuse and litter can cause water pollution. The debris is composed of dust and dirt and coarse materials including vegetation, rock, paper, wood, metal, and glass. The soluble portion of the dust and dirt fraction produces the most potential for biochemical oxygen demand, chemical oxygen demand, nitrogen, phosphorus, and bacteria (see Glossary for biochemical oxygen demand and chemical oxygen demand). If not removed, street litter may be washed off by storm runoff and enter the storm drain via street curb inlets.

Fujiwara (1973) sampled and analyzed the quality of urban storm water runoff collected from storm drains in residential, commercial, and industrial areas in Honolulu. Average water-quality concentrations from his study are presented in Table 7.11. Residential and commercial areas contributed high levels of organics as indicated by chemical oxygen demand. Fecal contamination was mainly derived from animals (fecal coliform/fecal streptococci ratio being less than 1.0 [Geldreich 1976; Fujioka 1983]). The high lead concentration presumably came from automobile exhaust emissions. The biochemical oxygen demand values are within the range (10 to 20 mg/l) best attainable in medium wastewater effluent after secondary treatment (McGauhey 1968).

Street sweeping and analysis of the dust and dirt fraction were done for both Honolulu and Kāne'ohe (Chun et al. 1972; Nakamura and Young 1974; Lau et al. 1976), and the results of all studies were evaluated by Young (1978). The studies indicate that the greatest potential pollutant contributions are suspended solids and bacteria. When compared with results reported in an earlier study for Chicago (American Public Works Association 1969), the Honolulu dust and dirt fraction contained more soluble material, higher phosphate but lower nitrogen levels, and higher biochemical oxygen demand but lower chemical oxygen demand levels.

In the Kāne'ohe investigation, solids from single- and multifamily residential and commercial streets were collected by sweepings. Only solids, nitrogen, phosphorus, heavy metals, and pesticides data were obtained from the fine fraction of collected material. There was no consistent pattern to the data, as was also noted in the earlier effort for Honolulu. Solids proved again to be an important quality factor. Nitrogen was present mostly in organic form, with the highest levels in samples from residential areas. Lead showed the greatest heavy metal concentration. The lead content was one to two orders of magnitude higher than that determined for stream sediments examined in this study. Levels of chromium and mercury, however, were within the range of values reported for the sediment and may reflect the street dust derived from natural soils.

Some modeling studies were also carried out to determine urban runoff loadings. Nakamura and Young (1974) attempted to correlate pollution associated with street litter for Honolulu streets (Chun et al. 1972) and with baseline quality data for Kalihi Stream (Matsushita and Young 1973). Correlations were made between street-sweeping quality and length of street, land use, street surface traits, and rainfall intensity to yield an estimate of the concentration of pollutants in urban storm water. Results were comparable with measured values in Kalihi Stream, but the study suffered from the limited database. Long-suspected contaminants were documented in a field survey in the storm drainage system of West Maui. They include oils and illicit discharges such as landscape debris, swimming pool filter water, cleaning fluids, and paints (Hawai'i Department of Health 1997).

A number of drainage basins have been investigated by upstream-downstream surveys. All of these basins drain both the leeward and windward slopes of the Ko'olau Mountains. The basins include all land-use types: conservation, urban, rural, and agricultural. Collectively and individually, these surveys provide a voluminous amount of data. One of the first efforts was a study of Kāne'ohe Bay

Table 7.11. Representative Urban Storm Water–Quality Data for Honolulu

| Parameter | Residential[a] | Commercial[b] | Industrial[c] |
|---|---|---|---|
| | (mg/l) | (mg/l) | (mg/l) |
| Total solids | 511 | 278 | 246 |
| Suspended solids | 252 | 142 | 12 |
| COD | 142 | 209 | 40 |
| $BOD_5$ | 10 | 19 | 7 |
| Dissolved oxygen | 7.1 | 5.7 | 6.7 |
| $NO_3$–N | 0.211 | 0.045 | 1.1 |
| Total Kjeldahl nitrogen | 0.381 | 0.272 | 2.70 |
| Total phosphorus | 0.57 | 0.53 | 2.17 |
| Orthophosphate | 0.27 | 0.19 | 1.27 |
| Grease | 2.8 | 1,919 | 2.2 |
| Lead | 0.407 | 0.987 | 1.657 |
| Chromiun | 0.013 | 0.021 | 0.013 |
| Zinc | 0.512 | 0.792 | 0.729 |
| Copper | 0.036 | 0.036 | 0.021 |
| Iron | 0.377 | 0.295 | 0.049 |
| | (no./100 ml) | (no./100 ml) | (no./100 ml) |
| Total coliform | 83,300 | 33,500 | 11,500 |
| Fecal coliform | 1,965 | 463 | 580 |
| Fecal streptococcus | 6,393 | 7,900 | 7,350 |

SOURCE: Fujiwara 1973.

NOTE: COD, chemical oxygen demand; $BOD_5$, 5-day biochemical oxygen demand.

[a]Storm water samples collected at Aupuni Street near Nuhelewai Stream.

[b]Storm water samples collected at Beretania Street between Maunakea and River Streets.

[c]Storm water samples collected near Iwilei and Pacific Streets.

and its tributary streams, wherein a full suite of water-quality parameters was monitored for 1 year to determine the concentration and loading (Cox et al. 1973). The pioneering work was followed by more-detailed studies of Kāne'ohe Bay, including investigations by the Hawai'i Environmental Simulation Laboratory (Dugan 1977) and Kāne'ohe Bay Urban Water Resources Study team (Young et al. 1975; Lau et al. 1976).

The more-sensitive parameters and their numerical ranges were identified. Some early leeward Ko'olau studies reported many first findings that were later confirmed and explained. The Kalihi Stream study showed that many water-quality parameters occurred in the same order of magnitude as those reported elsewhere in the mainland United States and other countries, that the concentration of some but not all water-quality parameters increased downstream of urban areas, that concentration hydrographs display the phenomenon of first flushing in the rising phase of flow during a rainstorm, and that pesticides were detected in the parts-per-trillion range (Matsushita and Young 1973).

More investigations confirmed that the concentra-

Table 7.12. Water-Quality Concentration and Loading in Streams Draining Basins of Different Land Use in Kāneʻohe, Oʻahu

| Water-Quality Parameter | Kamoʻoaliʻi Urban Residential | Luluku Conservation | Lower Waiheʻe Rural Agriculture |
|---|---|---|---|
| Concentration (median of monthly median) (mg/l) | | | |
| Nitrogen | | | |
| Total nitrogen | 0.450 | 0.401 | 0.723 |
| Organic nitrogen | 0.180 | 0.226 | 0.318 |
| $NH_4$–N | 0.02 | 0.01 | 0.17 |
| $NO_2 + NO_3$–N | 0.250 | 0.165 | 0.235 |
| Phosphorus | | | |
| Total phosphorus | 0.024 | 0.035 | 0.088 |
| Organic phosphorus | 0.006 | 0.008 | 0.009 |
| $PO_4$–P | 0.018 | 0.027 | 0.079 |
| Loading (1974–1975) (lb/ac/yr) | | | |
| Total nitrogen | 3.73 | 3.81 | 9.50 |
| Total phosphorus | 0.20 | 0.33 | 1.16 |
| Suspended solids | 223 | — | 555 |

SOURCES: Dugan 1977 for nitrogen and phosphorus series; Lau et al. 1973 for suspended solids.

NOTE: 1 lb/ac/yr = 1.12 kg/ha/yr.

tion and loading of nitrogen and phosphorus in urban-residential surface runoff are not unusually distinct from those generated under natural undisturbed conditions, as exhibited for Kamoʻoaliʻi and Luluku Streams (Table 7.12) (Dugan 1977). An interesting observation is that urban-residential areas contribute a somewhat lower concentration and load of nitrogen and phosphorus than rural-agricultural areas.

The effects of land use on runoff can be small enough to show dilution instead of increase of the original concentrations. This was observed in an urban basin (10.50 km$^2$ [4.05 mi$^2$]) drained by Kamoʻoaliʻi Stream into southern Kāneʻohe Bay. The total phosphorus concentration was reduced from 0.035 mg/l upstream to 0.024 mg/l downstream. Addition of water flow from urban land was considerable: the upstream discharge was 0.04 m$^3$/s (1.36 ft$^3$/s); the downstream discharge was 0.33 m$^3$/s (11.8 ft$^3$/s). Both groundwater seepage and overland flow contributed to the flow increase (Dugan 1977).

Hawaiʻi streams are not only flashy in flow but also in water quality in response to intense rainfall, as shown by water-quality hydrographs. The water-quality traits of a single storm event on Kamoʻoaliʻi Stream (January 31–February 2, 1975) are presented in Figure 7.8 (Dugan 1977). In general, the results revealed that turbidity is directly related to discharge, but conductivity, silica, and nitrite-nitrate nitrogen are inversely related—an apparent dilutional effect of the base flow. Ammonia nitrogen and phosphate stay fairly steady. The 1975 event represents a 2-year flow (i.e., the flow would be statistically expected to reach this magnitude once every 2 years).

Several other water-quality parameters appear to be widely distributed; among them are pesticides including $\alpha$ and $\gamma$ chlordane, dieldrin, DDT, and PCP in a few nanograms per liter (parts per trillion) in the water and one to two orders of magnitude higher in stream sediments (Lau et al. 1973). In a 1998 water-quality assessment, the first three of these chemicals that were analyzed contin-

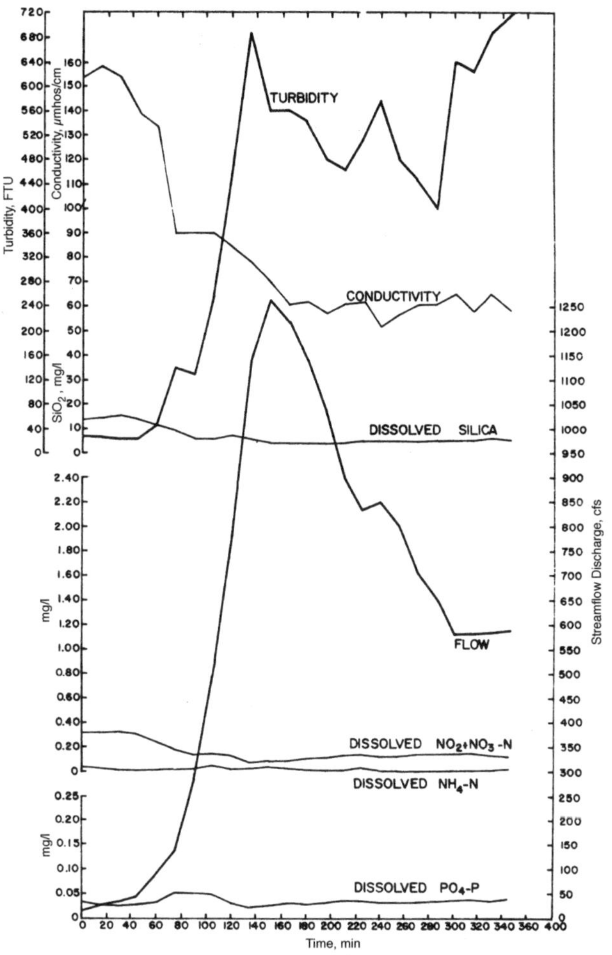

Figure 7.8. Water-quality and discharge hydrographs, Kamoʻoaliʻi Stream, Oʻahu, January 31 – February 2, 1975. USGS Station No. 2739. (Reprinted by permission from Dugan 1977)

ued to persist, as expected, in streambed sediments and also in fish tissues in several drainage basins in urban and agricultural use on Oʻahu (Brasher and Anthony 2000). Fecal coliform and fecal streptococci are common but not necessarily of human origin, as evidenced by the fecal coliform/fecal streptococci ratio analysis.

### *Agricultural land use*

For more than a century, sugarcane fields yielded excess surface water known as tailwater from the furrow method of irrigation. In the 1970s, a few investigative studies were conducted to identify the water-quality traits and pollution potentials of such practice. Today, the issue is less important because the plantations have been reduced to only one each on Kauaʻi and Maui and especially since the irrigation method was converted to drip irrigation, which does not produce any tailwater. However, during the early stage of growth when vegetal cover is sparse, the land surface is still susceptible to rainstorm-induced water-quality consequences. Pineapple fields are still plentiful in Hawaiʻi, but because they are drip irrigated, their water-quality issues have been with groundwater rather than with surface water (see Chapter 6). Other agricultural practices have not been known to induce serious water-quality consequences.

As part of a study of storm runoff and irrigation tailwater, samples were collected from a small (0.57-ha [1.4-ac]) plot in a gently sloping sugarcane field cultivated in Kunia silty clay soil at Kunia, Oʻahu, in 1973–1974, a relatively low rainfall year (1,194 mm [47 in.]) at the site (Yim and Dugan 1975). The medians of runoff water from twenty episodes indicated total suspended solids to be moderately high (46 mg/l); total dissolved solids, moderately low (85 mg/l); chloride, low (11 mg/l); chemical oxygen demand, low (18 mg/l); total nitrogen, moderately high (0.40 mg/l); and total phosphorus, high (0.42 mg/l). A stream containing nothing else but this overland flow would exceed water-quality standards, but if compared with the headwaters in ʻAihualama Stream in the rain forest of Mānoa basin, the Kunia sugarcane land would be considered a mild generator of water-quality parameters.

On a larger drainage basin scale, Anahulu and ʻŌpaeʻula Streams in Waialua, Oʻahu, were sampled manually biweekly and automatically during high-flow events. Streamflow was monitored at an upper station bordering the forest and at a lower station at the terminus of sugarcane agriculture from March 1992 to April 1993, a period of relatively low rainfall (55 to 69 percent of the long-term mean [DeVito et al. 1995]). The water passing the downstream stations had a higher concentration of turbidity, total nitrogen, and phosphorus than that passing the upstream stations. The sugarcane fields were mostly (90 percent) grown by drip irrigation, which results in smaller amounts and higher concentrations of many parameters in the return water. The downstream nitrogen, phosphorus, and suspended solid concentrations exceeded the 1992 water-quality standards, but so did the upstream nitrogen and phosphorus concentrations (Table 7.13). Statistical correlations were established between streamflow discharge during rainstorms and many water-quality parameters. For example, the concentrations of total suspended solids, turbidity, and total phosphorus increased with the discharge rate, some exponentially and some linearly. No causal explanations were offered, but soil erosion may have taken place. All nine organic chemicals were below the level of analytical detection for dry and wet weather. They include atrazine, ametryn, and diuron—all of which are applied as pesticides in sugarcane culture. The importance of the particulate form of nitrogen and phosphorus as nutrients in O'ahu stream water was confirmed in a later study of four streams in the windward Ko'olaus and 'Ōpae'ula and Poamoho Streams in the leeward Ko'olaus (Hoover 2002).

### *Rural land use*

Rural drainage basins may add large amounts of nitrogen, phosphorus, and suspended solids to surface runoff, thus raising the concentration of these parameters in the receiving streamflow. This is evidenced in the Waihe'e Stream data expressed as the median of monthly medians for the period from February 1974 to February 1975 (Dugan 1977). As the streamflow discharge increased from 0.14 $m^3/s$ (4.81 $ft^3/s$) at the upstream forest border to 0.27 $m^3/s$ (9.65 $ft^3/s$) when departing the rural area (3.44 $km^2$ [1.33 $mi^2$]) and emptying into Kāne'ohe Bay, the pickup of total nitrogen raised the concentration from 0.236 mg/l to 0.723 mg/l. The accrual of total phosphorus elevated the concentration from 0.034 mg/l to 0.088 mg/l. Rural and urban basins were compared in terms of loading of the water-quality parameters (Lau et al. 1976). The rural basins were in Waiāhole and Waikāne, which drain the northern part of Kāne'ohe Bay. On the basis of per-unit surface area of the basins, the stream loading of total nitrogen and phosphorus and suspended solids in these rural areas turned out to be actually higher than that from the urban basins in the southern part of the bay. During the wet-weather months (November 1975 to May 1976), these loadings were, respectively, 0.016, 0.0041, and 0.65 lb/ac/d for the rural basins vis-à-vis 0.015, 0.0015, and 0.46 lb/ac/d for the urban basins (1 lb/ac/d = 1.12 kg/ha/d).

## Lakes and Reservoirs

Few natural lakes exist in Hawai'i. The most unusual is Lake Waiau, an alpine lake at an altitude of 3,968 m (13,020 ft) on the island of Hawai'i. Other natural lakes include Waiākea Pond, Hawai'i; Kanahā Pond, Maui; and Halāli'i Lake and Halulu Lake, Ni'ihau. Large reservoirs are few in number also, but they are of major utility and include Wahiawā reservoir, Nu'uanu reservoir, and Ho'omaluhia reservoir on O'ahu, and others on Moloka'i and Kaua'i.

The ecological health of Wahiawā reservoir has been a subject of controversy and has been intensively investigated at various times (Young et al. 1975; Moore et al. 1981; Lum and Nakasone 1997). Damming the Kaukonahua Stream in Wahiawā in 1906 created a deep (24.4 m [80 ft]) and voluminous (11.4 million $m^3$ [3 billion gal]) water body. The three major uses—irrigation, fishing, and wastewater disposal—are not seasonally compatible, as occasionally manifested in major fish kills, such as when 90 percent of the total fish population expired in November 1962 during heavy draft for irrigation. Two more fish kills took place in November 1968 and September 1972.

The 1975 study examined hydrology, water quality, algal growth potential, fish bioassay, and sediment quality. The 1981 study modeled flow, dissolved oxygen, and temperature and tested the reservoir responses to several water-quality management options. Artificial aeration appeared to be the most effective strategy, as suggested by

Table 7.13. Quality of Stream Water Passing Sugarcane Fields, O'ahu, Hawai'i

| Parameter | Upper Anahulu Stream | | Lower Anahulu Stream | | Upper 'Ōpae'ula Stream | | Lower 'Ōpae'ula Stream | |
|---|---|---|---|---|---|---|---|---|
| Flow ($ft^3/s$)[a] | 40 | (55) | 21 | (54) | 18 | (78) | 2.1 | (55) |
| Median concentration | | | | | | | | |
| Total suspended solids (mg/l) | 4 | (56) | 7 | (52) | 8 | (66) | 7 | (54) |
| Turbidity (NTU) | 3.1 | (57) | 9.1 | (54) | 4.7 | (79) | 6.7 | (56) |
| Total dissolved solids (mg/l) | 42 | (55) | 72 | (52) | 34 | (66) | 107 | (54) |
| Conductivity (μmho/cm) | 52 | (56) | 78 | (53) | 52 | (77) | 81 | (55) |
| pH | 7.09 | (57) | 7.05 | (54) | 5.84 | (78) | 6.98 | (56) |
| Temperature (°C) | 23.0 | (27) | 23.0 | (28) | 23.0 | (31) | 22.5 | (32) |
| Total nitrogen (mg/l) | 0.55 | (35) | 0.75 | (28) | 0.73 | (40) | 1.25 | (27) |
| Total phosphorus (mg/l) | 0.064 | (39) | 0.080 | (38) | 0.039 | (47) | 0.176 | (36) |

SOURCE: DeVito et al. 1995.

NOTE: Basin areas 3,940 ha (9,736 ac) for Anahulu and 1,551 ha (3,834 ac) for 'Ōpae'ula. Numbers in parentheses are numbers of water samples analyzed over the monitoring period, March 1992–April 1993.

[a] 1 $ft^3/s = 0.02832\ m^3/s$.

the model. However, costs and benefits derived favored the alternative of tertiary treatment of the wastewater. Overgrowth of water hyacinth in the reservoir was reported in 1997, adding one more problem in managing this nutrient-enriched body of water.

## Stream Biota: Fishes

Gobies are freshwater fishes, called *'o'opu* in the Hawaiian language, present in some perennial streams. They are drab and small, with the male on the order of 10 cm (3.9 in.) in length and the female, 8.5 cm (3.3 in.). They are diadromous, that is, they are confined to freshwater as adults yet must spend their larval lives at sea as members of the marine zooplankton community. The pelvic fins of some gobies form a strong suction disk, enabling them to cling to and climb rocks. Their habitats span all reaches of perennial streams.

One of the species, *Lentipes concolor,* is endemic to Hawai'i (Plate 7.3). A statewide survey revealed that it is extinct on O'ahu but quite abundant on the other major islands. Physical water-quality parameters (dissolved oxygen, pH, and temperature) do not differ much between streams with *Lentipes* and streams without *Lentipes*. However, water turbidity is mostly absent to slight in streams with *Lentipes* but slight to high in streams without *Lentipes* (Timbol et al. 1980).

Many scientific observations have been made of this and two other goby species (*Awaous stamineus* and *Sicyopterus stempsoni*) in three sizable streams: Wainiha River on Kaua'i, Hanawī Stream on Maui, and Nānue Stream on Hawai'i Island. Analysis of the resulting data revealed preference and utilization of habitats as defined by parameters, including stream regime (pools, riffles, runs, falls), position (center, side, margin), velocity (incremental up to 1.39 m/s [4.6 ft/s]), water depth (incremental up to 1.37 m [4.5 ft]), and substratum (sand, gravel, cobble, boulder, bedrock) (Kinzie 1988).

Basically, the preference curves identify an optimal streamflow discharge for fish habitat and quantify habitat loss associated with incremental reduction of streamflow. Robust discharges also diminish habitat. However, habitat-utilization data are stream specific and cannot be reliably transferred from one stream to another (Kinzie et al. 1986). This conclusion was later disputed by Hoover (2002).

## Minimum Streamflows

Under the State Water Code (Hawai'i Revised Statutes 1987), the Commission on Water Resource Management is charged with the responsibility of establishing and administering a statewide in-stream use protection program as well as in-stream flow standards on a stream-by-stream basis "whenever necessary to protect the public interest." The standards established by the commission shall specify minimum flows or depths of water required in a stream at given times of the year to "protect fishery, wildlife, recreational aesthetic, scenic, and other beneficial in-stream uses."

According to the code, beneficial in-stream uses include but are not limited to:

1. Maintenance of fish and wildlife habitats
2. Outdoor recreational activities
3. Maintenance of ecosystems such as estuaries, wetlands, and stream vegetation
4. Aesthetic values such as waterfalls and scenic waterways
5. Navigation
6. In-stream hydropower generation
7. Maintenance of water quality
8. Conveyance of irrigation and domestic water supplies to downstream points of diversion
9. Protection of traditional and customary Hawaiian rights

The establishment of permanent in-stream standards is an enormous undertaking, although the commission has wide latitude in determining whether any given stream should require standards.

To preserve the environment in a perennial stream, some level of minimum flow is necessary. In establishing the minimum, flow characteristics need to be identified and ambient ecology understood. As a general rule, in-stream values are important only for perennial streams.

A perennial stream or reach of a stream needs to be identified. One approach is to choose a flow parameter that stipulates that there exists a lowest average flow over a specified period (low flow). The 14-day (consecutive) lowest average flow for the period of record means that in the entire period of record, the lowest average flow for any 14-day period is the listed value. If the value is zero, then in at least one interval of 14 consecutive days the stream had no running water. Such an index can be used to identify truly perennial streams.

A 14-day average minimum flow greater than zero probably would be required to sustain an ecology dependent on running water. However, continuation of such a low average flow indefinitely would likely result in ecological damage. The minimum flow necessary to sustain the biology of a stream undoubtedly exceeds the 14-day low yet is probably not as great as, say, the median flow. Perhaps a value in the range between the 14-day low and the base flow (usually taken as the 90 percentile exceedence) or, to be safer, the 75 percentile exceedence would be a reasonable compromise.

The average flow is usually about the same as the 25 percentile exceedence flow, which means that flows equal to or greater than the average occur only 90 days in a typical year. The other 275 days have lower flows. The average, therefore, is a poor parameter to employ in establishing minimum flows. Even the median is not reasonable. A minimum flow that assures continuation of a stream ecology is likely to lie between the 14-day average low flow and an exceedence percentile greater than the median but less than the 90 percentile exceedence.

Sensible use of the extensive streamflow data collected principally by the U.S. Geological Survey over the past 80 years provides the basis for establishing the physical parameters of minimum streamflow. The proposed stream classification system in the State Water Protection Plan may serve as the framework for utilizing the statistics (Yuen and Associates 1992).

A preliminary appraisal of statewide stream resources has provided an inventory of available information on

Hawai'i stream resources. Twenty-four streams were identified to possess three or more outstanding stream resources (Hawai'i Cooperative Park Service Unit 1990; Hawai'i Commission on Water Resource Management 1993). Eleven windward O'ahu Ko'olau streams were evaluated for various in-stream uses, including fauna (fish) habitat, waterbird habitat, aesthetic enjoyment, water-based recreation, and other criteria based on extant data and reconnaissance field observations (Wilson Okamoto and Associates 1983). Various aquatic surveys were recently conducted in several windward Ko'olau streams. Many more studies are needed, however, to lay the foundation for effective in-stream use management.

## Hawai'i Data and Models

### Data

Measurements of stream and diverted flows may have started with the ancient Hawaiians because their agriculture depended on an adequate quantum of water for irrigation. When the sugar industry began its rapid growth more than 100 years ago, the volume of water potentially available for irrigation had to be estimated before investment in diversion schemes could be justified. For example, measurements of all major streams in the Waipi'o Valley drainage basin, island of Hawai'i, were made by J. M. Lydgate in 1889 preparatory to planning the Lower Hāmākua Ditch. Similar instantaneous and short-term measurements were made of stream water sources on the other islands.

Continuous-flow recording stations were not established until 1909 when the U.S. Geological Survey (USGS) began its stream measurement program. Since then the USGS has amassed a large, extremely valuable database of both continuous and stage streamflow, as well as of other parameters. In parallel with the USGS, plantations and landowners maintained measuring stations at ditch intakes and at land boundaries to record the volume of water originating on private and government lands. The USGS data are the most complete. Table 7.14 presents a summary of the number of continuous-flow gaging stations maintained over the 80 some years of record by the USGS, along with the numbers in service in 1979, 1988, and 1994. Locations of gaging, water-quality, and partial-record stations on all major islands are provided in the USGS water resources data issued annually (Fontaine et al. 1997). Streamflow discharge is not continuously measured. Instead, it is obtained by stage-discharge relationship or rating curve (Appendix 7.1).

Hydrologic parameters become more stable with increasing length of record. The standard error is commonly used for the purpose of determining the length of record needed for the flow statistics to stabilize. The one proposed by Hardison (1969) is expressed as $100\ (Cv)/N^{0.5}$ where $Cv$ is the coefficient of variation, or the ratio of standard deviation to the mean, and $N$ is the number of events used in computing the statistics. This formulation shows that a combination of variability and a short record results in high standard error and that the standard error approaches an asymptotic value with a longer record. The asymptotic value is achieved with a 20-year record for annual discharges but requires a 50-year record for the 100-year flood statistics, as indicated in a study of selected O'ahu stream gaging stations (Fontaine 1996).

Records of surface water quality must have started with the first use of stream waters as sources for drinking water. Raw water quality is routinely checked for bacteria, salinity or chlorides, and turbidity. Databases are kept at the county departments of water supply and the State Department of Health. The list of parameters has been continually expanded as water-quality standards have become increasingly stringent.

In the late 1960s, both the University of Hawai'i and the USGS began their respective surface water-quality programs. In 1994 the USGS sites numbered nineteen, fourteen of them on O'ahu, two on Kaua'i, and one each on Moloka'i, Maui, and Hawai'i. The USGS publishes both flow and quality data annually (Fontaine et al. 1997). The University of Hawai'i programs are research projects focusing on the Ko'olau Mountain region, O'ahu; many of them are cited in this chapter. Water-quality data, which are relatively sparse for neighbor islands and

Table 7.14. Continuous Streamflow Gaging Stations in Hawai'i

| Period | Ni'ihau | Kaua'i | O'ahu | Moloka'i | Maui | Lana'i | Hawai'i | Total |
|---|---|---|---|---|---|---|---|---|
| Total 80 yr (1909–1988) | 0 | 137 | 88 | 23 | 98 | 0 | 60 | 406 |
| 1979 | 0 | 26 | 32 | 10 | 18 | 0 | 24 | 110 |
| 1988 | 0 | 19 | 29 | 8 | 7 | 0 | 15 | 78 |
| 1994 | 0 | 19 | 34 | 8 | 10 | 0 | 18 | 89 |

SOURCE: Fontaine et al. 1997.

the Wai'anae Mountain region on O'ahu, are especially needed for sites of emerging urban centers and diversified agriculture.

## Models

### Rainfall-Runoff Correlations: Annual Basis

The response of runoff to rainfall on annual and regional bases in Hawai'i is usually cast as a simple linear equation, RO = $a' + b'P$ (Giambelluca 1983; Shade and Nichols 1996), or in the form of a power equation, RO = $bP^n$ (Hirashima 1971), where $a'$, $b'$, $b$, and $n$ are constants based on correlation analysis of data. A different correlation equation using quadratic power appears to be an improvement (Yuen and Associates 1992).

Correlations can be made between rainfall and runoff by way of a variety of equation types. As a matter of last resort, a correlation can be expressed as an nth order polynomial. Every calculated correlation, however, is a statistical artifice whose descriptive worth is limited to the range of values employed in the computations, even though correlations are often used to make predictions outside the limits of the data range.

Although power and linear equations are not the only correlations that enjoy high regression coefficients, they are the ones most often employed in Hawai'i. The resulting correlation equation does not express actualities of the system under investigation. For example, in a power equation the exponent and multiplier constants have no concrete meaning and are merely numbers that force the correlation; they are not derived from a fundamental physical principle. The same is true for a linear equation. Both types of equations often have a positive intercept, which is impossible because without rain there can be no direct runoff, and groundwater will decay to zero unless its origin is outside the system.

Average direct runoff is expressed as a function of average annual rainfall in the Pearl Harbor region of southern O'ahu (Hirashima 1971). In this region, except near the coast where basal springs enter the streams, total runoff is equivalent to direct overland flow plus temporary bank storage. High-level aquifers that drain to streams do not occur on the leeward side of the Ko'olau Mountains in the Pearl Harbor sector, greatly simplifying the rainfall-runoff relationship.

Hirashima calculated the correlation of annual values (in inches) to be

$$\mathrm{RO} = 0.0266P^{1.37} \tag{7.2}$$

on the basis of the flow records for Hālawa, Kalauao, Waimalu, Waikele, Kīpapa, and Moanalua Streams. The average annual rainfall for these basins ranges from 1,727 to 4,318 mm (68 to 170 in.). Had he included the Kaukonahua Stream data in the very high rainfall region (average maximum 6,350 mm [250 in.] per year) between Wahiawā and the Ko'olau crest, the curve would not have been so flat at the high rainfall end.

The power expression calculated by Hirashima suggests that the rate of change of direct runoff with respect to the change in rainfall increases with some function of rainfall. At 2,540 mm (100 in.) average annual rainfall, the average direct runoff is 381 mm (15 in.), whereas at 5,080 mm (200 in.) it is 965 mm (38 in.), nearly two and a half times as great. If the correlation were cast as a straight line, runoff at 5,080 mm (200 in.) would be only twice that at 2,540 mm (100 in.). Still, the Hirashima curve is so gentle that at 6,350 mm (250 in.) average annual rainfall in the upper Kaukonahua drainage the computed direct runoff would be just 1,303 mm (51.3 in.). In fact, the Kaukonahua records show that average annual runoff is 3,810 mm (150 in.) where average annual rainfall is 6,350 mm (250 in.) and 2,159 mm (85 in.) where it is 5,080 mm (200 in.).

A different correlation for central and southern O'ahu yields the following expression for direct runoff (Yuen and Associates 1992):

$$\mathrm{RO} = 0.0021P^2 \tag{7.3}$$

It is based on stream records through 1988 for the following undiverted streams: North Fork Kaukonahua, Poamoho, Kīpapa, Waiawa, Waimalu, Kalauao, and North Hālawa. The correlation was forced to yield a quadratic expression, yet the regression coefficient is very high ($r^2 = 0.8750$; $r = 0.9354$).

The correlation is structured as a quadratic to satisfy a physical principle. By making the power exponent to be 2, the equation follows from the relationship in which the rate of change in runoff to the change in rainfall is proportional to the rainfall at the time the change takes place. This is intuitively logical and is expressed as follows:

$$d(\mathrm{RO})/dP = 2bP \tag{7.4}$$

which on solution becomes

$$\mathrm{RO} = bP^2 \tag{7.5}$$

The power equation, exponent 2, gives results similar to the Hirashima power equation and the Giambelluca-Shade linear equation for average annual rainfalls of less than about 2,540 mm (100 in.). For example, for average annual rainfall of 1,803 mm (71 in.) the direct runoff amounts to 267 mm (10.54 in.) by the linear equation, 235 mm (9.28 in.) by the Hirashima equation, and 275 mm (10.82 in.) by the quadratic power equation.

Figure 7.9 shows the quadratic power correlation equation for central and southern O'ahu. This basic relationship between direct runoff and rainfall on an annual basis is useful for estimating direct runoff in most natural Hawai'i terrains that are eroded to a surface consisting of soil and subsoil overlying a mantle of saprolite. The equation does not yield a good estimate for many regions on the island of Hawai'i where comparatively recent volcanic activity has not resulted in much soil-saprolite cover. In these regions little runoff takes place because the surface lavas have extremely high infiltration characteristics.

Virtually all perennial streams, except for headwater tributaries in high-rainfall areas, receive base flow from groundwater; thus the total flow consists of two components, direct runoff and base flow. The simple direct runoff equation is not applicable as a complete expression for these streams. Options for the total runoff (TRO) equations are

$$\mathrm{TRO} = B + bP^n \tag{7.6}$$

and

$$\mathrm{TRO} = cP + bP^n \tag{7.7}$$

The first equation assumes that the base flow, $B$, is constant; in the second equation the base flow term is a linear function of rainfall, with $c$ being a constant. The second may appear logical; however, decay from the 90 to a higher percentile is very slow, allowing base flow to be a constant for practical purposes.

Combining constant base flow with the equation for direct runoff gives

$$\mathrm{TRO} = B' + 0.0021P^2 \tag{7.8}$$

in which $B'$ is the 90 percentile flow. The equation is applicable to most terrains except the new lava landscapes on the island of Hawai'i.

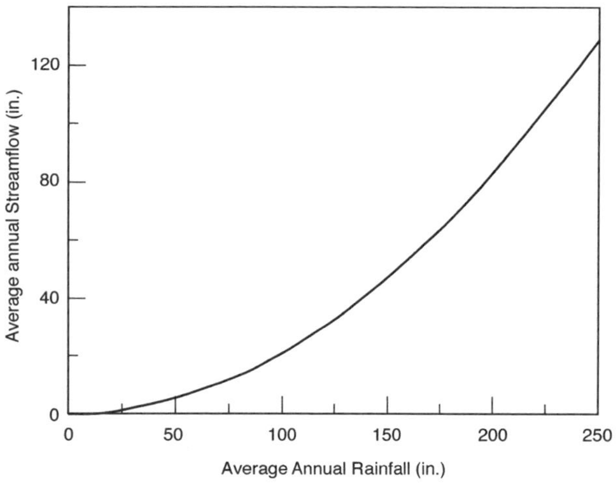


Figure 7.9. Correlation between direct runoff and rainfall on an annual basis for drainage basins in central and southern Oʻahu. (Adapted from Yuen and Associates 1992)

Testing the equation using streamflow records for Kahana and Punaluʻu Streams on Oʻahu by solving for the rainfall variable yields average annual rainfall in the drainages of 5,334 and 5,004 mm (210 and 197 in.), respectively, which is consistent with the isohyetal map. Quadratic equations have been developed with data of nondiverted streams in Kōloa and Nāpali Volcanics, island of Kauaʻi.

## Flow and Quality Frequency Analysis

Hawaiʻi's streamflow (mean daily flow) distribution is approximately lognormal, whereas ditch flow is approximately normally distributed (Yuen and Associates 1992). Stream water-quality parameters also appear to be lognormally distributed (Matsushita and Young 1973; Hawaiʻi Department of Health 1977). Stochastic modeling of ditch flow time series has been attempted on Maui (Fok and Miyasato 1976) and on the island of Hawaiʻi (Akinaka et al. 1975). Both studies used a Lag-1 Markov model and normal distribution for the random component.

## Design Floods

Virtually all computational methods for flood-flow occurrence are not original to Hawaiʻi, but they have been adapted to suit Hawaiʻi's geomorphic environment and climate. Two peak discharge formulas serve well as examples: the Rational formula and the envelope curve of maximum experience. Hawaiʻi usage of the Rational formula is restricted to small basins (40.5 ha [100 ac]), as was recommended by Chow (1966). This restriction has helped make the formula assumptions reasonable. Further improvements with a probabilistic approach and locally derived *C* values should be but have not yet been attempted. Generalized rain frequencies for the neighbor islands need to be updated. The envelope method introduced by Chow was adequate when the severe lack of hydrologic data made it inappropriate to use flood-frequency and other sophisticated methods.

Considerable investigations on flood frequencies incorporating geomorphic and climate parameters have been done in Hawaiʻi (Wu 1967; Jay 1978; Nakahara 1980; Lee 1984; Wong 1994). Many frequency distributions have been tested. The U.S. government designates the log-Pearson Type III method. It is worth noting that Oʻahu's coefficient of skewness is −0.14 rather than the −0.05 designated for use by the U.S. Water Resources Council. The Hawaiʻi experience suggests that the use of regional flood-frequency analyses should be limited to ungaged basins and basins with extremely short records only. Its development should include independent and statistically significant physical parameters instead of being limited to only one or two parameters simply for the reason of practicality of use. Finally, the recently advocated L-moments method should be investigated to realize possible statistical improvement (Hosking 1990).

## Time Distribution of Runoff Event

The linearity assumption in the unit-hydrograph theory was demonstrated to be applicable for numerous drainage basins on Oʻahu (Wang and Wu 1972). Kalihi basin appears to be an exception, having experienced an exceptionally high historical flood flow (Chong and Fok 1973).

The U.S. Natural Resources Conservation Service (formerly Soil Conservation Service) hydrograph method has been accepted for use in Hawaiʻi as it has been elsewhere, probably because of its ease of use. Disputes between critics (Pilgrim and Cordery 1993) and proponents (Ponce and Hawkins 1996) of the method are still unresolved. There appear reasons enough to evaluate critically its applicability and limitations in the Hawai'i-type environment. Perhaps, at a minimum, the curve number values should be calibrated against hydrologic records of some selected basins.

Simulation models of distributed runoff have been developed for two instrumented drainage basins: forested Moanalua Valley on O'ahu (Shade 1984) and urban St. Louis Heights, also on O'ahu (Phamwon and Fok 1977). However, they have yet to be transformed for practical applications.

The Standard Project Flood method provided the basis for designing the Kāne'ohe-Kailua flood-control reservoir now known as Ho'omaluhia Botanical Garden (U.S. Army Corps of Engineers 1972).

## Other Models

Among other models studied in Hawai'i are biohydraulic simulation (Kinzie et al. 1986) and spatial model (Parham 2002) applied to gobies; water-quality modeling of Wahiawā reservoir (Moore et al. 1981; Lum and Nakasone 1997); correlation between urban stormwater pollutant concentration and street-sweeping quality, street physical, and rainfall parameters (Nakamura and Young 1974); and statistical correlation between monthly rainfall and runoff in leeward Ko'olau basins (Anderson et al. 1966; Hirashima 1971; Giambelluca 1983). An emerging need in both data and models is urban stormwater pollution for low-frequency events. The volcanic terrain in Hawai'i offers interesting basic research opportunities for surface runoff and associated streamflow dynamics and geomorphology.

# Appendix 7.1. Streamflow Measurements

Streamflow is measured commonly by stage (elevation), velocity, and discharge (flow rate). Stage is the height of water surface above a datum, which is taken to be the mean sea level in Hawai'i. The stage is measured by means of a staff gage on which a scale is affixed so that a portion of it is immersed in the water at all times. This and other manual gages are suitable for slowly changing stages. For rapidly changing stages during a flood, a recording gage such as a float or bubble gage is required. A float gage records continuously the vertical position of a float resting on the water body in a stilling well that is connected to the stream. A bubble gage indicates the air pressure required to maintain a small flow of gas bubbles (usually carbon dioxide) from an orifice submerged in the stream. The air pressure is a measure of the depth of water over the orifice and is recorded by a manometer.

A crest gage provides a low-cost means to record the crest stage of a flood at locations where a recording gage is not justified and a manually read staff gage is inadequate. The U.S. Geological Survey crest gage consists of a length of pipe capped at the bottom and perforated with holes near the bottom to admit stream water. It contains a graduated stick and a small amount of ground cork. The cork floats as the water rises, and some adheres to the stick at the highest level reached by the water.

A stage record is transformed to a discharge record by calibration at stream controls where the stage-discharge relationship is strongly controlled by the physical features of the channels. For example, rapids or falls where the slope of the stream is steep and critical depth occurs are ideal controls.

The stage-discharge relationship, also known as a rating curve, is essentially a statistical regression, tempered by hydraulics, relating the three measured variables: stage, discharge, and slope of the water surface along the stream. The discharge data are obtained indirectly by taking the product of average water velocity and the area of a stream cross section. Velocity is measured with a Price current meter or a propeller-type current meter. A detailed discussion of streamflow measurements is provided in Linsley et al. (1982).

CHAPTER EIGHT

# Coastal Waters

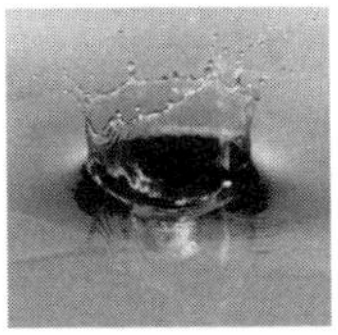

This chapter is a discussion of the effects of freshwater discharges on the quality of coastal waters in Hawai'i. To what extent do freshwater intrusions affect the quality parameters of water and sediment, and how influential are they in creating and altering the biota composed of microorganisms, fish, and coral and benthic communities? Added to the natural phenomena are the effects of human influences. The composite of natural processes and human activities results in a wide variety of coastal conditions, some immutably natural, others candidates for reclamation.

## Natural Controls

In Hawai'i the dynamics of ocean-atmosphere-earth interactions is powerful and pervasive in creating, maintaining, and destroying the shoreline and its associated biota. For example, lava flows and coral reefs create new shore environment in different ways and time frames, and waves from storm winds and hurricanes may substantially reduce coral covers.

### Coastal Water Quality

Surface water in the ocean surrounding the Hawaiian Islands is moderately warm, 24.0° to 26.1°C (75° to 79°F), and saline, 34.8‰ to 35.1‰ (parts per thousand). Both temperature and salinity vary seasonally within a small range (Seckel 1962; Flament et al. 1998). Surface water is low in nutrients: nitrogen and phosphorus. Overlain by the surface mixed layer, which is roughly 30 to 100 m (100 to 328 ft) thick and a thermocline, the deep ocean water is cold and high in nutrients. For example, the ocean off Keāhole Point of the Kona coast, Hawai'i, at a depth of 609 m (2,000 ft) is a cold 9°C (48°F), is about 200 times richer in nitrate nitrogen than the surface layer (545.6 versus 2.8 μg N/l), and is about twenty times richer in phosphate phosphorus than the surface layer (91.8 versus 5.0 μg P/l). Other water-quality parameters with highly contrasting deep water and surface water values include silica and dissolved oxygen (see Table 8.1).

The presence of the Islands alters the physicochemical quality of the ocean water. Hawai'i water-quality standards are promulgated to recognize three different water bodies: ocean water, open coastal water, and embayment water (Table 8.2). Embayments or estuaries of various sizes and shapes occur in the Hawaiian Islands (Cox and Gordon 1970). Terrestrial waters and ocean waters are detained in these relatively calm environments for an extended time, as determined by the embayment volume and the fluxes, to permit advection and dispersion of these waters and transformation of their water-quality parameters to take place and attain the characteristic quality of the embayment water. The Hawai'i coastline lacks a continental shelf but is surrounded by the slumps and debris avalanches from giant landslides (Moore et al. 1994). The bathymetry, with its ridges and canyons, provides an environment for robust mixing and transport of the open coastal water to attain its characteristic quality. The values listed in Table 8.2 are geometric

Table 8.1. Warm and Cold Ocean Water Quality off Keāhole Point, Hawai'i

| | Warm Seawater | Cold Seawater |
|---|---|---|
| Water-quality parameters | | |
| Temperature (°C) | 25.99 ± 0.93 | 8.91 ± 0.95 |
| Temperature (°F) | 78.79 ± 2.82 | 48.04 ± 5.12 |
| Salinity (‰) | 34.816 ± 0.172 | 34.298 ± 0.033 |
| pH | 8.227 ± 0.049 | 7.563 ± 0.040 |
| Alkalinity (mg/l as $CaCO_3$) | 115.9 ± 1.0 | 117.7 ± 1.05 |
| Dissolved oxygen (mg/l) | 6.98 ± 0.33 | 1.21 ± 0.19 |
| $NO_3 + NO_2$ (µg/l as N) | 2.8 ± 1.12 | 545.6 ± 16.7 |
| $NH_4$ (µg/l as N) | 5.04 ± 2.94 | 1.82 ± 2.8 |
| Dissolved organic N (µg/l as N) | 60.76 ± 9.94 | 24.92 ± 8.54 |
| $PO_4$ (µg/l as P) | 4.96 ± 1.24 | 91.76 ± 2.48 |
| Dissolved organic P (µg/l as P) | 7.68 ± 1.60 | 1.60 ± 1.92 |
| Total organic C (mg/l as C) | 0.77 ± 0.33 | 0.36 ± 0.14 |
| Particulate organic C (µg/l as C) | 34.56 ± 10.20 | 11.52 ± 4.20 |
| Si (mg/l as $SiO_2$) | 0.18 ± 0.09 | 4.48 ± 0.26 |
| Total suspended solids (mg/l) | 0.61 ± 0.52 | 0.25 ± 0.13 |
| Water intake locations | | |
| Sample depth below surface | 13.7 m (45 ft) | 586 m (1,925 ft) |
| Height above seafloor | 6.1 m (20 ft) | 21 m (70 ft) |
| Distance offshore | 92.4 m (303 ft) | 1,417 m (4,650 ft) |

SOURCE: Natural Energy Laboratory of Hawai'i 1986.

NOTE: Values for water-quality parameters are mean ± standard deviation of weekly samples, 1982–1986.

means; natural dispersion is recognized in the standards by the 10 percent exceedence and 2 percent exceedence probabilities.

## Coastal-Water Ecosystems

The definition of coastal water includes the water column, bottom sediment, and biota, which includes the benthic communities.

Coastal-water ecosystems of terrestrial-marine interest are those that can be adversely affected by freshwater discharges into the marine waters. They include all the shoreline ecosystems (Kay 1987) and some marine ecosystems (Maragos 1998). The most important ones are those associated with sandy beaches, estuaries, and coral communities.

Estuaries in Hawai'i include both bays receiving freshwater discharges and the tidal portions of streams (Cox and Gordon 1970). Endemic fishes such as *'o'opu nākea* (*Awaous stamineus*) and *āholehole* (*Kuhlia sandvicensis*) and endemic mollusks such as *hapawai* (*Neritina vespertina*) and *hīhīwai* (*Neritina granosa*) are found in many estuaries.

The coral communities include fringing reefs, barrier reefs, and lagoons. Fringing reefs are most extensive, accounting for one-half of the shoreline lengths of O'ahu, Kaua'i, Moloka'i, Lāna'i, and Maui, and the northwestern coast of Hawai'i. They are wide, shallow platforms ex-

Table 8.2. Select Parameters in Hawai'i Water-Quality Standards

| Water-Quality Parameters | Embayments | Open Coastal Waters | Ocean Waters |
|---|---|---|---|
| Total nitrogen | 200[a] | 150[c] | 50 |
| Total nitrogen as nitrogen (µg/l) | 150[b] | 110[d] | |
| Ammonia nitrogen | 6[a] | 3.5[c] | 1.0 |
| as nitrogen (µg/l) | 3.5[b] | 2.0[d] | |
| Nitrite and nitrate nitrogen | 8[a] | 5.00[c] | 1.50 |
| $NO_2^{-1} + NO_3^{-1}$–N as nitrogen (µg/l) | 5[b] | 3.5[d] | |
| Total phosphorus | 25[a] | 20[c] | 10 |
| Total phosphorus as phosphorus (µg/l) | 20[b] | 16[d] | |
| Light extinction coefficient | — | 0.2[c] | — |
| (k units) | — | 0.1[d] | |
| Chlorophyll *a* | 1.5[a] | 0.3[c] | 0.06 |
| (µg/l) | 0.5[b] | 0.15[d] | |
| Turbidity | 1.5[a] | 0.5[c] | 0.03 |
| (NTU) | 0.4[b] | 0.2[d] | |

SOURCE: Hawai'i Department of Health 2000.

NOTE: Values are geometric means. Embayments are land confined and physically protected marine waters with restricted openings to open coastal waters. Open coastal waters are marine waters bounded by the 183-m (600-ft) depth contour and the shoreline, excluding certain bays. Ocean waters are all other marine waters outside the 183-m depth contour.

[a]For embayments, wet criteria means the average freshwater flow from land per day equals or exceeds 1 percent of the embayment volume.

[b]For lesser flows, dry criteria apply.

[c]For open coastal waters, wet criteria apply when the waters receive more than 0.082 $m^3/s$ of freshwater discharge per shoreline kilometer (3 mgd/mi).

[d]Otherwise, dry criteria apply.

tending as far as 300 m (984 ft) seaward from shore and are subtidal at depths of 1 to 3 m (3 to 10 ft). They comprise sand, living coral veneers on fossil coral limestone rock, and coralline algae. Hawaiian reefs are neither as spectacularly developed nor as diverse as those off other Pacific islands deeper in the tropics. The Hawaiian Archipelago is geographically isolated and remote from the tropical reef zone, and its subtropical climate with cool winter ocean water prohibits colonization by coral varieties requiring continuous warm water throughout the year.

## Earth, Ocean, and Atmospheric Factors

Open coastlines that plunge steeply into the deep facilitate strong currents and clear water, such as the case near Keāhole Point in North Kona, island of Hawai'i. Huge submarine landslides off island coasts have been recently documented (Moore et al. 1994; Clague 1998), complicating former models of Hawaiian bathymetry. Conversely, estuaries and bays are most often relatively shallow and weakly indent coastlines. Water is detained and its movement and transport substantially restricted. South Kāne'ohe Bay and Pearl Harbor, both of which are

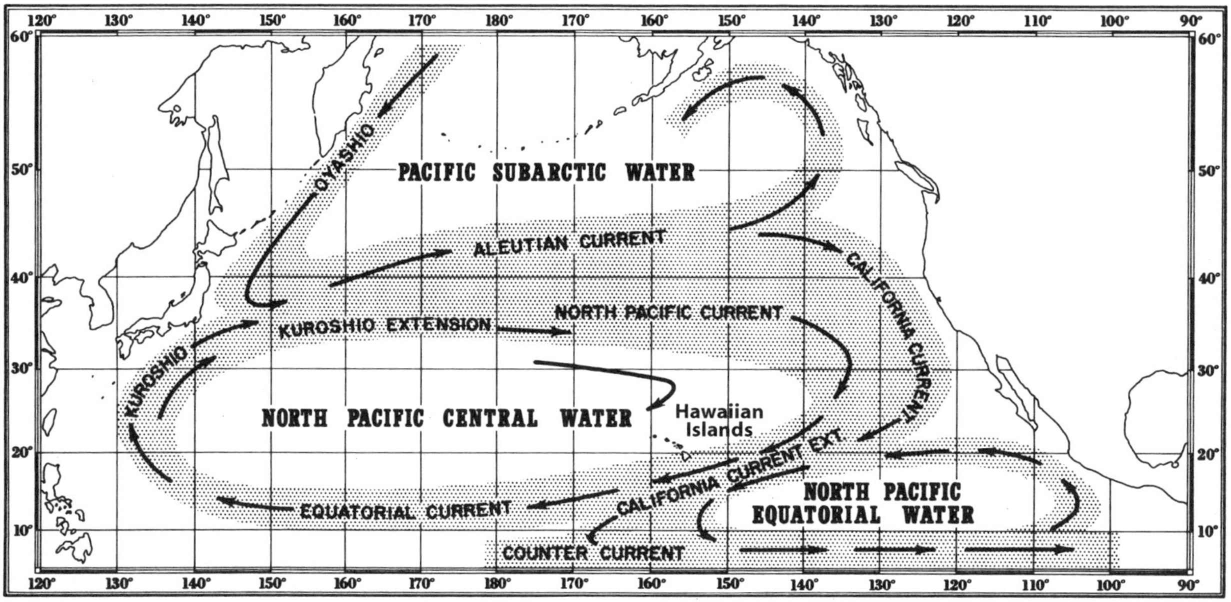


Figure 8.1. A model of the major North Pacific water types and currents. (Reprinted by permission from Seckel 1962)

drowned valleys, are good examples of large embayments. Less-visible freshwater discharges emerge as groundwater in the coastal zone along rocky shorelines. The more permeable rock formations permit greater releases of fresh or brackish groundwater, thus constantly diluting the salinity and marginally lowering the temperature of the coastal water. When these effects are strong, unique habitats for marine ecosystems are created.

In Hawai'i, basalts and fossil coral reefs are the most highly permeable rocks, and alluvium and most unlithified reef sediments are weakly permeable. The most copious measured discharges along a basaltic coastline are found in Waiākea Pond in Hilo (Fisher et al. 1966). Many other locations on the island of Hawai'i with notable groundwater discharge include the Ka'ū and Puna coastlines (Adams and Lepley 1968), the South Kohala coast from 'Anaeho'omalu to Lālāmilo (Kay et al. 1977), and Honokōhau on the North Kona coast (Bienfang 1980). On O'ahu the aquifers fronting on the Pearl Harbor estuary primarily discharge by means of terrestrial springs that extend from Honouliuli in 'Ewa to Kalauao near 'Aiea.

Three powerful physical processes drive the ocean waters that surround the Hawaiian Islands: regional currents, tides and tidal currents, and surface waves. Regional currents are created by the interactions of the Islands with large-scale ocean currents and with varying wind speeds. Ocean currents are generated by pressure variations of the water column, which in turn are caused by spatial variations of temperature and density of water. The earth's rotation causes the currents to rotate clockwise as a gyre. According to Seckel (1962), the Hawaiian Archipelago is predominantly bathed by North Pacific Central water, as indicated in his model of circulation (Figure 8.1). On the other hand, tides are generated by Earth's place in the solar system, but local bathymetry affects the range and phases of tides along the shore. The resulting tidal currents are often stronger than large-scale currents. Further, tidal currents oscillate within diurnal and semidiurnal periods, as indicated for Hawai'i in Plate 8.1 (Flament et al. 1998).

In Hawai'i, tidal ranges are limited; for example, less than 0.9 m (3 ft) off Honolulu. Although tidal ranges are routinely and accurately forecast, tidal currents are poorly known. According to a recent study of Māmala Bay, O'ahu (Colwell et al. 1996), tidal currents strongly influence the advection of the wastewater discharge (plume) from the Sand Island ocean outfall sewer so that a circulation mechanism exists, allowing the submerged plume to surface. The circulation of ocean water surrounding O'ahu based on measurements and computations is provided in the *Circulation Atlas for Oahu, Hawaii* (Bathen 1978).

Surface waves are generated by normal storm winds and in rare cases by hurricanes. Responding to nearshore bathymetry, they produce breaking waves and rip currents. These high-energy events exert transient effects on the water quality, but they can cause serious stresses and mortality on coral reefs on a global rather than local scale (Grigg and Dollar 1990). Two examples are cited: Māmala Bay off O'ahu and a single site along the South Kona coast of Hawai'i. Both have been sporadically surveyed since the 1970s. Māmala Bay water receives sewage effluent from ocean outfalls and total runoffs from highly developed land. The general area of the Kona site, located 10 km (6 mi) south of Kailua-Kona, experiences low levels of human disturbance. The waves of interest there are summer swells, winter storm swells, and local Kona storm waves. In addition, there are hurricane-force waves.

At the Kona site, several storms caused considerable destruction of the coral communities. A moderate storm in 1974 reduced the coral cover from 52 percent to 46 percent. In 1980 a Kona storm destroyed the coral zonation patterns almost entirely (10 percent of original cover). Twelve years later (1992) coral cover grew to 15 percent but only to be reduced again to 11 percent by Hurricane 'Iniki (Figure 8.2). In 1993 recovery was taking place, and by extrapolation the cover may return to 1973 conditions in 40 to 70 years (Dollar and Tribble 1993).

Before 1977, the Māmala Bay reefs were healthy except for a 4-km (2.5-mi) area surrounding the old Sand Island outfall. They were characterized by an average coral cover of 60 to 75 percent, representing mature communities of mostly *Porites compressa*. The 1982 ('Iwa) and 1992 ('Iniki) hurricane-wave events reduced these high-cover reefs to rubble fields, except for a few areas of high relief where *P. lobata* was the dominant species (Plate 8.2). Even these areas were significantly disturbed. The living coral cover was reduced to less than 20 percent (Figure 8.3). By 1994–1995, recovery had begun (Grigg 1995; Brock 1996).

During moderate wave events (height >2.0 m [6.6 ft], period >14 s), many chemical parameters in Māmala Bay water were unaffected and basically remained baseline in character (Gould 1995; Grigg 1995). This was true for salinity, turbidity, chlorophyll *a*, and silicate, although turbidity increased slightly. Two notable exceptions were nutrient concentrations: phosphate decreased below the baseline whereas nitrate exceeded it (Table 8.3). This set of water-quality measurements differentiates the effects of three different atmospheric-marine conditions: high waves, heavy rainfall, and normal (baseline) weather. The monitored water columns at eleven stations were located offshore in water 13 m (43 ft) deep and were sampled at the surface and the bottom (see Figure 8.3). The selection criterion for wave and rainfall days was low diurnal frequency (top 5 percent of the 1985 to 1993 record).

This data set generally met the Hawai'i water-quality standards. A few exceptions were, in particular, nitrates, turbidity, and chlorophyll *a* in surface water during rain events, and nitrates in both surface and bottom waters during wave events.

## Freshwater Intrusion

### Surface Runoff

Kahana Bay and the Kahana drainage basin are essentially pristine. According to a 2-year investigative study, the bay biota—coral, invertebrates, and micromollusks—show variations that are more related evidently to the oceanography and other traits of the bay than to the quality of water and sediments discharged from Kahana Stream. There was no evidence that the diversity of life in Kahana Bay was limited by the water-quality factors originating on land. The biological data are therefore subject to little interpretation in terms of response of living organisms to discharges from Kahana Stream (Lau et al. 1973).

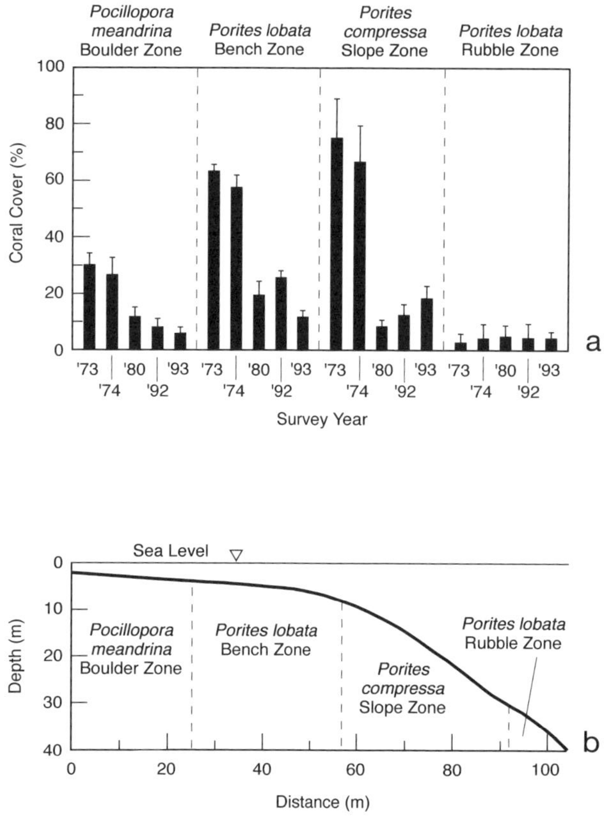


Figure 8.2. Destruction of coral communities by storms at a site along South Kona, island of Hawai'i: *a,* Mean and standard error of coral cover; *b,* depth profile and zonal boundary of coral species. (Adapted from Dollar and Tribble 1993)

In contrast, intensive rainstorm runoffs are most intrusive. Freshwater torrents, salinity, turbidity, and sediment instantly alter the receiving coastal water quality and aquatic habitats. These effects are more pronounced in embayments.

In Hawai'i uncommon events associated with voluminous surface runoff are known to seriously damage, even kill, reefs. Perhaps the only documented recent event is that of the New Year's Eve flood on December 31, 1987 (Jokiel et al. 1993). This is the same rare intensive rain cited in Chapter 3. The rainfall intensities had an annual recurrence interval of 100 to 200 years (frequency 1 percent to 0.5 percent). On that occasion, the floodwaters diluted the salinity of the surface waters in Kāne'ohe Bay to 15‰, resulting in massive mortality of coral reefs in shallow water. The salinity of some bay waters remained below 20‰ for as long as 5 days (previous studies suggest 15‰ to 20‰ sustained for 2 days as lethal to coral). A spectacular phytoplankton bloom occurred in the following weeks. The growth was apparently stimulated by the dissolved nutrients that were in part transported by flood runoff and in part derived from decomposed marine organisms killed by the flood. Within 2 weeks of the storm, the chlorophyll *a* concentration reached 40 mg/m$^3$, one of the highest values ever reported. For comparison, the Hawai'i water-quality standards stipulate a maximum limit of 1.5 mg/m$^3$ for an embayment. The plankton growth was followed by a dramatic decline as a result of depletion of nutrients. Water-quality parameters returned to normal after 2 to 3 months. Coral, echinoderms (sea urchins, sea cucumbers), and crustaceans suffered extremely high mortality in shallow water. Direct fish kill did not occur, however; reef fish apparently simply moved off to deeper water. Virtually all coral was killed to depths of 1 to 2 m (3 to 6 ft) in the western and southern portions of the bay. Elimination of coral species intolerant to lowered salinity led to later dominance by the coral *Porites compressa.* The rate of coral recovery observed as of 1992 was rapid, ranging from 40 percent to 83 percent of the coverage recorded in a 1986 preflood survey.

## Groundwater Discharges

Submarine groundwater discharge occurs where the aquifer is hydraulically connected to the ocean and the groundwater head is above sea level. The discharge is detectable by various means. Human senses easily detect submarine freshets gushing from a permeable basaltic shoreline. Waders in a lagoon in Waiulua Bay, Hawai'i, feel chilled water at the water surface. Scuba and snorkel divers' vision suddenly becomes blurred in the water as two waters of different density mix. With instruments, salinity, temperature, and nitrogen and phosphorus are the common indicating parameters of the discharge. Discharged groundwater is colder and contains higher

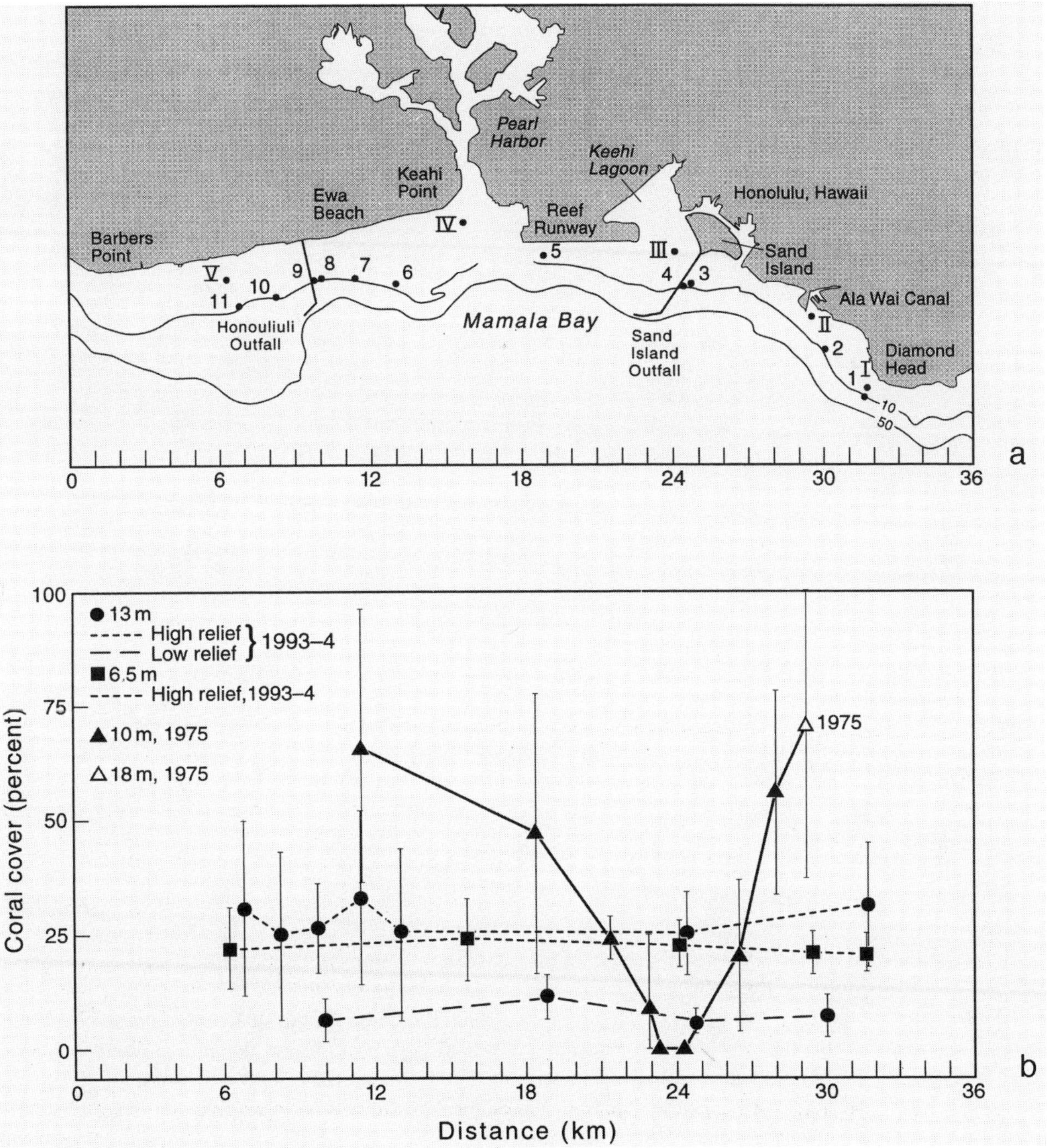


Figure 8.3. Coral covers in Māmala Bay at various times, 1975 – 1994: *a*, Location of stations. Stations I – V are at 6 m (20 ft) depth. Stations 1 – 11 are at 13 m (43 ft) depth; *b*, percentage coral cover. Error bars are standard deviations. (Reprinted from Grigg 1995 with permission from Springer Science and Business Media)

Table 8.3. Coastal Water Quality in Māmala Bay (offshore stations), 1993–1994

| Parameters | Surface Mean | Surface Range | Bottom Mean | Bottom Range |
|---|---|---|---|---|
| Baseline parameters | | | | |
| Salinity (‰) | 34.55 ± 0.03 | 34.48–34.60 | 34.55 ± 0.02 | 34.52–34.61 |
| Phosphate (μg/l as P) | 4.03 ± 0.31 | 3.41–4.65 | 4.03 ± 0.31 | 3.41–4.65 |
| Nitrate (μg/l as N) | 0.42 ± 0.28 | 0.14–0.98 | 0.56 ± 0.28 | 0.00–1.12 |
| Ammonium (μg/l as N) | 0.56 ± 0.28 | 0.00–0.98 | 0.70 ± 0.28 | 0.14–1.12 |
| Silicate (μg/l as SiO2) | 112.20 ± 28.80 | 64.80–160.80 | 105.60 ± 27.60 | 62.40–135.00 |
| Turbidity (NTU) | 0.29 ± 0.03 | 0.25–0.37 | 0.28 ± 0.02 | 0.25–0.32 |
| Chlorophyll (μg/l) | 0.16 ± 0.14 | 0.09–0.24 | 0.17 ± 0.06 | 0.07–0.24 |
| Wave event parameters | | | | |
| Salinity (‰) | 34.70 ± 0.07 | 34.59–34.83 | 34.72 ± 0.06 | 34.63–34.83 |
| Phosphate (μg/l as P) | 2.79 ± 0.62 | 1.86–3.41 | 2.48 ± 0.31 | 1.86–3.10 |
| Nitrate (μg/l as N) | 8.82 ± 12.04 | 0.28–47.32 | 10.08 ± 14.56 | 0.28–58.38 |
| Ammonium (μg/l as N) | 1.26 ± 0.70 | 0.14–3.08 | 1.40 ± 0.84 | 0.28–3.08 |
| Silicate (μg/l as SiO2) | 138.00 ± 33.00 | 75.00–178.80 | 125.40 ± 27.00 | 76.80–165.00 |
| Turbidity (NTU) | 0.37 ± 0.10 | 0.27–0.70 | 0.34 ± 0.06 | 0.27–0.48 |
| Chlorophyll (μg/l) | 0.18 ± 0.07 | 0.06–0.28 | 0.17 ± 0.07 | 0.01–0.30 |
| Rain event parameters | | | | |
| Salinity (‰) | 34.13 ± 0.88 | 31.53–34.41 | 34.51 ± 0.21 | 33.97–34.60 |
| Phosphate (μg/l as P) | 4.65 ± 2.79 | 3.10–13.02 | 3.72 ± 0.62 | 3.10–5.27 |
| Nitrate (μg/l as N) | 8.54 ± 19.04 | 0.14–72.10 | 1.54 ± 1.82 | 0.14–7.42 |
| Ammonium (μg/l as N) | 1.54 ± 0.70 | 0.56–3.66 | 1.54 ± 0.56 | 0.56–2.24 |
| Silicate (μg/l as SiO2) | 210.60 ± 128.40 | 126.60–573.00 | 151.20 ± 18.00 | 127.80–201.00 |
| Turbidity (NTU) | 0.93 ± 0.96 | 0.28–3.20 | 0.39 ± 0.16 | 0.29–0.77 |
| Chlorophyll (μg/l) | 0.32 ± 0.34 | 0.03–1.23 | 0.27 ± 0.20 | 0.13–0.89 |

SOURCES: Gould 1995; Grigg 1995.

concentrations of nitrogen and phosphorus than surface seawater.

Copious submarine groundwater discharges occur in Honokōhau Harbor in North Kona (Bienfang 1980). An aerial infrared photo taken from 609 m (2,000 ft) reveals the discharge as distinct plumes of bright image emerging from the harbor (lower right of Figure 8.4) for about 0.8 km (0.5 mi). This photo is a negative print so the brighter images are water colder than the surroundings. Infrared imaging was done along the Kaʻū and Kona coastlines in 1968 (Adams and Lepley 1968) following the first report with such remote sensing of Hilo Bay. Ground truth is always necessary to confirm submarine groundwater discharges. Quantification of submarine groundwater discharges by infrared imaging has not yet been successful.

At morning low tide in the Wainānāliʻi pond near Kīholo Bay, Hawaiʻi, the surface water was 24°C (75°F) and only 3 m (10 ft) below, the bottom water was 27.5°C (81°F). The associated salinity was 15‰ at the surface but doubled at the bottom. The dissolved oxygen gradient was also observed. The strong physicochemical gradients appeared to have affected the fauna in the uppermost 1.5 m (5 ft), where strongly select euryhaline organisms lived in an environment of widely varying salinity. Elsewhere,

Figure 8.4. Infrared photo (negative print) of groundwater discharge in the vicinity of Honokōhau Harbor, island of Hawai'i. (Courtesy of Hawai'i Institute of Geophysics Photography Laboratory, 1990)

the community composition appeared relatively consistent with the substrate rather than variably influenced by the differences in water quality (Kay et al. 1977).

Another natural intrusion is the transport of nitrogen and phosphorus in the basal groundwater discharge into the ocean. A shoreline spring in Waiulua Bay discharges nitrite-nitrate nitrogen at a concentration of 558 μg nitrogen/l and total phosphorus at 56 μg phosphorus/l. These values far exceed the Hawai'i water-quality standards of 8 and 25 μg/l, respectively. The source of nitrogen may be natural because the whole coast was virtually undeveloped at the time of the study in 1974–1975. A probable source of nitrogen is the lush growth of *kiawe* (*Prosopis pallida*), which is a nitrogen-fixing (leguminous) plant capable of imparting nitrates to the ground. The nearshore marine water fronting recreation resorts along the Waikōloa coast, of which Waiulua Bay is a part, was found to experience no statistical alterations in water quality, as reported in a long-term (9-year) study (Brock 1999).

## Human Effects: Land Use and Water Discharge

The Polynesians first arrived in the Hawaiian Islands about 200 to 300 A.D. (Kirch 1985; Franco 1995). Their survival depended on sensible use of land and freshwater. Initial settlements took place in the coastal areas where fishing was good and in the wet valleys. To support an increased population that reached hundreds of thousands of people at one time, they modified the environments: clearing with fire (slash and burn) the lowlands and valley slopes, converting shorelines to fishponds, and building barrage terraces, diversion dams, ditches, and pondfields.

During the 100 years before European contact in 1779, the *ahupuaʻa* was well suited to support a stratified social system and decreased available land. An *ahupuaʻa* is a large wedge of land extending from the mountain crest to the sea and bound on the sides by ridges and valleys. Within it is a gradation of climate zones and managed land and ocean resources (Figure 8.5). Within an *ahupuaʻa*, the ancient Hawaiians made the most of the natural resources in harmony with nature.

Since the 1850s, several events have demanded a great deal from the land, water, and natural environment: mechanized sugarcane and pineapple culture, population growth in Honolulu, World War II, statehood, and the tourism industry. In the 1960s numerous municipal and agricultural wastewaters of unacceptable nature were discharged into coastal waters (Dugan and Young 1973). Abatement of waste discharge by regulations (see Chapter 7: Water-Quality Standards) has since reversed the trend. Water conservation and water reuse can further reduce the marine discharges. However, the coastal waters surrounding the Islands will continue to be the ultimate sink for wastewaters, either directly or indirectly (McGauhey and Lau 1975). Transport of conservative water-quality parameters in receiving water is relatively tractable. But identifying subtle (sublethal) and ultimate impact requires protracted biological monitoring, which has been in progress (Moravcik 2000).

## Point Sources

A point source is a source of water-quality parameters that is discharged from the end of a conduit at a known location, such as a wastewater treatment plant, a sugar mill, or an electricity power plant. Nonpoint-source pollution refers to alteration of water quality by contaminants that originate from developed land surface. Wet weather washes out the contaminants to cause storm runoff pollution. Because they are diffuse, these sources are difficult to eliminate.

### Deep Outfalls

Māmala Bay, which extends from Diamond Head to Barbers Point, receives two large point-source discharges from ocean sewer outfalls totaling 4.22 $m^3/s$ (149 $ft^3/s$) or about one-third of the total freshwater discharge into the bay (Stevenson et al. 1996). These submarine outfalls are Sand Island (3.17 $m^3/s$ [112 $ft^3/s$]) and Barbers Point (Honouliuli) (1.05 $m^3/s$ [37 $ft^3/s$]) (see Figure 8.3*a*). Nonpoint-source discharges from land accounts for the remaining two-thirds of the discharge: Pearl Harbor (6.13 $m^3/s$ [216 $ft^3/s$]), Keʻehi Lagoon/Honolulu Harbor (1.98 $m^3/s$ [70 $ft^3/s$]), and Ala Wai Canal (0.92 $m^3/s$ [32 $ft^3/s$]).

The outfall discharges occur in deep areas (61 to 72 m [200 to 240 ft]) distant from shore (Sand Island: 3,811 m [12,500 ft]; Barbers Point: 3,203 m [10,500 ft]). Each outfall is connected to a long diffuser pipe (Sand Island: 1,031 m [3,382 ft]; Barbers Point: 533 m [1,750 ft]) that is perforated with numerous ports and laid on the edge of a steeply sloping bed. The permitted discharges are conditioned upon acceptable performance by monitoring.

The permitted discharges have resulted in no deficiency of dissolved oxygen in the water column surrounding the outfalls. Also, observed biostimulatory effects on phytoplankton were minor (Colwell et al. 1996; Laws et al. 1999). Results of a 10-year monitoring program indicate that no organic matter was accumulated on the bottom, even at the stations nearest the outfall diffuser (Nelson et al. 1995). The data set lends support to an earlier nutrient budget study using benthic respirometry and particle

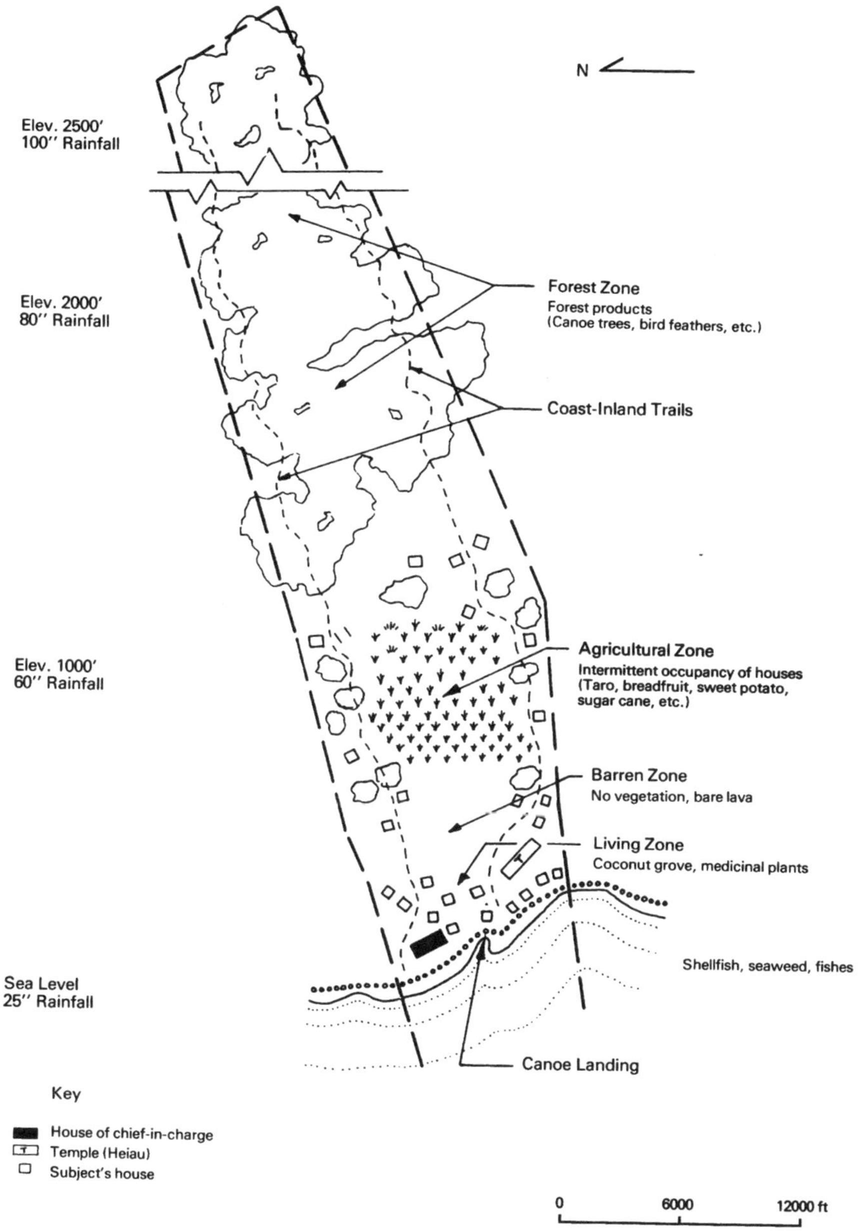


Figure 8.5. The *ahupuaʻa,* basic unit of land division in ancient Hawaiʻi. 1 ft = 0.305 m, 1 in. = 25.4 mm. (Based on Kirch 1985; reprinted with permission from Bishop Museum Press)

trap that found that only about 1 percent of the effluent reached the sediment surface (Dollar 1986).

All data have indicated that the swimming beach waters in Māmala Bay are hygienic and acceptable for contact recreation. There have not been any epidemiological evidences at any Māmala Bay swimming beaches. Microbiological monitoring statistics invariably show very low frequencies of fecal indicators. Most swimming beach waters meet the Hawaiʻi marine recreation water standards of 7 colony-forming units (CFU) enterococci/100 ml, which is stricter than the U.S. Environmental Protection Agency standard of 33 to 35 CFU/100 ml (Fujioka and Loh 1996). The sources and pathway of pathogens and fecal indicators that are sporadically recovered remain a matter of incomplete certainty. Pathogens (*Salmonella, Cryptosporidium,* and *Giardia*) and fecal indicators (fecal coliform, *E. coli,* enterococci, and *Clostridium perfringens*) were present in Sand Island outfall discharges (100 percent). However, in the Ala Wai Canal, which discharges at the shoreline in midbay, some of these microbes also occurred in high frequency (17 to 77 percent). No observed evidences indicate that outfall discharge contributed to the pathogens present in the Māmala Bay swimming beach waters (Colwell et al. 1996).

Multiyear regulatory monitoring and special investigations have indicated that Sand Island and Barbers Point discharges have resulted in no measurable impact on coral and benthic communities. These include shallow reef communities shoreward of the outfalls (Colwell et al. 1996); benthic and fish communities including micromollusks and macroinvertebrates, Mollusca, Annelida, and Echinodermata (Nelson et al. 1995; Brock 1996); coral calcification, growth, species composition, density, or community structure (Grigg 1995); and species abundance and growth rates of coral stands (Brock 1996).

The discharges from the two deep-ocean outfalls appear to have resulted in minimal and acceptable environmental and health impacts in Māmala Bay. Initial dilution followed by ocean current transport apparently provide robust advection and dispersion of the discharges. In addition, neutral buoyancy results in submergence of the plume most of the time. Even so, the predischarge treatment was recommended to be upgraded to chemically enhanced primary treatment and disinfected with ultraviolet irradiation (Colwell et al. 1996).

## Shallow Outfalls

Not all outfalls that operated in Māmala Bay delivered acceptable performance. Before 1977, the outfall in service since the early 1940s was a short (1,100 m [3,600 ft]) pipe without a diffuser, discharging untreated sewage up to 2.4 $m^3/s$ (85 $ft^3/s$) at shallow-water depth (12 m [40 ft]). Besides domestic wastewater, the discharge contained wastes from light industry and pineapple canneries. Infiltration of brackish groundwater into sewers was high (over 40 percent), making water reuse infeasible. The plume was visible from boats and high grounds in Honolulu. The quality of the receiving water in the vicinity of the outfall exceeded all applicable water-quality standards (Dugan and Young 1973). Benthic surveys made before and after the replacement of the "old" outfall provide amazing contrasts (Grigg 1975; Dollar 1979). The impacted zone in 1975 extended 2 km (1.2 mi) to the east and 4 km (2.4 mi) to the west, reflecting the prevailing currents. Within the zone, corals were either absent or suddenly depressed in abundance. The bottom area within 1 km (0.6 mi) of the outfall was completely dominated by *Chaetopterus,* a tube-building polychaete that builds thick mounds. Within the mounds, nematodes up to $10^5/m^2$ ($10^4/ft^2$) were found. Other favored species within the zone were the alga *Ulva* sp., sponges, and the urchin *Echinothrix diadema*. Urchins were exceedingly abundant and succeeded in bioeroding and excavating virtually all living corals within the 6 km (3.6 mi) of coastline affected. In 1978, 1 year after the diversion into deeper water, Grigg made another survey that was followed by the one made by Dollar in 1979. They observed that all of the dominant species previously present had become absent or rare. The polychaete mounds had vanished, urchins were rare, and sponges and *Ulva* were absent. The bottom was a hardpan barren limestone substratum with abundant cobbles and rubble and layers of sand. No coral recruitment was observed.

The Waiʻanae and Mōkapu ocean outfalls are also shallow varieties (33 m [110 ft] and 28 m [92 ft], respectively). The discharges from the outfalls appear to be biologically compatible with the natural ecology. For example, the Wai'anae outfall discharging primary effluent in 1996 actually enhanced the coral and fish assemblages (Russo 1995). Specifically, at the diffuser, fish abundance and species richness increased over time, and corals actually grew on the diffuser pipe (Plate 8.3). The armor rock over the pipeline acted to recruit corals and fishes. The water was clear (10 to 16 m [32 to 52 ft] horizontal visibility), and the surrounding sediments were clean and white. The outfall discharged 0.13 $m^3/s$ (4.5 $ft^3/s$) flow and is located 1.8 km (6,000 ft) offshore with a diffuser length of 162 m (531 ft). Even so, the predischarge treatment was required to be upgraded to secondary treatment later in 1996.

The experience with ocean outfalls around O'ahu offers criteria as to what is acceptable. Important considerations are as follows:

1. Locate along an open coast, as in the case of the Wai'anae and Mōkapu outfalls, and definitely not in an embayment, as in the case of the abandoned Kāne'ohe Bay outfall
2. Treat the predischarge effluent at least to the advanced primary stage
3. Choose marine conditions that are robust and offshore, vectored away from protected waters
4. Construct diffuser that enhances high initial mixing
5. Periodically monitor water, sediment, and biota

Stringent regulatory monitoring requires autopsies and examination of liver tissues of fishes sampled in the vicinity of all major ocean outfalls on O'ahu (Brock 1998).

## Mill Discharges and Injection Wells

Waste discharges from sugarcane mills were a serious marine environmental problem in Hawai'i for about 30 years (from the late 1940s to the late 1970s). The impacts of these point discharges proved to be readily reversible, as indicated by investigations conducted along the Hāmākua coast off the island of Hawai'i and off Kīlauea, Kaua'i.

A 1983 survey revealed depletion of the coral reef ecosystem within 1.6 km (1 mi) of mill discharge points at Pepe'ekeo, 'O'ōkala, and Haina along the Hāmākua coast (Grigg 1985). The degree of depletion correlated positively with the thickness of the deposited sediment and approximately coincided with the size of the surface plume of total suspended solids. The nearby streams also discharged considerable sediment into the marine environment, totaling 180,000 tons (1 ton = 0.907 metric ton) in an average year and 560,000 tons during a wet year compared with 459,000 tons from the three mills in 1978, the year before U.S. Environmental Protection Agency regulations went into effect. The observed recovery of the coral assemblage at Pāpa'aloa, where a mill was shut down 17 years earlier, indicated the effects of mill discharge to be temporary and reversible (Grigg 1972). Besides corals, fishes were also surveyed for species diversity and abundance. Bagasse, the fibrous residue of the mechanically crushed sugarcane stalks, was eliminated from the discharges as a result of the regulations. The effects of bagasse removal on fishes and fishing were mixed and inconclusive.

The effects of the discharge of sugarmill wastes on the marine environment (water, sediment, and biota) were investigated before and after the demise of the 94-year-old Kīlauea Sugar Company, Kaua'i, in 1971 (Lau et al. 1972; Lau et al. 1973). The untreated mill waste carried coliforms, sediments, leaf trash, and bagasse to the coastal waters. Sediments rather than water harbored most of the nutrients, heavy metals, and chlorinated hydrocarbons (used as herbicides) in the wastes. Coral depletion occurred within 1.6 km (1 mi) of the discharges, similar to the change along the Hāmākua coast (see Figure 8.6).

Aesthetics, coliforms, and sediments are evidently the major problems of mill waste discharge. A striking aesthetic improvement of the coastal water and the beach occurred quickly (just one winter) after the cessation of mill waste discharge. The coastal water quality also improved, meeting contemporary standards. Part of the improvement was attributable to heavy sea, which provides strong dilution. However, beach and ocean sediments continued to harbor plenty of nutrients. Fishes reappeared quickly.

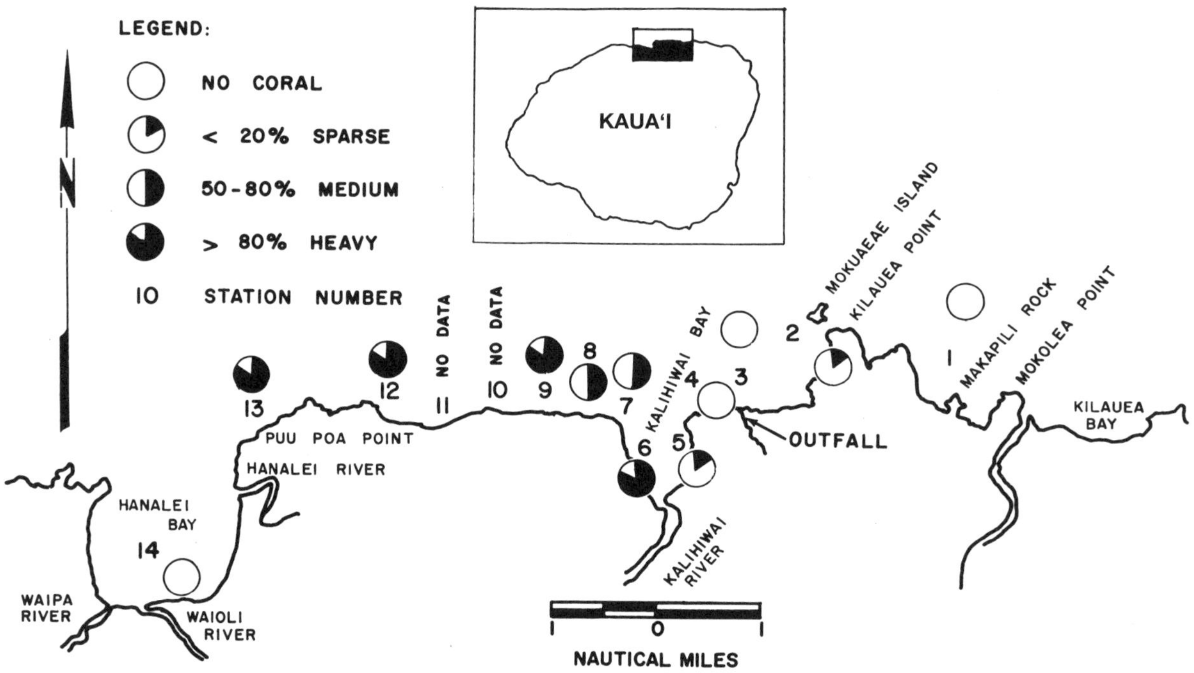


Figure 8.6. Coral coverage offshore between Kīlauea Bay and Hanalei Bay, July 15, 1971. 1 nautical mile = 1.15 mi = 1.85 km. (Reprinted by permission from Lau et al. 1972)

From numerous investigations, the chief conclusions are (1) there is no evidence of eutrophication (increased algal production due to excess nutrients), (2) the adverse effects of the discharges are mostly transitory, and (3) the epibenthic community (plants and animals living on the sea bottom between the low-tide level and 182 m [100 fathoms]) is more influenced by waves, currents, and coastal topography than by mill waste discharge.

An indirect discharge into coastal water may be transported by way of injection wells in the coastal area. The concerns of wastewater disposal and control of seawater intrusion by injection date back to the 1950s (Harder et al. 1953; Lau et al. 1958). Both practices can be successful, provided precautions are undertaken to minimize clogging of the wells, but the threat of coastal pollution persists. In Hawai'i wastewater injection is permitted only in brackish groundwater seaward of the underground injection line (Hawai'i Department of Health 1992). The only serious attempt to make measurements in coastal waters downgradient of injection wells was at Honokōwai on the west coast of West Maui (Tetra Tech Inc. 1994). A steady-state discharge of secondary-treated effluent at a substantial rate of 132 liters/second (2,100 gallons/minute) from an injection well located at a distance of 609 m (2,000 ft) from the shoreline was traced for 2 months with a dye, Rhodamine WT. Extensive nearshore monitoring in a large water surface (7.8 km² [3.0 mi²]) area nearby failed to detect the presence of the dye tracer, except for a few sporadic samples at slightly above detectable limit.

## Nonpoint Sources

### Urban Recreation

Hanauma Bay is a flooded volcanic crater with a fringing reef that protects a sandy beach (Plate 8.4). Radiocarbon dating suggests that the reef started about 7,000 years ago (Easton and Olson 1976). Because the bay has been designated a marine life conservation district, the coral cover, fish species, and fish abundance are expectedly high (Grigg 1995). Many fish species that usually dart away from human observers have become remarkably tame there, adding to the attributes of Hanauma as a popular tourist attraction. The annual visitor count rose from 200,000 in 1970 to over 3 million in 1990, leading to recent regulations in an attempt to control overuse. Sewage has been pumped out from the beach park since 1992. The microbial water quality at the beach has become a matter requiring periodic monitoring and observation. In a 1993 study, the water was deemed to meet the stringent state marine recreation standard of 7 CFU enterococci/100 ml. The sources of these indicator bacteria were determined to be bird (pigeon) droppings on beach sands and possibly infrequent runoff (Fujioka and Roll 1993).

## Urban Residential Land Use

Hawai'i Kai marina presents an excellent opportunity to assess what effects urban residential land use may have on water quality. A general trend of improvement in water quality from the marina to the nearby ocean bay waters was indicated for nitrogen, phosphorus, turbidity, and suspended solids during 2 years of monitoring (Lau et al. 1973). DDT, dieldrin, and PCP (the latter a termiticide and all of them then permitted for use) were present at low levels (in the nanograms-per-liter [parts-per-trillion] range) in the water but at much higher levels in the sediments, reflecting the intensive urban activities associated with the then relatively new and growing residential area.

Another urban residential effect on coastal water quality is fecal contamination, as indicated in the case of Enchanted Lake, which is connected to popular Kailua Beach by Ka'elepulu Stream (Fujioka et al. 1997; Roll and Fujioka 1997). The fecal indicators in the CFU in the stream waters greatly exceeded the three U.S. Environmental Protection Agency recreational water-quality standards of 200 CFU fecal coliform/100 ml, 126 CFU *E. coli*/100 ml, and 33 CFU enterococci/100 ml). Primary sources of pollution were determined to be nonpoint sources, including tributary streams, storm drains, duck feces, soil, and surface runoff directly from land. Ka'elepulu Stream water affects the Kailua Beach water only when the sandbar at the stream mouth is breached, allowing the stream water to discharge freely into Kailua Bay.

## Agriculture: Sugarcane

The use of land for mechanized sugarcane culture in Hawai'i has had few permanent effects on coastal water quality when irrigation tailwater and mill wastewaters are properly managed. For example, at the former McBryde Sugar Plantation on Kaua'i (Lau et al. 1973), the sediment-laden irrigation tailwater from furrows was captured in some seventy small ponds located along the coastline. The ponds served as settling and evaporation ponds. The pond sediment was periodically dredged and recycled to the fields. The only discharge to the ocean was occasional overflows during heavy rains. The mill waste discharge was totally retained on land with waste and sediment separated and recycled, thus eliminating three major problems caused by mill waste discharge: aesthetics, coliform organisms, and sediment in the coastal waters.

The most common herbicides used in sugarcane culture, such as atrazine and ametryn, did not appear to be a factor in water quality. In studies focused on Kaiaka Bay near Waialua and West Loch of Pearl Harbor and their drainage basins (Green et al. 1977), these two chemicals dissipated rapidly in field soils, and neither was found normally in stream and estuarine sediments. However, diuron, being more persistent, appeared in nearly all sampled stream and estuarine sediments. West Loch estuarine water contained 0.1 to 1.0 μg/l of diuron, and concentrations in stream runoff water were several times higher. It is not known if these chemicals at chronic low levels adversely affect the ecological balance of the estuary.

## Ala Wai Canal

The Ala Wai Canal is a prominent landmark on the Honolulu coastal plain. Hydrologically, the canal drains the surface water from Diamond Head to Punchbowl, including Mānoa and Pālolo Valleys as the two main drainage basins, for a total area of 42.2 $km^2$ (16.3 $mi^2$). Constructed by dredging in the 1920s, it originally drained a sizable wetland (duck ponds, fishponds, taro patches, and rice fields). Acting as a surrogate for the original wetland, the canal detains, settles, and filters surface-water contaminants. It was dredged three times in recent years (1966, 1978, and 2002). Its design capacity is 5,200 $ft^3/s$ (10-year flood); however, a 1968 rainstorm resulted in about 19,247 $ft^3/s$, an estimated 50-year flood.

The Ala Wai Canal has been identified as a major source of fecal indicators (Fujioka and Loh 1996) and pathogens (Gerba et al. 1996) to Māmala Bay. There are no point-source sewage effluent discharges into the surface waters of the Ala Wai Canal drainage basin. Nonpoint sources are the origin of the pollutants. In any event, the fecal indicators are fecal coliforms, *E. coli,* and enterococci. Their concentrations were high in all samples taken from the upper freshwater layer of the canal water.

Four pathogens present in the canal water are *Salmonella,* enterovirus, *Cryptosporidium,* and *Giardia.* They were found in high frequency (17 to 77 percent of the time) in the daytime samples collected at the Ala Moana Bridge (Gerba and Pepper 1996). In comparison, they were present in all samples of the Sand Island outfall discharge. A real concern is that they have been detected in low incidences at Waikīkī beaches and close inshore marine waters in Māmala Bay (Gerba and Pepper 1996).

Suspended solids and turbidity in the Ala Wai Canal are high, not surprisingly reflecting the traits of its chief contributor, the Mānoa drainage basin (see Chapter 7). The canal collects bulk sediment averaging annually 3,100 tons (2,812 metric tons) (McMurtry et al. 1995a).

Certain heavy metals in the canal sediments are noticeably greater in concentrations than the natural baselines — namely, Zn, Cu, and Cd and, over a limited period, Pb and Hg. Enrichment of the first three has been attributed to the particles of worn brake pads of automobiles (Zn and Cu) and particles of worn vulcanized tires (Zn and Cd), all transported by runoff from urban roads (McMurtry et al. 1995a; DeCarlo and Anthony 2002). Pb enrichment was evident in cored samples of sediment layers from the canal and has been attributed to leaded gasoline, which has been banned since 1975 (DeCarlo and Spencer 1995). Enriched Hg was attributed to antifouling paints used in the Ala Wai Yacht Harbor; they were first used in the 1950s and then banned in the early 1970s (Raine et al. 1995). Fallout from volcanic eruptions on the island of Hawai'i added measurable spikes of Hg in the canal sediments. No organisms (shrimps, worms, fishes) sampled in the canal in 1974 exceeded the contemporary federal standard of 500 ppb Hg in edible tissue (Luoma 1974). Further, Luoma's laboratory experiments with $^{203}Hg$-labeled canal sediments indicated that sediment-bound Hg is essentially not biologically available to the detritus feeders. Also, Hg in the tissue of feeders at any one time is but a snapshot because these feeders cleanse themselves.

A best management practice was proposed for the canal drainage basin in an attempt to improve the canal water quality as a joint state and county program (Dashiell 1997). Also, alternative measures to control nonpoint sources of contamination from the Ala Wai Canal particularly during and immediately after intense rainstorm events were recommended by the Māmala Bay Commission (Colwell et al. 1996).

## Global Situations

Māmala Bay and Kāne'ohe Bay represent two global cases of a mixture of coastal land uses and point-source discharges. They are sufficiently different environmentally that they have become the subjects of numerous studies. Uses of land adjoining Māmala Bay are urban (recreational, residential, commercial), military, industrial, and agricultural, whereas uses of land surrounding Kāne'ohe Bay are urban residential and rural with some agriculture.

### Māmala Bay

Pollution should be viewed from both local and global perspectives. Local effects are manifested in the immediate area receiving pollutants, whether point or nonpoint, whereas global effects are distributed over a much larger field. The Sand Island and Barbers Point (Honouliuli) ocean outfalls are the two major point sources of pollution in Māmala Bay. In addition, four drainage basins, which include Ala Wai Canal, Pearl Harbor, Ke'ehi Lagoon–Honolulu Harbor, and Kewalo Basin contribute substantially to the pollution. Each source impacts its respective nearshore coastal waters, which behave as a buffer zone against global pollution (Laws et al. 1999). Some contaminants may be retained for a time in the buffer zone to allow their transformation from nutrients to phytoplankton, whereas some settle and bind with sediments.

Under normal weather (baseline) conditions, global effects are subtle and much less discernible than local effects, but supporting data are rare. Māmala Bay data (Gould 1995; Grigg 1995) were probably the only suitable set for this type of assessment: the monitored water columns were located offshore from the coast in 13 m (43 ft) deep water and were sampled at the surface and bottom of the water column during normal weather (trade winds 10 to 20 mph [4.5 to 8.9 m/s], waves <2.0 m [6.6 ft], and diurnal rainfall <1.25 cm [0.5 in.]). The data (Table 8.3) indicate that the water met the nitrogen, phosphorus, chlorophyll *a*, and turbidity standards (Hawai'i water-quality standards for open coastal water, wet criteria 1993). The only exception was nitrate in the surface-water samples. All water-quality parameters remained relatively uniform throughout the bay. Sedimentation was also measured offshore of the Ala Wai Canal at water stations 6 m (20 ft) deep under baseline condition in one summer week in 1994 and reported in terms of volume flux and dry weight. All values were considered to be exceedingly low (Grigg 1995). In sum, the global effects of nonpoint-source pollution appear to be minor, at least in terms of mean values.

Local pollution effects on coastal water have been evident at Pearl Harbor and the Ala Wai Canal during runoff from high-rainfall events. The concentrations of nitrates and phosphates were elevated at the surface off the Ala Wai outlet in 1993–1994 (Figure 8.7). But they remained relatively moderate off Pearl Harbor, evidently resulting from their transformation to chlorophyll, which was high in the surface water (Grigg 1995). Reduced recruitment of some benthic organisms was observed in the vicinity of Pearl Harbor and near the Ala Wai Canal (Colwell et al. 1996). See the preceding section on the Ala Wai Canal for other local effects.

To what extent does a point source like the Sand Island deep-ocean outfall impact Māmala Bay water quality on a global scale? The assessment has focused on pathogens and coral communities in nearshore waters. These issues have been addressed in a previous section on deep-ocean outfalls. Loadings of pathogenic microorganisms, although substantial at the point sources, were so reduced by dilution that the concentrations were far below detectable levels except near the outfall (Colwell et al. 1996). Before 1977, the discharge of raw sewage into Māmala Bay via a short outfall caused serious but localized environmental impacts. The wastewater is currently treated at the advanced primary level and discharged into deep water about 4 km (2.49 mi) offshore; it poses no observable and measurable environmental threat to the reef ecosystems.

## Kāne'ohe Bay

Another example of global pollution conditions is the water in Kāne'ohe Bay. It drew attention in the early 1970s when the shallow outfall was installed in South Kāne'ohe Bay. Increased discharge from population pressure eventually generated unacceptable environmental conditions leading to the diversion of the discharge to Mōkapu Point outside the bay in 1978. The aquatic environmental recovery has been studied and documented (Smith et al. 1981; Maragos et al. 1985). In fact, this case together with two other examples, Lake Washington and Lake Erie, has become a textbook example in *Aquatic Pollution* (Laws 1993) as cultural eutrophication.

The environmental issue with Kāne'ohe Bay water was primarily the deterioration of the coral-reef community as a result of sewage discharge and possibly sedimentation. The bay has a surface area of 46 km$^2$ (18 mi$^2$) and a mean depth of 6 m (20 ft) but is twice as deep in its southern portion. Normal water salinity ranges from 33‰ to 35‰, and water temperature varies seasonally between 20° and 27°C (68° and 81°F). The water residence time

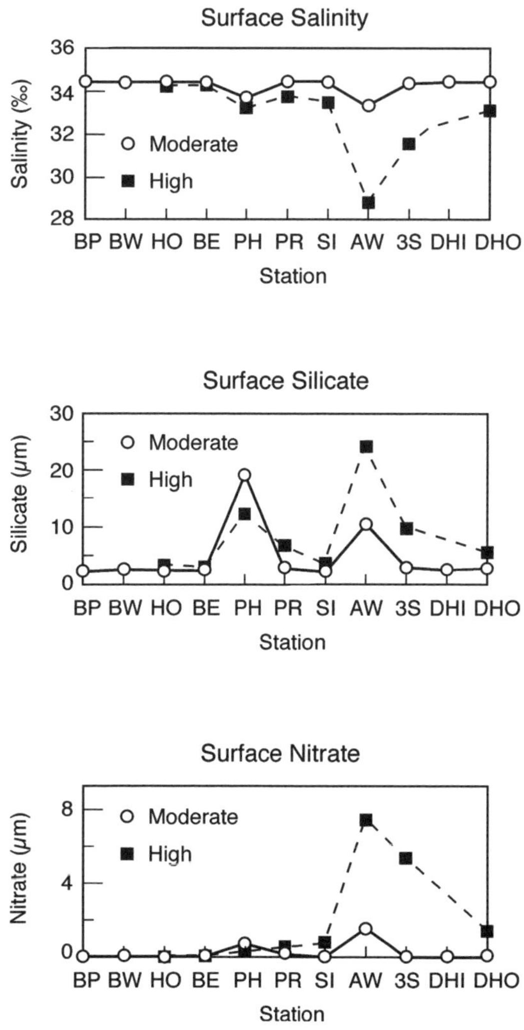


Figure 8.7. Surface water quality in Māmala Bay during moderate and high diurnal rainfall days in 1993 – 1994. Moderate rain depth ~5 cm (2 in.), high rain depth ~13 cm (5.1 in.); Stations BP, BW, and DHI were not sampled during high-rainfall events; 1 μm nitrate = 14 μg/l as N, 1 μm silicate = 60 μg/l as $SiO_2$. Station names refer to the station arabic numbers in Figure 8.3*a*. (Adapted from Gould 1995)

is about 12 days but is twice as long in South Kāne'ohe Bay. The bay receives runoff from eleven small perennial streams with a combined area that is slightly larger than the bay. Many species of coral thrive in the bay. Across the mouth of the bay is a barrier reef that restricts circulation. Rural and agriculture land uses prevailed in the drainage basin before World War II when taro, rice, sugarcane, pineapple, and bananas were principal crops. Dredging and filling of the bay for military and other purposes, followed by urban and residential developments in Kāne'ohe and Kailua, have substantially altered utilization of its environmental attributes. An early, highly important impact on the bay's environment was the loss of stream runoff, which was diverted to leeward O'ahu via the Waiāhole Tunnel system starting in 1913. Streamflow into the bay was reduced by 42 percent (Cox and Gordon 1970).

Cesspools had been the wastewater disposal practice until after 1951 when the Kāne'ohe Marine Corps Air Station constructed its sewage-treatment plant and discharged the effluent into the south part of the bay. For the 1963–1978 period, the municipal-operated Kailua regional sewage-treatment plant also discharged the effluent into the south part of the bay. A third but smaller discharge into the bay was a tertiary effluent via 'Āhuimanu Stream, but the practice has been discontinued.

During the period that sewage effluent was discharged into the bay, eutrophication occurred. However, within six years after the diversion of the effluent to the open ocean, a remarkable recovery of coral was observed (Maragos et al. 1985). Current uses of Kāne'ohe Bay water include recreational water sports and capture of tuna baitfish, *nehu*. Observations on recovery have been reported in great detail (Smith et al. 1981).

## Models and Data

Few investigations employing quantitative models have attempted to simulate Hawai'i coastal water quality. A notable early example is a model simulating the hydrodynamic and water-quality behavior of Pearl Harbor water as driven by tides and tributary streams (Orlob 1978). It is a link-node estuarial model (pseudo two-dimensional) based on one-dimensional equations of motion, continuity, and mass conservation. Simulated water-quality parameters include temperature, salinity, dissolved oxygen,

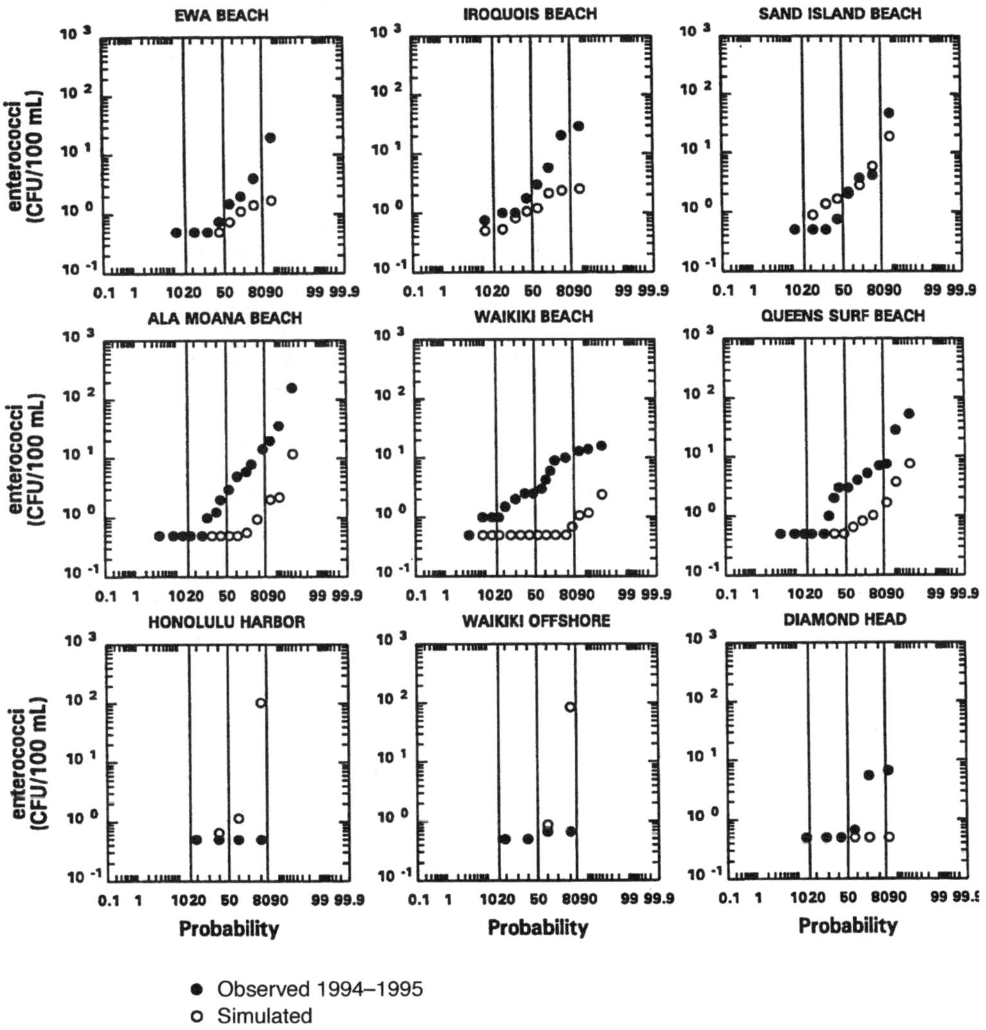


Figure 8.8. Observed and simulated enterococci cumulative frequency distributions, Māmala Bay. Marine recreation water standards: 7/100 ml, Hawai'i; 35/100 ml, U.S. Environmental Protection Agency. (Adapted from Blumberg and Connolly 1996)

biochemical oxygen demand, coliforms, nutrients, and algal biomass (chlorophyll *a*). Calibration of the model resulted in a reasonable representation of net tidal circulation, although stratification associated with groundwater accretions could not be confirmed. Actual water-quality data varied more widely than did the simulated results, although the mean values were generally in good agreement. Data for calibration included the East Loch water temperature, which exceeded water-quality standards. The exceedence was apparently a result from the large permitted discharge (22.43 $m^3/s$ [792 $ft^3/s$]) of expended cooling water from the power-generation plant at Waiau. Application of the calibrated model for management of waste load discharges indicated that the water-quality violation could be substantially reduced by selective consolidation, elimination, and upgrading of wastewater discharges into the harbor.

Since 1996 the U.S. Navy–operated Fort Kamehameha sewage-treatment plant located at the eastern outer shore of the harbor has been the only outfall of permitted sewage effluent discharge (0.30 $m^3/s$ [10.67 $ft^3/s$]) (Stevenson et al. 1996), whereas at one time the harbor received discharges from 105 sewer outfalls (Dugan and Young 1973).

Māmala Bay was simulated with two models designed for the Māmala Bay Study. First, the initial mixing and the near- and far-field conditions resulting from the two huge ocean-outfall discharges (Sand Island and Barbers Point [Honouliuli]) were simulated with the Roberts, Snyder, and Baumgartner (RSB) model (Roberts 1996). The model was developed with the traditional similitude principle and laboratory experiments with towed diffusers in a density-stratified water tank (Roberts et al. 1989). For use in the Māmala Bay Study, the input oceanographic data were measured with current and thermistor strings near the diffusers in 1994–1995. Based on some 20,000 near-field simulations, the frequency of surfacing of the submerged Sand Island plume was predicted to be 22 percent with very high dilution. During the summer months, the plume would be deeply submerged but with a reduced dilution of about 600. Little effects of the discharges would be expected beyond about 5 km (3 mi) from the diffusers.

The second investigative modeling effort attempted to simulate the transport and fate of pathogenic organisms in the entire Māmala Bay (Blumberg and Connolly 1996). The transport part was assessed with a numerical hydrodynamic circulation model that is three-dimensional, unsteady, advective, and dispersive and involves vertical mixing and open boundaries. It was calibrated and compared with the measured water-surface elevation, current velocity, water temperature, salinity of the bay, and other observed data near the Ala Wai Canal mouth. The fate part was evaluated with mathematical formulation of a conceptual model specifically intended for Māmala Bay, based on conservation of mass and four loss processes. The most important loss was phototoxicity (die-off in sunlight), which is modeled as a second-order kinetic process dependent on irradiance and abundance of organisms. The rate constants were 0.80 $m^2/MJ$ for fecal coliform, 0.45 $m^2/MJ$ for enterococci, and zero for *C. perfringens,* based on data by Fujioka and Loh (1996). The other three loss processes were predation, starvation, and settling. The model was applied to the three microorganisms, total suspended solids, and biochemical oxygen demand. The bacterial data used for calibration included both the STORET data (1975–1993) and the 1993–1995 project data. Comparison of simulation and data was made with frequency distributions, as indicated in Figure 8.8 for enterococci. The model is biased toward low levels for enterococci, suggesting that the loss of this organism may be overestimated. Also, the model tends to overpredict *Giardia* and *Salmonella* and is most sensitive to the rate of phototoxicity.

The coastal-water database for Hawai'i is by no means sufficient to address most problems. Physical aspects of the database include tides, currents, and freshwater inputs. The water-quality aspects include the water column, sediment, and biota. The data and information appear in diverse sources, such as City and County of Honolulu (1972).

The Hawai'i data are generated for two broad purposes. The first is regulatory monitoring such as those performed by National Pollutant Discharge Elimination System permittees to comply with data-gathering requirements associated with wastewater discharges, and other supplemental monitoring usually performed by government agencies. The data are commonly subject to statistical processing. The repository of regulatory monitoring is known as STORET, which is maintained by the U.S. Environmental Protection Agency, Region IX, in San Francisco. The second is research inquiries, usually by higher education and research institutions (e.g., the University of Hawai'i) and certain government agencies and consultants. The information generated is traditionally published as journal articles and technical reports. In Hawai'i, the Water Resources Research Center, Sea Grant College Program, Hawai'i Institute of Marine Biology, and School of Earth Science and Technology are among the major contributors.

# GLOSSARY

**Algae** A large and diverse group of aquatic plants in salt and fresh waters. They make their own food by photosynthesis and range in size from microscopic single-celled plants to huge brown kelps.

**Anchialine ponds** Pools located in recent coastal lavas and elevated fossil reefs that have only a subsurface connection with the ocean and that respond to tidal fluctuations.

**Aquiclude** Poorly permeable formation that virtually precludes the flow of water and that serves as an impermeable boundary of an aquifer.

**Aquifer** Geologic formation that is composed of permeable and porous rock and that permits a substantial amount of water to be extracted.

**Aquitard** Poorly permeable formation in which the flow of water is highly restrained.

**Bacteria** One-celled microscopic (plant) organisms that are capable of free living for growth and reproduction, even in the absence of free oxygen. Many bacteria are a major cause of human diseases.

**Basal groundwater** Fresh groundwater resting on seawater in the subsurface. The name "basal" denotes the relatively low elevation of the upper surface of the groundwater; the highest measured in Hawaiʻi is 12.8 m (42 ft) above sea level.

**Basalt** A primitive volcanic rock formed from molten magma that effused from volcanic vents. The most common rock in Hawaiʻi.

**Biochemical oxygen demand (BOD)** The amount of oxygen needed by microorganisms to reduce biodegradable organic matter to stable compounds under aerobic conditions.

**Breccia** Cemented angular fragments of older rocks. Masses of angular blocks are called volcanic breccia.

**Caldera** A large, approximately oval, and sunken crater with steep cliffs and nearly flat floor.

**Chemical oxygen demand (COD)** The amount of oxygen required for oxidizing all organic matter to carbon dioxide and water. COD is always greater than BOD.

**Coral reef** Biogenic or organic mass of rock consisting of corals and other calcium carbonate–secreting animals; such reefs grow in warm, shallow marine waters.

**Effective porosity (specific yield)** A parameter that measures the storage of water in the saturated zone of an aquifer; it is the ratio of water volume drained, after saturation, under gravity to the bulk volume of an unconfined aquifer; the value ranges from a few percent to 30 percent; for confined aquifers see Chapter 6.

**Fungus** A wide variety of plants that cannot make their own food by photosynthesis. They have a simple structure with no roots or leaves and range in size from single-celled yeasts to mushrooms.

**Hydraulic conductivity** A parameter that measures the permeability of rock types; generally speaking, a value of 1 m/d is considered permeable to allow economic exploitation.

**Hydrolysis** Nucleophilic substitution in which water acts as a nucleophile and attacks the bond of an organic chemical. There is no change in the oxidation state of the organic during the transformation. The reaction may be abiotic or biotic. It is typically a second-order or pseudo-first-order reaction.

**Ion exchange** A sorptive process causing attenuation (retardation) of cations and certain radionuclides. It occurs between ions or ion complexes and charged solid particles such as clay.

**Isostatic adjustment** The sinking of the earth's crust caused by the great weight of volcanic mountains.

**Kinetics** Speed of chemical reactions. It is proportional to the concentration of the reactants raised by one, two, or three powers, thus known as first, second, or third order. The proportional constant is the rate in unit of reciprocal time.

**Lava** Molten rock that flows out of a volcano or fissures in the earth.

**Magma** Molten rock beneath the Earth's surface.

**Nephelinitic basalt** A fine-grained, dark-colored rock resembling basalt but containing nepheline in place of feldspar.

**Orographic lifting** Elevation of wind-driven air mass up the mountain slopes.

**Photolysis** Destruction of a cell by light.

**Pleistocene Epoch** Geological time period beginning about 2 million years ago and succeeded by the present Holocene Epoch around 10,000 years ago.

**Pliocene Epoch** Geological time period beginning about 12 million years ago and succeeded by the Pleistocene Epoch.

**Precipitation (chemical)** Separation of chemical ions from a solution as a solid.

**Protozoans** One-celled animals (some parasitic) in marine or fresh water. They are motile.

**Reduction** Reduction occurs when a chemical gains (accepts) electrons from a reducing agent (electron doner). Both abiota and biota reductions are common. In anaerobic environment, hydrogen sulfide, which causes a serious odor problem, is present primarily due to microbial reduction of sulfate.

**Rift zone** A belt of fissures and craters extending from the summit of a volcano to its base and providing passageway for the rising magma.

**Sorption** Attachment of dissolved molecules to solids or solid surfaces. It is influenced by the fluid motion that transports the molecules, charges of the molecule and surfaces, and other chemical and thermodynamic properties of the water, molecule, and solid. Adsorption is attachment of molecules to the solid surface. Absorption is the incorporation of the molecules within the structure of a solid.

**Still-stand** A period of stability of land not undergoing depression or elevation.

**Strike (bearing or trend)** Compass direction of the intersecting line between a dike and a horizontal plane.

**Temperature inversion layer** An extensive air layer in the lower troposphere within which the temperature increases instead of decreases with altitude. The tradewind inversion profoundly affects the climate and weather of the Hawaiian Islands.

**Tertiary Period** Geological time period beginning about 70 million years ago and consisting of five epochs, the last of which is the Pliocene Epoch.

**Transpiration** Evaporation of water through the leaves of a plant.

**Unconformity** A buried erosional surface that marks the absence of intermediate-age rocks, which either have never been deposited at that locality or have been eroded away after they were deposited and before the area was covered by younger rocks.

**Valence** Oxidation number of certain chemical elements. It is determined by the number of electrons that can be taken on, given up, or shared with other atoms.

**Virus** Submicroscopic infectious organism that can grow and reproduce only when it enters another cell. The host's metabolism is subverted to favor viral reproduction. Harsh measures such as desiccation, heat, and ultraviolet light are required to kill.

## REFERENCES

References marked with a star (*) have been placed in the Water Resources Research Center Library at the University of Hawai'i at Mānoa, Honolulu. The collection of references is identified as for Lau and Mink, *Hydrology of the Hawaiian Islands.*


Adams, W. M., and L. K. Lepley. 1968. Infrared images of the Kau and Puna coastlines on Hawaii. Technical Report No. 26, Water Resources Research Center, University of Hawai'i at Mānoa, Honolulu. 15 p.

Ahuja, L. R. 1983. Modeling infiltration into crusted soils by the Green-Ampt approach. *Soil Science Society of America Journal* 47:412–418.

Ahuja, L. R., E. W. Dangler, and S. A. El-Swaify. 1976a. Predicting runoff initiation times under field conditions in tropical (Hawaii) soils. *Soil Science Society of America Journal* 40:777–779.

Ahuja, L. R., and S. A. El-Swaify. 1979. Determining soil hydrologic characteristics on a remote forest watershed by continuous monitoring of soil-water pressure, rainfall and runoff. *Journal of Hydrology* 44:135–147.

Ahuja, L. R., S. A. El-Swaify, and A. Rahman. 1976b. Measuring hydrologic properties of soil with a double-ring infiltrometer and multiple-depth tensiometers. *Soil Science Society of America Journal* 40:494–499.

Ahuja, L. R., R. E. Green, S.-K. Chong, and D. R. Nielsen. 1980. A simplified functions approach for determining soil hydraulic conductivities and water characteristics in situ. *Water Resources Research* 16 (5): 947–953.

Ahuja, L. R., and D. R. Nielsen. 1990. Field soil-water relations. In *Irrigation of agricultural crops*, 143–190. Agronomy Monograph No. 30, American Society of Agronomy, Madison, Wisconsin.

*Akinaka, A. Y., R. Y. Akinaka, S. P. Bowles, and J. F. Mink. 1975. Kohala water resources management and development plan. Phase III. Unpublished report. 77 p.

American Public Works Association. 1969. *Water pollution aspects of urban runoff.* Federal Water Pollution Control Administration, Washington, D.C. 272 p.

American Society of Agronomy and the Soil Science Society of America. 1986. Methods of soil analysis, Part 1, Physical and mineralogical methods. Agronomy Monograph No. 9. Madison, Wisconsin.

Anderson, H. W., P. D. Duffy, and T. Yamamoto. 1966. Rainfall and streamflow from small tree-covered and fern-covered and burned watersheds in Hawaii. Research Paper PSW-34, U.S. Forest Service, Berkeley, California. 10 p.

Anderson, M. P. 1995. Groundwater modeling in the 21st century. In *Groundwater models for resources analysis and management*, ed. A. I. El-Kadi, 79–93. Boca Raton: Lewis Publishers.

Anderson, M. P., D. S. Ward, E. G. Lappala, and T. A. Prickett. 1993. Computer models for subsurface water. In *Handbook of hydrology*, ed. D. R. Maidment, 22.1–22.34. New York: McGraw-Hill.

Andrews, C. B. 1909. The structure of the southeastern portion of the island of Oahu. M.S. thesis, Rose Polytechnic Institute, Terre Haute, Indiana. 18 p.

Anthony, S. S. 1997. Evaluation of the U.S. Geological Survey ground-water data-collection program in Hawaii. Water-Resources Investigations Report 97-4232, U.S. Geological Survey, Honolulu, Hawai'i. 76 p.

Back, W., and B. B. Hanshaw. 1965. Chemical geohydrology. In *Advances in hydroscience*, vol. 2, ed. V. T. Chow, 49–109. New York: Academic Press.

Badon-Ghyben, W. 1888/1889. Nota in verband met de voorgenomen putboring nabij Amsterdam [Notes on the probable results of well drilling near Amsterdam]. *Tijdschrift van het Koninklijk Instituut van Ingenieurs* (The Hague), 8–22.

Barcelona, M., J. F. Keely, W. A. Pettyjohn, and A. Wehrmann. 1987. Handbook of ground water. Center for Environmental Research Information, U.S. Environmental Protection Agency, Cincinnati, Ohio. 212 p.

Barry, R. G., and R. J. Chorley. 1992. *Atmosphere, weather and climate*, 6th ed. New York: Routledge.

Bathen, K. H. 1978. Circulation atlas for Oahu, Hawaii. MR-78-05, University of Hawai'i Sea Grant College Program, Honolulu. 94 p.

Bean, C., J. O. Juvik, and D. Nullet. 1994. Mountain evaporation profiles on the island of Hawaii. *Journal of Hydrology* 156:181–192.

Bear, J. 1972. *Dynamics of fluids in porous media*. New York: American Elsevier. 764 p.

———. 1979. *Hydraulics of groundwater*. New York: McGraw-Hill. 567 p.

Bear, J., A. H.-D. Cheng, S. Sorek, D. Ouazar, and I. Herrera, eds. 1999. *Seawater intrusion in coastal aquifers: Concepts, methods, and practices*. Norwell, Massachusetts: Kluwer Academic Publishers.

Bear, J., and A. Verruijt. 1987. *Modeling groundwater flow and pollution*. Dordrecht, Holland: D. Reidel Publishing Co. 414 p.

Beckwith, M. W. 1951. *The Kumulipo: A Hawaiian creation chant*, 331–333. Chicago: University of Chicago Press.

Bennett, H. H. 1939. *Soil conservation*. New York: McGraw-Hill.

Bienfang, P. 1980. Water quality characteristics of Honokohau Harbor: A subtropical embayment affected by groundwater intrusion. *Pacific Science* 34:279–291.

Biggar, J. W., and D. R. Nielsen. 1976. Spatial variability of the leaching characteristics of a field soil. *Water Resources Research* 12 (1): 78–84.

Biswas, A. K. 1972. *History of hydrology*. Amsterdam: North-Holland Publishing Co.

Black, T. A. 1979. Evapotranspiration from Douglas fir stands exposed to soil water deficits. *Water Resources Research* 15 (1): 164–197.

Blaney, H. F., and W. O. Criddle. 1950. Determining water requirements in irrigated areas from climatological and irrigation data. TP-96, Division of Irrigation and Water Conservation, Soil Conservation Service, U.S. Department of Agriculture, Washington, D.C.

*Blumberg, A. F., and J. P. Connolly. 1996. Modeling transport and fate of pathogenic organisms in Mamala Bay. In Mamala Bay Study, Final Report, Vol. I. Honolulu. 122 p.

Blumenstock, D. I., and S. Price. 1967. Climates of the states: Hawaii. In *Climatology of the United States*. No. 60-51, U.S. Department of Commerce, Washington, D.C. 27 p.

Boucher, R. J., and R. J. Newcomb. 1962. Synoptic interpretation of some TIROS vortex patterns: A preliminary cyclone model. *Journal of Applied Meteorology* 1:122–136.

Brakensiek, D. L., H. B. Osborn, and W. J. Rawls, eds. 1979. Field manual for research in agricultural hydrology. Science and Education Administration, U.S. Department of Agriculture, Washington, D.C. 547 p.

Brasher, A. M., and S. S. Anthony. 2000. Occurrence of organochlorine pesticides in streambed sediment and fish from selected streams on the island of O'ahu, Hawai'i, 1998. U.S. Geological Survey, Honolulu, Hawai'i.

Bresler, E., and R. E. Green. 1982. Soil parameters and sampling scheme for characterizing soil hydraulic properties of a watershed. Technical Report No. 148, Water Resources Research Center, University of Hawai'i at Mānoa, Honolulu. 42 p.

Brock, J. A. 1998. Necropsy and liver histopathology for fish sampled in the vicinity of Waianae ocean outfall, Oahu, Hawaii. September–June 1998. Project Report PR-99-03, Water Resources Research Center, University of Hawai'i at Mānoa, Honolulu. 12 p.

Brock, R. E. 1996. A study of the impact of hurricane Iniki on coral communities at selected sites in Mamala Bay, Oahu, Hawaii. Project Report PR-96-09, Water Resources Research Center, University of Hawai'i at Mānoa, Honolulu. 23 p.

———. 1999. Annual review of the water quality monitoring program for Waikoloa, 1998. Environmental Assessment Company, 1820 Kihi Street, Honolulu, Hawai'i. 77 p.

Brooks, N. H. 1960. Diffusion of sewage effluent in an ocean current. In *Waste disposal in the marine environment*, ed. E. A. Pearson, 246–267. New York: Pergamon Press.

Brutseart, W., and H. Stricker. 1979. An advection-aridity approach to estimate actual regional evapotranspiration. *Water Resources Research* 15 (2): 443–450.

Buxton, D. S., and R. E. Green. 1992. Desorption and leachability of sorbed DBCP residues in Hawaii soils. Technical Report No. 180. Water Resources Research Center, University of Hawai'i at Mānoa, Honolulu. 72 p.

Cabelli, V. J. 1983. Health effects criteria for marine recreational waters. EPA 600/1-80-031, U.S. Environmental Protection Agency, Washington, D.C.

Cagauan, B. G., L. S. Lau, R. E. Green, and G. Uehara. 1968. Solute dispersion in two Hawaiian soils under saturated flow. In Transactions, 9th International Congress of Soil Science, Paper 20. University of Adelaide, Adelaide, Australia. Vol. 1, p. 185–194.

Calder, I. R. 1993. Hydrologic effects of land-use changes. In *Handbook of hydrology*, ed. D. R. Maidment, 13.1–13.50. New York: McGraw-Hill.

Chang, J. H. 1993. Hydrology in humid tropical Asia. In *Hydrology and water management in the humid tropics*, ed. M. Bonell, M. M. Hufschmidt, and J. S. Gladwell, 55–66. Cambridge: Cambridge University Press.

Chang, J. H., R. B. Campbell, H. W. Brodie, and L. D. Baver. 1967. Evapotranspiration research at the HSPA Experimental Station. *Proceedings of the International Society of Sugar Cane Technologists* 12:10–24.

Chang, J. H., and L. S. Lau. 1993. Definition of the humid tropics. In *Hydrology and water management in the humid tropics*, ed. M. Bonell, M. M. Hufschmidt, and

J. S. Gladwell, 571–574. Cambridge: Cambridge University Press.

Chang, J. H., and G. Okimoto. 1970. Global water balance according to the Penman approach. *Geographical Analysis* 2 (1): 55–67.

Chen, C., R. E. Green, D. M. Thomas, and J. A. Knuteson. 1995. Modeling 1,3-dichloropropene fumigant volatilization with vapor-phase advection in the soil profile. *Environmental Science and Technology* 29 (7): 1816–1821.

Chen, C., D. M. Thomas, R. E. Green, and R. J. Wagenet. 1993. Two-domain estimation of hydraulic properties in macropore soils. *Soil Science Society of America Journal* 57:680–686.

Chen, Y.-L., and A. J. Nash. 1994. Diurnal variation of surface airflow and rainfall frequencies on the island of Hawaii. *Monthly Weather Review* 122:34–56.

Cheng, A. H.-D., and D. Ouazar. 1999. Analytical solutions. In *Seawater intrusion in coastal aquifers: Concepts, methods, and practices*, ed. J. Bear, A. H.-D. Cheng, S. Sorek, D. Ouazar, and I. Herrera, 163–191. Norwell, Massachusetts: Kluwer Academic Publishers.

Cheng, E. D. H. 1992. Debris flow prediction for an urbanized watershed. *Proceedings*, 6th IAHR International Symposium on Stochastic Hydraulics, Taipei, 213–220.

*Chong, S.-K., and Y. S. Fok. 1973. Linear and nonlinear watershed rainfall response, Oahu, Hawaii. (Paper presented at the 1973 winter meeting, American Society of Agricultural Engineers, Conrad Hilton Hotel, Chicago, Illinois, December 11–14, 1973, 10 p.)

Chong, S.-K., R. E. Green, and L. R. Ahuja. 1981. Simple in-situ determination of hydraulic conductivity by power function descriptions of drainage. *Water Resources Research* 17 (4): 1109–1114.

Chow, V. T. 1954. The log-probability law and its engineering applications. *Proceedings American Society of Civil Engineers* 80:1–25.

———. 1959. *Open-channel hydraulics.* New York: McGraw-Hill.

———. 1964a. Statistical and probability analysis of hydrologic data. Part I. Frequency analysis. In *Handbook of applied hydrology*, ed. V. T. Chow, 8-1 to 8-42. New York: McGraw-Hill.

———, ed. 1964b. *Handbook of applied hydrology.* New York: McGraw-Hill.

*———. 1966. An investigation of the drainage problems of the City and County of Honolulu. Unpublished report. 45 p.

Chow, V. T., D. R. Maidment, and L. W. Mays. 1988. *Applied hydrology.* New York: McGraw-Hill. 572 p.

Chu, P.-S. 1989. Hawaiian drought and the Southern Oscillation. *International Journal of Climatology* 9:619–631.

———. 1995. Hawaii rainfall anomalies and El Nino. *Journal of Climate* 8 (6): 1697–1703.

———. 1999. El Nino and La Nina effects on rainfall in Hawaii. *Proceedings,* Universities Council on Water Resources 1999 Annual Meeting, Kamuela, Hawai'i, June 29–July 2, p. 209–212.

Chu, P.-S., and Y. He. 1994. Long-range prediction of Hawaiian winter rainfall using canonical correlation analysis. *International Journal of Climatology* 14:659–669.

Chu, P.-S., A. J. Nash, and F.-Y. Porter. 1993. Diagnostic studies of two contrasting rainfall episodes in Hawaii: Dry 1981 and wet 1982. *Journal of Climate* 6 (7): 1457–1462.

Chun, M. J., R. H. F. Young, and G. K. Anderson. 1972. Wastewater effluents and surface runoff quality. Technical Report No. 63, Water Resources Research Center, University of Hawai'i at Mānoa, Honolulu. 67 p.

City and County of Honolulu. 1972. Water quality program for Oahu with special emphasis on waste disposal. Final Report. Department of Public Works, Honolulu, Hawai'i.

———. 1988. Storm drainage standards. Department of Public Works, Honolulu, Hawai'i. 41 p.

———. 1996. Soil erosion: Standards and guidelines. Supplement I and 1975 document. Department of Public Works, Honolulu, Hawai'i.

Clague, D. A. 1998. Geology. In *Atlas of Hawai'i*, 3rd ed., ed. S. P. Juvik and J. O. Juvik, 37–46. Honolulu: University of Hawai'i Press.

Clague, D. A., and G. B. Dalrymple. 1988. Geologic evolution. In *Volcanism in Hawaii,* vol. 1, ed. R. W. Decker, T. L. Wright, and P. H. Stauffer, 1–54. U.S. Geological Survey Professional Paper 1350, U.S. Government Printing Office, Washington, D.C.

Clark, T. L. 1979. Numerical simulations with a three-dimensional cloud model: Lateral boundary condition experiments and multicellular severe storm simulations. *Journal of Atmospheric Sciences* 36:2191–2215.

Clark, T. L., and R. Farley. 1984. Severe downslope windstorm calculations in two and three spatial dimensions using an elastic interactive grid nesting: A possible mechanism for gustiness. *Journal of Atmospheric Sciences* 41:329–350.

Cline, M. G., ed. 1955. *Soil survey*, Territory of Hawaii, Islands of Hawaii, Kauai, Lanai, Maui, Molokai, and Oahu. Soil Survey Series 1939, No. 25. U.S. Soil Conservation Service, Washington, D.C.

*Colwell, R. R., G. T. Orlob, and J. R. Schubel. 1996. Executive summary. In Mamala Bay Study, Final Report, Vol. I. Honolulu. 56 p.

Cooley, K. R., and L. J. Lane. 1982. Modified runoff curve

numbers for sugarcane and pineapple fields in Hawaii. *Journal of Soil and Water Conservation* 37 (5): 295–298.

Cornell, C. A. 1972. First-order analysis of model and parameter uncertainty. In *Proceedings, International Symposium on Uncertainties in Hydrologic and Water Resources Systems*, Tucson, Arizona, vol. 3, 1245–1274.

County of Maui. 1995. Rules for the design of storm drainage facilities in the County of Maui. Title MC-15, Chapter 4. Department of Public Works and Waste Management, Wailuku. 12 p.

———. 1998. Soil erosion and sedimentation control. Ordinance No. 2684. Department of Public Works and Waste Management, Wailuku.

Cox, D. C. 1981. Groundwater technology in Hawaii. In *Groundwater in Hawaii: A century of progress*, ed. F. N. Fujimura and W. B. C. Chang, 51–79. Water Resources Research Center, University of Hawai'i at Mānoa, Honolulu.

Cox, D. C., P.-F. Fan, K. E. Chave, R. I. Clutter, K. R. Gundersen, N. C. Burbank Jr., L. S. Lau, and J. R. Davidson. 1973. Estuarine pollution in the state of Hawaii — Volume II: Kaneohe Bay study. Technical Report No. 31, Water Resources Research Center, University of Hawai'i at Mānoa, Honolulu. 44 p.

Cox, D. C., and L. C. Gordon Jr. 1970. Estuarine pollution in the state of Hawaii. Vol. 1: Statewide study. Technical Report No. 31, Water Resources Research Center, University of Hawai'i at Mānoa, Honolulu. 131 p.

Cox, R. H. 1981. Groundwater technology in Hawaii. In *Groundwater in Hawaii: A century of progress*, ed. F. N. Fujimura and W. B. C. Chang, 15–30. Water Resources Research Center, University of Hawai'i at Mānoa, Honolulu.

Dagan, G., and J. Bear. 1968. Solving the problem of interface upconing in a coastal aquifer by the method of small perturbation. *Journal of Hydraulic Research* 6 (1): 15–44.

Dagan, G., and D. G. Zeitoun. 1999. Steady interface in stratified aquifers of random permeability distribution. In *Seawater intrusion in coastal aquifers: Concepts, methods, and practices*, ed. J. Bear, A. H.-D. Cheng, S. Sorek, D. Ouazar, and I. Herrera, 193–211. Norwell, Massachusetts: Kluwer Academic Publishers.

Dale, R. H. 1974. Relationship of ground-water tides to ocean tides: A digital simulation model. Ph.D. dissertation, University of Hawai'i at Mānoa, Honolulu. 150 p.

Dashiell, E. P. 1997. Watershed problems and best management practices for Ala Wai Canal watershed. City and County of Honolulu, Hawai'i.

Davis, G. H., J. H. Green, F. H. Olmsted, and D. W. Brown. 1959. Ground-water conditions and storage capacity in the San Joaquin Valley, California. U.S. Geological Survey Water Supply Paper 1469:164–168.

Davis, S. N. 1969. Porosity and permeability of natural materials. In *Flow through porous media*, ed. R. J. M. De Wiest, 54–89. New York: Academic Press.

DeCarlo, E. H., and S. S. Anthony. 2002. Spatial and temporal variability of trace element concentrations in an urban subtropical watershed, Honolulu, Hawaii. *Applied Geochemistry* 17:475–492.

DeCarlo, E. H., and K. J. Spencer. 1995. Records of lead and other heavy metal inputs to sediments of the Ala Wai Canal, O'ahu, Hawai'i. *Pacific Science* 49:471–491.

de Marsily, G. 1986. *Quantitative hydrogeology: Groundwater hydrology for engineers*. San Diego: Academic Press. 464 p.

DeVito, P. A., G. L. Dugan, E. T. Murabayashi, and H. K. Gee. 1995. Nonpoint source pollution within the Anahulu River and Opaeula Stream drainage areas, Kaiaka-Waialua Bay, hydrologic unit area, Oahu, Hawaii. Project Report PR-95-11, Water Resources Research Center, University of Hawai'i at Mānoa, Honolulu. 81 p.

Diaz, G. E., O. Elbadawy, J. C. Hughes, and J. D. Salas. 1995. In search of hydrologic similarity: A case study on Molokai, Hawaii. *Proceedings*, 1995 annual summer symposium, American Water Resources Association, 227–236.

Dingman, S. L. 1994. *Physical hydrology*. Englewood Cliffs, New Jersey: Prentice-Hall. 575 p.

Dollar, S. J. 1979. Ecological response to relaxation of sewage stress off Sand Island, Oahu, Hawaii. Technical Report No. 124, Water Resources Research Center, University of Hawai'i at Mānoa, Honolulu. 78 p.

———. 1986. Response of benthic ecosystem to deep ocean sewage outfalls in Hawaii: Nutrient fluxes at the sediment-water interface. Ph.D. dissertation, University of Hawai'i at Mānoa, Honolulu. 217 p.

Dollar, S. J., and G. W. Tribble. 1993. Recurrent storm disturbance and recovery: A long-term study of coral communities in Hawaii. *Coral Reefs* 12:223–233.

Dooge, J. C. I. 1984. The waters of the Earth. *Hydrological Sciences Journal* 29 (2): 149–176.

Doorenbos, J., and W. O. Pruitt. 1977. Guidelines for predicting water requirements. Irrigation and Drainage Paper 24. Food and Agricultural Organization of the United Nations, Rome, Italy. 144 p.

Dorman, C. E., and R. H. Bourke. 1979. Precipitation over the Pacific Ocean, 30° S to 60° N. *Monthly Weather Review* 107:896–910.

Doty, R. D., H. B. Wood, and R. A. Merriam. 1981. Suspended sediment production from forested watersheds on Oahu, Hawaii. *Water Resources Bulletin* 17 (3): 399–405.

Dracup, J. A., E. D. H. Cheng, J. M. Nigg, and T. A. Schroeder. 1991. *The New Year's Eve flood on Oahu, Hawaii.* National Academy Press, Washington, D.C. 72 p.

Drazin, P. G. 1961. On the steady flow of a fluid of variable density past an obstacle. *Tellus* 13:239–251.

Dugan, G. L. 1977. Water quality of normal and storm-induced surface water runoff: Kaneohe Bay watershed, Oahu, Hawaii, February 1974 to March 1975. Technical Report No. 106, Water Resources Research Center, University of Hawai'i at Mānoa, Honolulu. 74 p.

Dugan, G. L., and P. C. Ekern. 1984. Chemical constituents of rainfall at different locations on Oahu, Hawaii. Technical Report No. 160, Water Resources Research Center, University of Hawai'i at Mānoa, Honolulu. 25 p.

Dugan, G. L., H. K. Gee, and L. S. Lau. 1984. Decontamination of chromium-contaminated soil in Hawaii. Technical Report No. 159, Water Resources Research Center, University of Hawai'i at Mānoa, Honolulu. 23 p.

Dugan, G. L., and L. S. Lau. 1981. Sewage irrigation and recharge consequences, Oahu. *Journal of the Environmental Engineering Division, American Society of Civil Engineers* 107 (EE4): 699–711.

Dugan, G. L., and R. H. F. Young. 1973. Effects of coastal waste disposal in Hawaii. *Journal of the Environmental Engineering Division, American Society of Civil Engineers* 99 (EE5): 691–701.

Dupuit, J. 1863. *Etudes théoriques et pratiques sur le mouvement des eaux dans les canaux découverts et á travers les terrains perméables,* 2nd ed. Dunod, Paris. 304 p.

Eagleson, P. S. 1970. *Dynamic hydrology.* New York: McGraw-Hill. 462 p.

Easton, W. H., and E. A. Olson. 1976. Radiocarbon profile of Hanauma Reef, Oahu, Hawaii. *Geological Society of America Bulletin* 87:711–719.

Einstein, H. A. 1950. The bed load function for sediment transportation in open channel flow. U.S. Department of Agriculture Technical Bulletin 1026.

Ekern, P. C. 1950. Raindrop impact as the force initiating soil erosion. *Soil Science Society of America Journal* 15:7–10.

———. 1964. Direct interception of cloud water on Lanaihale, Hawaii. *Soil Science Society of America Journal* 28:419–421.

———. 1965a. Disposition of net radiation by a free water surface in Hawaii. *Journal of Geophysical Research* 70 (4): 798–800.

———. 1965b. Evapotranspiration of pineapple in Hawaii. *Plant Physiology* 40:736–739.

———. 1966a. Evapotranspiration by Bermudagrass sod, *Cynodon dactylon* L. Pers., in Hawaii. *Agronomy Journal* 58:387–390.

———. 1966b. Evapotranspiration from bare low humic latosol in Hawaii. *Journal of Applied Meteorology* 5:431–435.

———. 1972. Use of water by sugarcane in Hawaii measured by hydraulic lysimeter. *Proceedings of the International Society of Sugar Cane Technologists* 14:805–812.

———. 1976. Turbidity and sediment-rating curves for streams on Oahu, Hawaii. In *Soil erosion: Prediction and control,* 242–254. Soil Society of America, Ankeny, Iowa.

———. 1977. Drip irrigation of sugarcane measured by hydraulic lysimeters, Kunia, Oahu. Technical Report No. 109, Water Resources Research Center, University of Hawai'i at Mānoa, Honolulu. 99 p.

———. 1983. Measured evaporation in high rainfall areas, leeward Koolau Range, Oahu, Hawaii. Technical Report No. 156, Water Resources Research Center, University of Hawai'i at Mānoa, Honolulu. 60 p.

———. 1993. Evaporation along a transect across southern Oahu, Hawaii. Project Report PR-94-01, Water Resources Research Center, University of Hawai'i at Mānoa, Honolulu. 23 p.

Ekern, P. C., and J. H. Chang. 1985. Pan evaporation: State of Hawaii, 1893–1983. Report R74, Division of Water and Land Development, Department of Land and Natural Resources, State of Hawai'i, Honolulu. 172 p.

El-Kadi, A. I. 1993. Core sampling versus field-measured data for modeling drainage in field soils: Validity of the equivalent media approach. *Advances in Water Resources* 16:153–162.

Ellen, S. D., R. K. Mark, S. H. Cannon, and D. L. Knifong. 1993. Map of debris-flow hazard in the Honolulu District of Oahu, Hawaii. Open-File Report 93-213. U.S. Geological Survey, Menlo Park, California. 25 p.

Elliott, W. P., and R. K. Reed. 1984. A climatological estimate of precipitation for the world ocean. *Journal of Applied Meteorology* 23:434–439.

El-Swaify, S. A., and K. R. Cooley. 1980. Sediment losses from agricultural watersheds in Hawaii. Agricultural Reviews and Manuals ARM-W-17. U.S. Department of Agriculture, Oakland, California. 124 p.

Engman, E. T. 1993. Remote sensing. In *Handbook of hydrology,* ed. D. R. Maidment, 24.1–24.23. New York: McGraw-Hill.

Engman, E. T., and R. J. Gurney. 1991. *Remote sensing in hydrology.* London: Chapman and Hall. 225 p.

Essaid, H. I. 1986. A comparison of the coupled freshwater-saltwater flow and the Ghyben-Herzberg sharp interface approaches to modeling of transient behavior in coastal aquifer systems. *Journal of Hydrology* 86:169–193.

———. 1990. The computer model Sharp, a quasi-three-dimensional finite-difference model to simulate freshwater and saltwater flow in coastal aquifer systems. Water-Resources Investigations Report 90-4130. U.S. Geological Survey, Menlo Park, California. 181 p.

Eto, M. A., N. C. Burbank Jr., H. W. Klemmer, and L. S. Lau. 1967. Behavior of selected pesticides with percolating water in Oahu soils. Technical Report No. 9, Water Resources Research Center, University of Hawai'i at Mānoa, Honolulu. 35 p.

Evans, R. A., S. J. Turnbull, and D. R. Richards. 1995. The use of the DOD groundwater modeling system (GMS) to model the hydrodynamics of central Oahu, Hawaii. *Proceedings*, Water resources and environmental hazards symposium, Honolulu, 257–266. American Water Resources Association.

Eyre, P. R. 1985. Simulation of ground-water flow in southeastern Oahu, Hawaii. *Ground Water* 23 (3): 325–330.

Falta, R. W., K. Priess, S. Finsterle, and A. Battirtelli. 1995. T2VOC user's guide. LBL-36400, Lawrence Berkeley Laboratory, Berkeley, California.

Federal Emergency Management Agency. 1988. Interagency flood hazard mitigation report in response to the January 8, 1988 disaster declaration covering the City and County of Honolulu, Oahu, Hawaii. Honolulu. 34 p.

Fiering, M. B. 1967. *Streamflow synthesis.* Cambridge: Harvard University Press. 139 p.

Fiering, M. B., and B. B. Jackson. 1971. *Synthetic streamflow.* American Geophysical Union, Washington, D.C. 98 p.

Fischer, H. B., E. J. List, R. C. Y. Koh, J. Imberger, and N. H. Brooks. 1979. *Mixing in inland and coastal waters.* New York: Academic Press. 483 p.

Fisher, W. A., D. A. Davis, and T. M. Souza. 1966. Fresh-water springs of Hawaii from infrared images. U.S. Geological Survey, Hydrologic Investigation Atlas HA-218. Washington, D.C.

Flament, P., S. Kennan, R. Lumpkin, M. Sawyer, and E. Stroup. 1998. The ocean. In *Atlas of Hawai'i*, 3rd ed., ed. S. P. Juvik and J. O. Juvik, 82–86. Honolulu: University of Hawai'i Press.

Fok, Y.-S. 1967. Preliminary report on flood hydrology and urban water resources, Oahu, Hawaii. Technical Report No. 64, Water Resources Research Center, University of Hawai'i at Mānoa, Honolulu. 47 p.

———. 1975. A comparison of the Green-Ampt and Philip two-term infiltration equations. *Transactions of the American Society of Agricultural Engineers,* Paper No. 74-2574, pp. 1073–1075.

Fok, Y.-S., and C. S. Miyasato. 1976. Low-flow frequency and stochastic analysis of irrigation ditch flows for central Maui, Hawaii. Technical Report No. 99, Water Resources Research Center, University of Hawai'i at Mānoa, Honolulu. 32 p.

Fok, Y.-S., E. T. Murabayashi, and S. Phamwon. 1977. Digital simulation and evaluation of storm drainage system: St. Louis Heights watershed model. Technical Report No. 114, Water Resources Research Center, University of Hawai'i at Mānoa, Honolulu. 138 p.

Fontaine, R. A. 1996. Evaluation of the surface-water quantity, surface-water quality, and rainfall data-collection programs in Hawaii. U.S. Geological Survey, Honolulu. 125 p.

Fontaine, R. A., R. I. Taogoshi, V. E. Kunishige, and W. S. Shibata. 1997. Water resources data, Hawaii, Water Year 1995. U.S. Geological Survey, Honolulu. 458 p.

Foote, D. E., E. L. Hill, S. Nakamura, and F. Stephens. 1972. Soil survey of islands of Kauai, Oahu, Maui, Molokai, and Lanai, State of Hawaii. U.S. Soil Conservation Service, Washington, D.C.

Franco, R. W. 1995. Water: Its meaning and management in pre-contact Hawaii. Project Report PR-96-02, Water Resources Research Center, University of Hawai'i at Mānoa, Honolulu. 76 p.

Fread, D. L. 1998. A perspective on hydrologic prediction trends. *Proceedings*, Special Symposium on Hydrology (January 11–16, 1998, Phoenix, Arizona), J-1 to J-6. American Meteorological Society, Boston, Massachusetts.

Freeze, R. A., and J. A. Cherry. 1979. *Groundwater.* Englewood Cliffs, New Jersey: Prentice-Hall. 604 p.

Fujioka, R. S. 1983. Stream water quality assessment based on fecal coliform and fecal streptococcus analysis. Technical Memorandum Report No. 70, Water Resources Research Center, University of Hawai'i at Mānoa, Honolulu. 37 p.

———. 1996. Indicators of marine recreational water quality. In *Manual of environmental microbiology*, ed. C. Hurst, G. Knudson, M. McInerny, L. Stetzenback, and M. Walter, 176–183. Washington, D.C.: American Society of Microbiology Press.

Fujioka, R. S., and L. S. Lau. 1987. A virus monitoring program for irrigating sugarcane with Mililani sewage effluent. Special Report 01:15:87, Water Resources Research Center, University of Hawai'i at Mānoa, Honolulu.

*Fujioka, R. S., and P. C. Loh. 1996. Characterization of the microbiological quality of water in Mamala Bay. In Mamala Bay Study, Final Report, Vol. I. Honolulu. 72 p.

Fujioka, R. S., and K. P. Roll. 1993. Assessing the effects of natural events and management practices on the microbial quality of recreational waters. Project Completion Report 04:08:93, Water Resources Research Center, University of Hawai'i at Mānoa, Honolulu. 71 p.

Fujioka, R., B. Roll, and M. Byappanahalli. 1997. Appropriate recreational water quality standards for Hawaii and other tropical regions based on concentrations of *Clostridium perfringens*. *Proceedings, Water Environment Federation 1997* 4:405–411.

Fujioka, R. S., and L. K. Shizumura. 1985. *Clostridium perfringens,* a reliable indicator of stream water quality. *Journal of Water Pollution Control Federation* 57:986–992.

Fujiwara, T. O. 1973. Characterization of urban stormwater discharges from separate storm sewers. M.S. thesis, University of Hawai'i at Mānoa, Honolulu.

Fullerton, C. M., and D. J. Raymond. 1973. Rainfall intensity instrument and measurements. Technical Report No. 67, Water Resources Research Center, University of Hawai'i at Mānoa, Honolulu. 32 p.

Fullerton, C. M., and S. K. Wilson. 1974. Some characteristics of Hawaiian high-intensity rainfall. Technical Report No. 78, Water Resources Research Center, University of Hawai'i at Mānoa, Honolulu. 47 p.

Gavenda, R., C. Smith, and N. Vollrath. 1998. Soils. In *Atlas of Hawai'i*, 3rd ed., ed. S. P. Juvik and J. O. Juvik, 92–96. Honolulu: University of Hawai'i Press.

Geldreich, E. E. 1976. Fecal coliform and fecal streptococcus density relationship in waste discharges and receiving waters. In *CRC Critical Reviews in Environmental Control,* 349–369. Boca Raton, Florida: CRC Press.

Gelhar, L. W. 1986. Stochastic subsurface hydrology from theory to applications. *Water Resources Research* 22 (9): 135S–145S.

———. 1993. *Stochastic subsurface hydrology.* Englewood Cliffs, New Jersey: Prentice-Hall. 390 p.

Gelhar, L. W., C. Welty, and K. D. Rehfeldt. 1992. A critical review of data on field-scale dispersion in aquifer. *Water Resources Research* 28 (7): 1955–1974.

*Gerba, C. P., C. N. Haas, and J. B. Rose. 1996. Risk of swimming—acquired illnesses in Mamala Bay. In Mamala Bay Study, Final Report, Vol. I. Honolulu. 23 p.

*Gerba, C. P., and I. L. Pepper. 1996. Overall impact of Sand Island outfall on the incidence of pathogens in Mamala Bay. In Mamala Bay Study, Final Report, Vol. I. Honolulu. 106 p.

Giambelluca, T. W. 1983. Water balance of the Pearl Harbor–Honolulu Basin, Hawaii 1946–1975. Technical Report No. 151, Water Resources Research Center, University of Hawai'i at Mānoa, Honolulu. 151 p.

Giambelluca, T. W., L. S. Lau, Y.-S. Fok, and T. A. Schroeder. 1984. Rainfall frequency study for Oahu. Report R73. Division of Water and Land Development, Department of Land and Natural Resources, State of Hawai'i, Honolulu. 232 p.

Giambelluca, T. W., D. L. McKenna, and P. C. Ekern. 1992. An automated recording atmometer: 1. Calibration and testing. *Agricultural and Forest Meteorology* 62:109–125.

Giambelluca, T. W., and D. Nullet. 1991. Influence of the trade-wind inversion on the climate of a leeward mountain slope in Hawaii. *Climate Research* 1:207–216.

———. 1992a. Evaporation at high elevations in Hawaii. *Journal of Hydrology* 136:219–235.

———. 1992b. An automated recording atmometer: 2. Evaporation measurement on a high elevation transect in Hawaii. *Agricultural and Forest Meteorology* 62:127–138.

Giambelluca, T. W., D. Nullet, M. A. Ridgley, P. R. Eyre, J. E. T. Moncur, and S. Price. 1991. Drought in Hawaii. Report R88, Commission on Water Resource Management, State of Hawai'i, Honolulu. 232 p.

Giambelluca, T. W., D. Nullet, and T. A. Schroeder. 1986. Rainfall atlas of Hawaii. Report R76, Division of Water and Land Development, Department of Land and Natural Resources, State of Hawai'i, Honolulu. 267 p.

Giambelluca, T. W., and M. Sanderson. 1993. The water balance and climatic classification. In *Prevailing trade winds: Weather and climate in Hawai'i,* ed. M. Sanderson, 56–72. Honolulu: University of Hawai'i Press.

Gilman, C. S. 1964. Rainfall. In *Handbook of hydrology*, ed. V. T. Chow, 9-1 to 9-68. New York: McGraw-Hill.

Gould, B. M. 1995. Water quality of a nearshore Hawaiian embayment: Variability and forcing processes. M.S. thesis, University of Hawai'i at Mānoa, Honolulu. 49 p.

Green, R. E., L. R. Ahuja, S.-K. Chong, and L. S. Lau. 1982. Water conduction in Hawaii oxic soils. Technical Report No. 143, Water Resources Research Center, University of Hawai'i at Mānoa, Honolulu. 122 p.

Green, R. E., K. P. Goswami, M. Mukhtar, and H. Y. Young. 1977. Herbicides from cropped watersheds in stream and estuarine sediments in Hawaii. *Journal of Environmental Quality* 6 (2): 145–154.

Green, R. E., and C. W. Guernsey. 1981. Soil-water relations and physical properties of irrigated soils in the Kula area, island of Maui, Hawaii. Research Bulletin 173, Hawai'i Agricultural Experiment Station, University of Hawai'i at Mānoa, Honolulu. 55 p.

Green, R. E., C. C. K. Liu, and N. Tamrakar. 1986. Modeling of pesticide movement in the unsaturated zone of Hawaii soils under agriculture use. In *Evaluation of pesticides in ground water*, ed. W. Y. Garner, R. C. Honeycutt, and H. N. Nigg, 366–383. American Chemical Society Symposium Series 315, Washington, D.C.

Green, T. R., R. E. Green, and M. E. Miller. 1993. Impacts of vertical heterogeneity on simulated water flow in Hawaiian basaltic saprolites relation to recharge. In *Water flow and solute transport in soils*, ed. D. Russo and G. Dagan, 233–246. New York: Springer-Verlag.

Green, W. H., and G. A. Ampt. 1911. Studies on soil physics, part 1, flow of air and water through soils. *Journal of Agricultural Science* 4 (1): 1–24.

Grigg, R. W. 1972. Some ecological effects of discharged sugar mill wastes on marine life along the Hamakua coast, Hawaii. Seminar Series 2:27–45, Water Resources Research Center, University of Hawaiʻi at Mānoa, Honolulu.

———. 1975. The impact of raw sewage effluent on shallow reef corals off Honolulu, Hawaii [abstract]. Pacific Science Congress, Vancouver, Canada. pp. 125–126.

———. 1985. Hamakua coast sugar mill ocean discharges: Before and after EPA compliance. UNIHI-SEAGRANT-TR-85-02. Sea Grant College Program, University of Hawaiʻi, Honolulu. 25 p.

———. 1995. Coral reefs in an urban embayment in Hawaii: A complex case history controlled by natural and anthropogenic stress. *Coral Reefs* 14:253–266.

Grigg, R. W., and S. J. Dollar. 1990. Natural and anthropogenic disturbance on coral reefs. In *Coral reefs*, ed. Z. Dubinsky, 439–452. Amsterdam: Elsevier Science Publishers.

Gupta, S. K., L. S. Lau, and P. S. Moravcik. 1994a. Groundwater tracing with injected helium. *Ground Water* 32 (1): 96–102.

Gupta, S. K., L. S. Lau, P. S. Moravcik, and A. El-Kadi. 1990. Injected helium: A new hydrological tracer. Special Report 06.01.90, Water Resources Research Center, University of Hawaiʻi at Mānoa, Honolulu. 94 p.

Gupta, S. K., P. S. Moravcik, and L. S. Lau. 1994b. Use of injected helium as a hydrological tracer. *Hydrological Sciences Journal* 39 (2): 109–119.

Handley, L. L., and P. C. Ekern. 1984. Effluent irrigation of para grass: Water, nitrogen, and biomass budgets. *Water Resources Bulletin* 20 (5): 669–677.

Haraguchi, P., and P. Matsunaga. 1985. The El Nino relationship to Oahu rainfall. Circular C112, Division of Water and Land Development, Department of Land and Natural Resources, State of Hawaiʻi, Honolulu.

Harder, J. A., L. S. Lau, T. R. Simpson, F. L. Hotes, and P. H. McGauhey. 1953. Laboratory research on sea water intrusion into fresh ground water sources and methods of its prevention. Final report. Institute of Engineering Research, University of California, Berkeley. 68 p.

Harding, D., and J. M. Miller. 1982. The influence on rain chemistry of Hawaiian volcano Kilauea. *Journal of Geophysical Research* 87 (C2): 1225–1230.

Harding Lawson Associates. 1996. Final remedial investigation report for operable unit 2, Schofield Army Barracks, island of Oahu, Hawaii. Denver, Colorado.

———. 1997. Final interim long-term groundwater monitoring report for operable units 2 and 4, Schofield Army Barracks, island of Oahu, Hawaii. Denver, Colorado.

Hardison, C. H. 1969. Accuracy of streamflow characteristics. Professional Paper 650-D, D210–D214. U.S. Geological Survey, Washington, D.C.

Hawaiʻi Commission on Water Resource Management. 1992. Rainfall trends in the Hawaiian islands: Through March 1992. Department of Land and Natural Resources, State of Hawaiʻi, Honolulu. 4 p.

———. 1993. Hawaii streams: Nākahawai o Hawaiʻi. State of Hawaiʻi, Honolulu. 40 p.

Hawaiʻi Cooperative Park Service Unit. 1990. Hawaii stream assessment: A preliminary appraisal of Hawaii's stream resources. Report R84, National Park Service, Honolulu. 294 p.

Hawaiʻi Department of Health. 1977. An ecosystem approach to water quality standards: Report of the Technical Committee on water quality standards. State of Hawaiʻi, Honolulu.

*———. 1989. Lead contamination in water cisterns on the island of Hawaii. Paper presented at the 4th international conference on rain water cistern systems, Manila, Philippines, Aug. 2–4. State of Hawaiʻi, Honolulu.

———. 1991a. Potable water systems. Chapter 20. Hawaiʻi Administrative Rules. State of Hawaiʻi, Honolulu.

———. 1991b. Wastewater systems. Chapter 62. Hawaiʻi Administrative Rules. State of Hawaiʻi, Honolulu.

———. 1992. Underground injection control. Chapter 23. Hawaiʻi Administrative Rules. State of Hawaiʻi, Honolulu.

———. 1994. Remediation and environmental monitoring conference. *Proceedings*, Honolulu, September 12–16. State of Hawaiʻi, Honolulu.

———. 1995a. Risk-based corrective action (RBCA) and decision making at sites with contaminated soil and groundwater. December 15, 1995. State of Hawaiʻi, Honolulu.

———. 1995b. Hawaii wellhead protection program plan. State of Hawaiʻi, Honolulu.

———. 1996. Updated groundwater contamination maps for the state of Hawaii. State of Hawaiʻi, Honolulu.

———. 1997. West Maui watershed owners manual. State of Hawaiʻi, Honolulu. 104 p.

———. 2000. Water quality standards. Chapter 54. Hawaiʻi Administrative Rules. State of Hawaiʻi, Honolulu.

———. 2002. Guidelines for the treatment and use of reclaimed water. State of Hawaiʻi, Honolulu. 110 p.

Hawaiʻi Department of Land and Natural Resources. 1977.

Hawaii water resources plan: Hawaii water resources regional study. State of Hawai'i, Honolulu. 247 p.
———. 1990. Hawaii stream assessment: A preliminary appraisal of Hawaii's stream resources. Report R84, State of Hawai'i, Honolulu. 294 p.
Hawai'i Department of Planning and Economic Development. 1982. Geothermal power development in Hawaii, vol. 1, 11–24. State of Hawai'i, Honolulu.
Hawai'i Division of Water and Land Development. 1970. Flood frequencies for selected streams in Hawaii. Report R36. Department of Land and Natural Resources, State of Hawai'i, Honolulu. 120 p.
———. 1973a. Climatologic stations in Hawaii. Report R42. Department of Land and Natural Resources, State of Hawai'i, Honolulu. 187 p.
———. 1973b. Pan evaporation in Hawaii: 1894–1970. Report R51. Department of Land and Natural Resources, State of Hawai'i, Honolulu. 82 p.
———. 1982. Median rainfall: State of Hawai'i. Circular 88, Department of Land and Natural Resources, State of Hawai'i, Honolulu. 45 p.
Hawai'i Revised Statutes. 1985. Chapter 205. Land Use Commission. State of Hawai'i, Honolulu.
———. 1987. Chapter 174C. State Water Code. State of Hawai'i, Honolulu.
Herzberg, A. 1901. Die wasserversorgung einiger Nordseebaden [The water supply on parts of the North Sea coast in Germany]. *Zeitschrift für Gasbeleuchtung und Wasserversorgung* 44:815–819, 824–844.
Hill, B. R. 1996. Streamflow and suspended-sediment loads before and during highway construction, North Halawa, Haiku, and Kamooalii drainage basin, Oahu, Hawaii, 1983–91. Water-Resources Investigations Report 96-4259, U.S. Geological Survey, Honolulu, Hawai'i. 34 p.
Hirashima, G. T. 1971. Availability of streamflow for recharge of the basal aquifer in the Pearl Harbor area, Hawaii. U.S. Geological Survey Water Supply Paper 1999-B. 26 p.
Hoover, D. J. 2002. Fluvial nitrogen and phosphorus in Hawaii: Storm runoff, land use, and impacts on coastal waters. Ph.D. dissertation, University of Hawai'i at Mānoa, Honolulu. 437 p.
Horel, J. D., and J. M. Wallace. 1981. Planetary scale atmospheric phenomena associated with the Southern Oscillation. *Monthly Weather Review* 109:819–829.
Hosking, J. R. M. 1990. L-moments: Analysis and estimation of distributions using linear combinations of order statistics. *Journal of the Royal Statistical Society*, Series *B* 52 (1): 105–124.
How, K. T. S. 1978. Solar radiation in Hawaii, 1932–1975. Report R57, Department of Land and Natural Resources, State of Hawai'i, Honolulu. 99 p.
Hubbert, M. K. 1940. The theory of ground-water motion. *Journal of Geology* 48 (8): 785–944.
Huber, R. D., and W. M. Adams. 1971. Density logs from underground gravity surveys in Hawaii. Technical Report No. 45, Water Resources Research Center, University of Hawai'i at Mānoa, Honolulu. 39 p.
Hudlow, M. D. 1988. Technological developments in real-time operational hydrologic forecasting in the United States. *Journal of Hydrology* 102:69–92.
Hufen, T. H. 1974. A geohydrologic investigation of Honolulu's basal waters based on isotope and chemical analyses of water samples. Ph.D. dissertation, University of Hawai'i at Mānoa, Honolulu. 160 p.
Hufen, T. H., R. W. Buddemeier, and L. S. Lau. 1972. Tritium and radiocarbon in Hawaiian natural waters: Part I. Technical Report No. 53, Water Resources Research Center, University of Hawai'i at Mānoa, Honolulu. 54 p.
Hufen, T. H., P. R. Eyre, and W. McConachie. 1980. Underground residence times and chemical quality of basal groundwater in Pearl Harbor and Honolulu aquifers, Oahu, Hawaii. Technical Report No. 129, Water Resources Research Center, University of Hawai'i at Mānoa, Honolulu. 75 p.
Hufen, T. H., L. S. Lau, and R. W. Buddemeier. 1974. Radiocarbon, $^{13}C$, and tritium in water samples from basaltic aquifers and carbonate aquifers on the island of Oahu, Hawaii. In *Proceedings of International Atomic Energy Agency Symposium, Isotope Techniques in Groundwater Hydrology*, Vienna, Austria, vol. 2, 111–127.
Hunt, C. R., and W. H. Snyder. 1980. Experiments on stable and neutrally stratified flow over a model three-dimensional hill. *Journal of Fluid Mechanics* 96:671–704.
Hurst, H. E. 1951. Long term storage capacities of reservoirs. *Transactions, American Society of Civil Engineers* 116:776–808.
Ishizaki, K., N. C. Burbank Jr., and L. S. Lau. 1967. Effects of soluble organics on flow through thin cracks of basaltic lava. Technical Report No. 16, Water Resources Research Center, University of Hawai'i at Mānoa, Honolulu. 56 p.
Jay, D. 1978. 100-year flood inundation maps for Hawaii. Seminar Series 2: 7–11, Water Resources Research Center, University of Hawai'i at Mānoa, Honolulu.
Jokiel, P. L., C. L. Hunter, S. Taguchi, and L. Watarai. 1993. Ecological impact of a fresh-water "reef kill" in Kaneohe Bay, Oahu, Hawaii. *Coral Reefs* 12:177–184.
Jones, B. L., and C. J. Ewart. 1973. Hydrology and sediment

transport: Moanalua Valley, Oahu, Hawaii. Report No. HI-HWY-71-1-I1. U.S. Geological Survey, Honolulu, Hawai'i. 111 p.

Jones, B. L., R. H. Nakahara, and S. S. Chin. 1971. Reconnaisance study of sediment transported by streams, island of Oahu. Circular C33, U.S. Geological Survey, Honolulu, Hawai'i. 44 p.

Jones, C. A. 1980. A review of evapotranspiration studies in irrigated sugarcane in Hawaii. *Hawaiian Planters' Record* 59 (9): 195–214.

Jury, W. A. 1982. Simulation of solute transport using a transfer function. *Water Resources Research* 18 (2): 363–368.

Juvik, J. O., and P. C. Ekern. 1978. A climatological mountain fog on Mauna Loa, Hawaii island. Technical Report No. 118, Water Resources Research Center, University of Hawai'i at Mānoa, Honolulu. 63 p.

Juvik, J. O., and D. Nullet. 1994. A climate transect through tropical montane rain forest in Hawaii. *Journal of Applied Meteorology* 33:1304–1312.

———. 1995. Relationship between rainfall, cloud-water interception, and canopy throughfall in a Hawaiian montane forest. In *Tropical montane cloud forests*, ed. L. S. Hamilton, J. O. Juvik, and F. N. Scatena, 165–182. New York: Springer-Verlag.

Juvik, J. O., D. C. Singleton, and G. G. Clarke. 1978. Climate and water balance on the island of Hawaii. In *Mauna Loa Observatory*, 20th Anniversary Report, ed. J. Miller, 129–139. NOAA Environmental Research Laboratory, U.S. Department of Commerce, Washington, D.C.

Kanehiro, B. Y., and F. L. Peterson. 1977. Groundwater recharge and coastal discharge for the northwest coast of the island of Hawaii: A computerized water budget approach. Technical Report No. 110, Water Resources Research Center, University of Hawai'i at Mānoa, Honolulu. 83 p.

Kay, E. A. 1987. Marine ecosystems in the Hawaiian islands. In *Reef and shore fauna of Hawaii*, ed. D. M. Devaney and L. G. Eldredge, 1–9. Bernice P. Bishop Museum Special Publication 64.

Kay, E. A., L. S. Lau, E. D. Stroup, S. J. Dollar, D. P. Fellows, and R. H. F. Young. 1977. Hydrologic and ecologic inventories of the coastal waters of west Hawaii. Technical Report No. 105, Water Resources Research Center, University of Hawai'i at Mānoa, Honolulu. 94 p.

Khan, M. A. 1979. Computer simulation of soil processes relevant to irrigation and fertilizer management. Ph.D. dissertation, University of Hawai'i at Mānoa, Honolulu. 289 p.

Khan, M. A., T. Liang, P. S. C. Rao, and R. E. Green. 1986. Use of an interactive computer graphics and mapping system to assess the potential for groundwater contamination with pesticides. EOS 67 (16): 278.

Kilonsky, B. J., and C. S. Ramage. 1976. A technique for estimating tropical open-ocean rainfall from satellite observations. *Journal of Applied Meteorology* 15:972–975.

Kinzie, R. A. 1988. Habitat utilization by Hawaiian stream fishes with reference to community structure in oceanic island streams. *Environmental Biology of Fishes* 22 (3): 179–192.

Kinzie, R. A., J. I. Ford, A. R. Yuen, and S. J. L. Chow. 1986. Habitat modeling of Hawaiian streams. Technical Report No. 171, Water Resources Research Center, University of Hawai'i at Mānoa, Honolulu. 126 p.

Kirch, P. V. 1985. *Feathered gods and fishhooks: An introduction to Hawaiian archaeology and prehistory.* Honolulu: University of Hawai'i Press. 349 p.

Kong, L.-J. 1996. A real-time automated control system to minimize the potential contamination of groundwater. M.S. thesis, University of Hawai'i at Mānoa, Honolulu. 166 p.

Konikow, L. F., and J. D. Bredehoeft. 1978. Computer model of two-dimensional solute transport and dispersion in groundwater. In *Techniques of water resources investigation*, book 7, chap. C2. Washington, D.C.: U.S. Geological Survey. 90 p.

Kroopnick, P. M., D. Thomas, L. S. Lau, R. W. Buddemeier, and D. Bills. 1980. Geochemical techniques in geothermal research—the Hawaii example. *Tectonophysics* 62:87–97.

Lamoureux, C. H. 1973. Plants. In *Atlas of Hawaii*, ed. R. W. Armstrong, 63–66. Honolulu: University Press of Hawai'i.

Landsberg, L. B. 1951. Statistical investigations into the climatology of rainfall on Oahu, T.H. *Meteorological Monographs* 1 (3): 7–23.

Lane, L. J., and M. A. Nearing, eds. 1989. USDA-Water erosion prediction project: Hillslope profile model documentation. U.S. Department of Agriculture, Agricultural Research Service, National Soil Erosion Research Laboratory Report No. 2, Lafayette, Indiana.

Lane, W. L., and D. K. Frevert. 1989. Applied stochastic techniques. Personal Computer version 5.2, User's manual, Earth Sciences Division, U.S. Bureau of Reclamation, Denver, Colorado.

Lao, C. 1995. Converging on sustainable yield for the Honolulu aquifer. In *Proceedings of American Water Resources Association Annual Summer Symposium, Water Resources and Environmental Hazards*, 217–226. Herndon, Virginia: American Water Resources Association.

Lao, C., F. L. Peterson, and D. C. Cox. 1969. Application of electric well logging and other well logging methods in Hawaii. Technical Report No. 21, Water Resources Research Center, University of Hawai'i at Mānoa, Honolulu. 108 p.

Lappala, E. G., R. W. Healy, and E. P. Weeks. 1987. Documentation of computer program VS2D to solve the equations of fluid flow in variably saturated porous media. Water-Resources Investigations Report 83-4009, U.S. Geological Survey, Reston, Virginia.

Lau, L. S. 1962. Water development of Kalauao basal springs: Hydraulic model studies. University of Hawai'i at Mānoa, Honolulu. 102 p.

———. 1967. Dynamic and static studies of seawater intrusion. Technical Report No. 3, Water Resources Research Center, University of Hawai'i at Mānoa, Honolulu. 31 p.

———. 1982. Professor Van Te Chow's contributions to Hawaii's storm drainage standards. *Water International* 7:102–108.

———. 1994. Artificial recharge of groundwater for all purposes. *Proceedings,* 2nd International Symposium on Artificial Recharge of Ground Water, 761–769. Walt Disney World Swan, Orlando, Florida, July 17–22. New York: American Society of Civil Engineers.

Lau, L. S., D. S. Buxton, R. S. Fujioka, H. K. Gee, T. W. Giambelluca, R. E. Green, K. M. Loague, M. E. Miller, J. F. Mink, E. T. Murabayashi, D. S. Oki, S. Orr, and F. L. Peterson. 1987. Organic chemical contamination of Oahu groundwater. Technical Report No. 181, Water Resources Research Center, University of Hawai'i at Mānoa, Honolulu. 153 p.

Lau, L. S., D. C. Cox, P. F. Fan, J. W. Hylin, H. W. Klemmer, P. H. McGauhey, E. D. Stroup, et al. 1972. The quality of coastal waters: First annual progress report. Technical Report No. 60, Water Resources Research Center, University of Hawai'i at Mānoa, Honolulu. 213 p.

Lau, L. S., G. L. Dugan, P. C. Ekern, R. H. F. Young, and N. C. Burbank Jr. 1975. Water recycling of sewage effluent by irrigation: A field study on Oahu — Final progress report for August 1971 to June 1975. Technical Report No. 94, Water Resources Research Center, University of Hawai'i at Mānoa, Honolulu. 151 p.

Lau, L. S., P. C. Ekern, P. C. S. Loh, R. H. F. Young, G. L. Dugan, R. S. Fujioka, and K. T. S. How. 1980. Recycling of sewage effluent by sugarcane irrigation: A dilution study, October 1976 to October 1978. Phase II-A. Technical Report No. 130, Water Resources Research Center, University of Hawai'i at Mānoa, Honolulu. 107 p.

*Lau, L. S., and A. El-Kadi. 1995. Injected helium technology for near-shore environmental water monitoring [abstract]. Paper presented at the American Geophysical Union 1995 fall meeting, San Francisco. 25 p.

Lau, L. S., W. R. Hardy, H. K. Gee, P. S. Moravcik, and G. L. Dugan. 1989. Groundwater recharge with Honouliuli wastewater irrigation: Ewa plain, southern Oahu, Hawaii. Technical Report No. 182, Water Resources Research Center, University of Hawai'i at Mānoa, Honolulu, 121 p.

Lau, L. S., W. J. Kaufman, G. T. Orlob, and D. K. Todd. 1958. Studies of dispersion in a radial flow system. Progress Report No. 4, Sanitary Engineering Research Laboratory, University of California, Berkeley. 67 p.

Lau, L. S., E. A. Kay, S. N. Luoma, J. F. Mink, S. A. Reed, A. R. Russo, H. Yamauchi, et al. 1973. The quality of coastal waters: Second annual progress report. Technical Report No. 77, Water Resources Research Center, University of Hawai'i at Mānoa, Honolulu. 275 p.

Lau, L. S., and J. F. Mink. 1967. A step in optimizing the development of the basal water lens of southern Oahu, Hawaii. International Association of Scientific Hydrology, Publication No. 72, 500–508. Wallingford, Oxfordshire, United Kingdom.

———. 1987. Organic contamination of groundwater: A learning experience. *American Water Works Association Journal* 79 (8): 37–42.

———. 1995a. Groundwater modeling in Hawaii: A historical perspective. In *Groundwater models for resources analysis and management*, ed. A. I. El-Kadi, 253–274. Boca Raton, Florida: CRC Press.

———. 1995b. Organic contamination of groundwater in Hawaii: A learning experience revisited. *Acta Universitatis Carolinae Geologica* (Prague, Czech Republic) 39:213–220.

Lau, L. S., S. Price, P. C. Ekern, R. Pulfrey, and F. L. Peterson. 1971. Hydrologic systems in Hawaii. In *Systems approach to hydrology*, ed. V. Yevjevich, 187–201. Water Resources Publications, Fort Collins, Colorado.

Lau, L. S., R. H. F. Young, S. K. Konno, R. J. Oldnall, and H. H. Lee. 1976. Wet-weather water quality monitoring: Kaneohe Bay, Oahu, Hawaii. Technical Report No. 100, Water Resources Research Center, University of Hawai'i at Mānoa, Honolulu. 126 p.

Lavoie, R. L. 1974. A numerical model of trade wind weather on Oahu. *Monthly Weather Review* 102:630–639.

Laws, E. A. 1993. *Aquatic pollution*, 2nd ed. New York: John Wiley and Sons. 611 p.

Laws, E. A., D. Zieman, and D. Schulman. 1999. Coastal water quality in Hawaii: The importance of buffer zone and dilution. *Marine Environmental Research* 48:1–21.

Lee, C.-C., R. E. Green, and W. J. Apt. 1986. Transformation and adsorption of fenamiphos, f. sulfoxide and f. sufone in Molokai soil and simulated movement with irrigation. *Journal of Contaminant Hydrology* 1:211–225.

Lee, R. 1984. Generalized skew coefficient for flood frequency

computations for the state of Hawaii. Water-Resources Investigations Report 84-4027, U.S. Geological Survey, Honolulu, Hawai'i. 21 p.
Leopold, L. B. 1949. The interaction of trade wind and sea breeze, Hawaii. *Journal of Meteorology* 6 (5): 312–320.
———. 1997. *Water, rivers and creeks.* Sausalito: Science Books. 185 p.
Leopold, L. B., M. G. Wolman, and J. P. Miller. 1964. *Fluvial processes in geomorphology.* New York: W. H. Freeman. 522 p.
Li, Y.-H. 1988. Denudation rates of the Hawaiian islands by rivers and groundwaters. *Pacific Science* 42:253–266.
Libby, W. F. 1962. Tritium geophysics: Recent data and results. In *Tritium in the physical and biological sciences,* Vol. 1, 5–32. Vienna, Austria: International Atomic Energy Agency.
Lin, P. 1993. Two-dimensional modeling analysis of volatile and non-volatile nematicides in Hawaiian soils. M.S. thesis, University of Hawai'i at Mānoa, Honolulu. 118 p.
Lin, P., C. C. K. Liu, R. E. Green, and R. Schneider. 1995. Simulation of 1, 3-dichloropropene in topsoil with pseudo first-order kinetics. *Journal of Contaminant Hydrology* 18:307–317.
Ling, G., and A. I. El-Kadi. 1998. A lumped parameter model for nitrogen transformation in the unsaturated zone. *Water Resources Research* 34 (2): 203–212.
Linsley, R. K., Jr., M. A. Kohler, and J. L. H. Paulhus. 1958. *Hydrology for engineers.* New York: McGraw-Hill. 340 p.
———. 1982. *Hydrology for engineers,* 3rd ed. New York: McGraw-Hill. 508 p.
Liu, C. C. K. 1988. Solute transport modeling in heterogeneous soils: Conjunctive application of physically based and systems approaches. *Journal of Contaminant Hydrology* 3:97–111.
Liu, C. C. K., C. Ewart, and Q. Huang. 1991a. Response of a basal-water body to forced draft. *Proceedings,* Symposium Ground Water in the Pacific Rim, 36–42. Honolulu, Hawai'i, July 22–24. New York: American Society of Civil Engineers.
Liu, C. C. K., R. E. Green, and P. Lin. 1993. Fumigant DBCP mixing and transport in soils. *Proceedings,* Joint CSCE-ASCE National Conference on Environment Engineering, 1233–1240. Montreal, Canada, July 12–14. New York: American Society of Civil Engineers.
Liu, C. C. K., P. Lin, and A. Firman. 1994. Modeling chemical transport in top soil with a gamma distribution function. *Water Science and Technology* 30 (2): 121–129.
Liu, C. C. K., K. Loague, and J.-S. Feng. 1991b. Fluid flow and solute transport processes in unsaturated heterogeneous soils. *Journal of Contaminant Hydrology* 7:161–283.
Liu, C. C. K., N. K. Tamrakar, and R. E. Green. 1987. Biodegradation and adsorption of DBCP and the mathematical simulation of its transport in tropical soils. *Toxicity Assessment* 2:239–252.
Lo, A., S. A. Swaify, and C. W. Rose. 1988. Analysis of erodibility of two tropical soils using a process model. *Soil Science Society of America Journal* 52:781–784.
Loague, K., T. W. Giambelluca, R. E. Green, C. C. K. Liu, T. C. Liang, and D. S. Oki. 1989. Simulation of organic chemical movement in Hawaii soils with PRZM: 2. Predicting deep penetration of DBCP, EDB, and TCP. *Pacific Science* 43:362–383.
Loague, K., R. E. Green, T. W. Giambelluca, T. C. Liang, and R. S. Yost. 1990. Impact of uncertainty in soil, climatic and chemical information in a pesticide leaching assessment. *Journal of Contaminant Hydrology* 5:171–194.
Loague, K., D. Lloyd, T. W. Giambelluca, A. Nguyen, and B. Sakata. 1996. Land misuse and hydrologic response: Kaho'olawe, Hawai'i. *Pacific Science* 50:1–35.
Loague, K., R. N. Miyahara, R. E. Green, D. S. Oki, T. W. Giambelluca, and R. C. Schneider. 1995. Chemical leaching near the Waiawa Shaft, Oahu, Hawaii: 2. Modeling results. *Ground Water* 33 (1): 124–138.
Loveridge, E. H. 1924. Diurnal variation of precipitation at Honolulu, Hawaii. *Monthly Weather Review* 52:584–585.
*Lum, L. W. K., and M. Nakasone. 1997. Water quality standards for wastewater disposal, central Oahu. Report prepared for U.S. Army Pacific. R. M. Towill Corporation, Honolulu, Hawai'i. 238 p.
Luoma, S. N. 1974. Mercury cycling in a small Hawaiian estuary. Technical Memorandum Report No. 42, Water Resources Research Center, University of Hawai'i at Mānoa, Honolulu. 38 p.
Lyman, R. T., T. A. Schroeder, and G. M. Barnes. 2005. The heavy rain event of 29 October 2000 in Hana, Maui. *Weather and Forecasting* 20 (4): 397–414.
Lyons, S. W. 1982. Empirical orthogonal function analysis of Hawaiian rainfall. *Journal of Applied Meteorology* 21:1713–1729.
Macdonald, G. A., and A. T. Abbott. 1970. *Volcanoes in the sea: The geology of Hawaii.* Honolulu: University Press of Hawai'i. 441 p.
Macdonald, G. A., A. T. Abbott, and F. L. Peterson. 1983. *Volcanoes in the sea: The geology of Hawaii,* 2nd ed. Honolulu: University of Hawai'i Press. 517 p.
Macdonald, G. A., D. A. Davis, and D. C. Cox. 1960. Geology

and ground-water resources of the island of Kauai, Hawaii. Bulletin No. 13, Hawai'i Division of Hydrography, Honolulu. 207 p.

*Maciolek, J. 1978. Environmental and biological restraints in storm drainage management. (In lecture notes for a workshop on storm drainage management in Hawaii, pp. II-E, 1–6. Water Resources Research Center, University of Hawai'i at Mānoa, Honolulu, January 17–18, 1978, Pagoda Hotel, Honolulu.)

Maidment, D. R., ed. 1993. *Handbook of hydrology.* New York: McGraw-Hill.

Makkind, G. F. 1957. Testing the Penman formula by means of lysimeters. *Journal of the Institution of Water Engineers* 11 (3): 277–288.

Maragos, J. E. 1998. Marine ecosystems. In *Atlas of Hawai'i,* 3rd ed., ed. S. P. Juvik and J. O. Juvik, 111–120. Honolulu: University of Hawai'i Press.

Maragos, J. E., C. Evans, and P. Holthus. 1985. Reef corals in Kaneohe Bay six years before and after termination of sewage discharges (Oahu, Hawaiian Archipelago). In *Proceedings of the 5th International Coral Reef Congress,* Tahiti, vol. 4, 189–194.

Margolis, S. 1988. *Adventuring in the Pacific.* San Francisco: Sierra Club Books. 406 p.

Matalas, N. C., and B. Jacobs. 1964. A correlation procedure for augmenting hydrologic data. U.S. Geological Survey Professional Paper 434-E:E1–E7, Washington, D.C.

Mather, J. R. 1978. *The climatic water budget in environmental analysis.* Lexington, Massachusetts: Lexington Books.

Matheron, G. 1965. *Les variables régionalisées et leur estimation.* Paris: Masson.

Matsushita, G. K., and R. H. F. Young. 1973. Baseline quality data for Kalihi Stream. Technical Report No. 71, Water Resources Research Center, University of Hawai'i at Mānoa, Honolulu. 61 p.

McCall, W. W. 1973. Soil classification in Hawaii. Circular 476. Cooperative Extension Service, University of Hawai'i at Mānoa, Honolulu.

McCutcheon, S. C., J. L. Martin, and T. O. Barnwell. 1993. Water quality. In *Handbook of hydrology,* ed. D. R. Maidment, 11.1–11.73. New York: McGraw-Hill.

McGauhey, P. H. 1968. *Engineering management of water quality.* New York: McGraw-Hill. 295 p.

McGauhey, P. H., and L. S. Lau. 1975. Management of Hawaii's coastal zone for water quality objectives. Technical Supplement No. 2, Hawai'i Coastal Zone Management Program, State of Hawai'i, Honolulu. 47 p.

McKnight, J. H., and J. O. Juvik. 1975. Methodological approaches in Hawaiian fog research. Technical Report No. 85, Water Resources Research Center, University of Hawai'i at Mānoa, Honolulu. 34 p.

McMurtry, G. M., A. Snidvongs, and C. R. Glenn. 1995a. Modeling sediment accumulation and soil erosion with $^{137}Cs$ and $^{210}Pb$ in the Ala Wai Canal and central Honolulu watershed, Hawai'i. *Pacific Science* 49:412–451.

McMurtry, G. M., J. C. Wiltshire, and J. P. Kauahikaua. 1995b. Heavy metal anomalies in coastal sediments of O'ahu, Hawai'i. *Pacific Science* 49:452–470.

Meinzer, O. E. 1923. The occurrence of ground water in the United States. U.S. Geological Survey Water Supply Paper 489. 321 p.

———, ed. 1942. *Hydrology.* New York: Dover Publications. 712 p.

Meisner, B. N. 1976. A study of Hawaiian and Line Islands rainfall. UHMET-76-04, Department of Meteorology, University of Hawai'i at Mānoa, Honolulu. 83 p.

———. 1979. Ridge regression—Time extrapolation applied to Hawaiian rainfall normals. *Journal of Applied Meteorology* 18:904–912.

Mercado, A. 1984. Utilization of the buffering properties of the unsaturated zone in groundwater management on a regional scale: The coastal aquifer of Israel as a case study. In *Pollutants in porous media: The unsaturated zone between soil surface and groundwater,* ed. B. Yaron, D. Dagan, and J. Goldshmid, 275–291. New York: Springer-Verlag.

Mercer, J. W., S. P. Larson, and C. R. Faust. 1980. Simulation of salt-water interface motion. *Ground Water* 18 (4): 374–385.

Meyers, C. K., D. C. Kleinecke, D. K. Todd, and L. E. Ewing. 1974. Mathematical modeling of fresh water aquifers having salt water bottoms. Technical Report GE74 TEM-43, Center for Advanced Studies, TEMPO, General Electric Company, report to Office of Water Resources Research, U.S. Department of the Interior. 176 p.

Miike, L. 1996. Updated groundwater contamination maps for the state of Hawaii. Department of Health, State of Hawai'i, Honolulu.

Miller, M. E., R. E. Green, F. L. Peterson, R. C. Jones, and K. Loague. 1988. Hydrogeologic characteristics of subsoil and saprolite and their relation to contaminant transport, central Oahu, Hawaii. Technical Report No. 178, Water Resources Research Center, University of Hawai'i at Mānoa, Honolulu. 76 p.

Mink, J. F. 1960. Distribution patterns of rainfall in the leeward Koolau mountains, Oahu, Hawaii. *Journal of Geophysical Research* 65 (9): 2869–2876.

———. 1962a. Excessive irrigation and the soils and groundwater of Oahu, Hawaii. *Science* 135:672–673.
———. 1962b. Rainfall and runoff in the leeward Koolau mountains, Oahu, Hawaii. *Pacific Science* 16:147–159.
———. 1964. Groundwater temperature in a tropical island environment. *Journal of Geophysical Research* 69 (24): 5225–5230.
———. 1978. Waianae water development study. Honolulu Board of Water Supply. Unpublished report (available at U.S. Geological Survey, Honolulu). 109 p.
———. 1980. State of the groundwater resources of southern Oahu. Board of Water Supply, City and County of Honolulu, Hawai'i. 83 p.
*———. 1982. DBCP and EDB in soil and water, Kunia, Oahu, Hawaii. Report prepared for Del Monte Corporation, Honolulu, Hawai'i. 32 p.
*———. 1997. Wailoa Ditch at Kamole Weir: Analysis of daily record for proposed storage requirements to supply County of Maui, Department of Water Supply. Unpublished report. 12 p.
———. 1998. Equivalency of global results for RAM and SHARP. In *The Pearl Harbor aquifer systems Waipahu-Waiawa and Waimalu: Proposed sustainable yields.* Mink and Yuen, Inc., Honolulu.
Mink, J. F., and L. S. Lau. 1980. Hawaii groundwater geology, hydrology, and early mathematical models. Technical Memorandum Report No. 62, Water Resources Research Center, University of Hawai'i at Mānoa, Honolulu. 74 p.
———. 1990a. Aquifer identification and classification for Oahu: Groundwater protection strategy for Hawaii. Technical Report No. 179, Water Resources Research Center, University of Hawai'i at Mānoa, Honolulu. 28 p.
———. 1990b. Aquifer identification and classification for Maui: Groundwater protection strategy for Hawaii. Technical Report No. 185, Water Resources Research Center, University of Hawai'i at Mānoa, Honolulu. 47 p.
———. 1992a. Aquifer identification and classification for Kauai: Groundwater protection strategy for Hawaii. Technical Report No. 186, Water Resources Research Center, University of Hawai'i at Mānoa, Honolulu. 37 p.
———. 1992b. Aquifer identification and classification for Molokai: Groundwater protection strategy for Hawaii. Technical Report No. 187, Water Resources Research Center, University of Hawai'i at Mānoa, Honolulu. 31 p.
———. 1993a. Aquifer identification and classification for Lanai: Groundwater protection strategy for Hawaii. Technical Report No. 190, Water Resources Research Center, University of Hawai'i at Mānoa, Honolulu. 20 p.
———. 1993b. Aquifer identification and classification for the island of Hawaii: Groundwater protection strategy for Hawaii. Technical Report No. 191, Water Resources Research Center, University of Hawai'i at Mānoa, Honolulu. 108 p.
———. 1997. Conservation zoning for groundwater source protection. *Proceedings,* Conference of water resources management, 121–134. August 10–13, Seattle, Washington. American Water Works Association, Denver.
Mink, J. F., E. T. Murabayashi, and L. S. Lau. 1993. Five-year review of state land use districts. Technical Report No. 188, Water Resources Research Center, University of Hawai'i at Mānoa, Honolulu. 32 p.
Mink, J. F., and S. T. Sumida. 1984. Aquifer classifications, State of Hawaii. Technical Memorandum Report No. 75, Water Resources Research Center, University of Hawai'i at Mānoa, Honolulu. 34 p.
Moberly, R., Jr. 1963. Rate of denudation in Hawaii. *Journal of Geology* 71 (3): 371–375.
Monteith, J. L. 1965. Evaporation and environment. In *The state of movement of water in living organisms.* Proceedings, Symposium of the Society for Experimental Biology 19:205–234.
Moore, J. G., W. R. Normark, and R. T. Holcomb. 1994. Giant Hawaiian underwater slides. *Science* 264:46–47.
Moore, S. F., G. S. Lowry, G. P. Young, and R. H. F. Young. 1981. Water quality simulation of Wahiawa Reservoir, Oahu, Hawaii. Technical Report No. 138, Water Resources Research Center, University of Hawai'i at Mānoa, Honolulu. 76 p.
Moravcik, P. S. 2000. Ten years of biological monitoring. Information Transfer Brochure 1002-01, Water Resources Research Center, University of Hawai'i at Mānoa, Honolulu. 11 p.
Murabayashi, E. T., and Y.-S. Fok. 1979. Urbanization induced impacts on infiltration capacity and on rainfall-runoff relation in an Hawaiian urban area. Technical Report No. 127, Water Resources Research Center, University of Hawai'i at Mānoa, Honolulu. 48 p.
Nakahara, R. H. 1980. An analysis of the magnitude and frequency of floods on Oahu, Hawaii. Water-Resources Investigations Report 80-45, U.S. Geological Survey, Honolulu, Hawai'i. 20 p.
Nakamura, M. N., and R. H. F. Young. 1974. Estimation of urban stormwater quality in Kalihi Stream drainage basin, Oahu, Hawaii. Technical Memorandum Report No. 45, Water Resources Research Center, University of Hawai'i at Mānoa, Honolulu. 39 p.
National Research Council. 1991a. *Opportunities in the hydrologic sciences.* National Academies Press, Washington, D.C. 348 p.

———. 1991b. *The New Year's Eve flood on Oahu, Hawaii: December 31, 1987–January 1, 1988*. National Academies Press, Washington, D.C. 72 p.

———. 1993. *Ground water vulnerability assessment: Predicting relative contamination potential under conditions of uncertainty*. National Academies Press, Washington, D.C. 204 p.

Natural Energy Laboratory of Hawai'i. 1986. Annual report for the comprehensive environmental monitoring program 1991–1992. Honolulu.

Nelson, W. G., J. H. Bailey-Brock, W. J. Cooke, and E. A. Kay. 1995. Assessment of effects of deep-water sewage effluent discharge on a nearshore, tropical benthic community, Honolulu, Hawaii. In *Challenges of our changing global environment,* vol. 2, 907–917. Ocean '95 MTS/IEEE Conference Proceedings, October 9–12, San Diego, California.

Neuman, S. P. 1990. Universal scaling of hydraulic conductivities and dispersivities in geologic media. *Water Resources Research* 26 (8): 1749–1758.

Nielsen, D. R., M. Th. Van Genuchten, and J. W. Biggar. 1986. Water flow and solute transport processes in the unsaturated zone. *Water Resources Research* 22 (9): 89S–108S.

Nishimura, J. K., and L. S. Lau. 1978. An automatic hydraulic structure to open sand plugs at stream mouths on Oahu. Technical Report No. 117, Water Resources Research Center, University of Hawai'i at Mānoa, Honolulu. 71 p.

Noguchi, Y. 1982. Water balance and exposure: Their implications to vegetation asymmetry under the tradewind regime. Ph.D. dissertation, University of Hawai'i at Mānoa, Honolulu.

Nullet, D. 1987. Energy sources for evaporation on tropical islands. *Physical Geography* 8 (1): 36–45.

Nullet, D., and T. W. Giambelluca. 1990. Winter evaporation on a mountain slope, Hawaii. *Journal of Hydrology* 112:257–265.

Nullet, D., and J. O. Juvik. 1994. Generalized mountain evaporation profiles for tropical and subtropical latitudes. *Singapore Journal of Tropical Geography* 15 (1): 17–24.

Nullet, D., and M. Sanderson. 1993. Radiation and energy balances and air temperature. In *Prevailing trade winds: Weather and climate in Hawai'i*, ed. M. Sanderson, 37–55. Honolulu: University of Hawai'i Press.

Ogata, A., and L. S. Lau. 1990. Vertical movement of saltwater-freshwater interface in a thick groundwater system. Technical Memorandum Report No. 82, Water Resources Research Center, University of Hawai'i at Mānoa, Honolulu. 9 p.

Oki, D. S. 1997. Modeling the effects of pumping, barometric pressure, and ocean tides on groundwater levels in northern Oahu, Hawaii. Ph.D. dissertation, University of Hawai'i at Mānoa, Honolulu. 321 p.

———. 1998. Geohydrology of the central Oahu, Hawaii, groundwater flow system and numerical simulation of the effects of additional pumping. Water-Resources Investigations Report 97-4276, U.S. Geological Survey. Honolulu, Hawai'i. 132 p.

———. 1999. Geohydrology and numerical simulation of the ground-water flow system of Kona, island of Hawaii. Water-Resources Investigations Report 99-4073, U.S. Geological Survey, Honolulu, Hawai'i. 70 p.

———. 2000. Site selections for a deep monitor well, Kualapuu, Molokai, Hawaii. Water-Resources Investigations Report 99-4291, U.S. Geological Survey, Honolulu, Hawai'i. 50 p.

Oki, D. S., L. S. Lau, and R. E. Green. 1991. Impact of proposed urbanization on groundwater quality: A case study in Hawaii. In *Proceedings* of 1991 Annual Conference of American Water Works Association, Philadelphia, Pennsylvania, 587–599.

Oreskes, N., K. Shrader–Frechette, and K. Belitz. 1994. Verification, validation, and confirmation of numerical models in the earth sciences. *Science* 263:641–646.

*Orlob, G. T. 1978. Mathematical modeling of Pearl Harbor. (A seminar presented at University of Hawai'i, Honolulu, June 17.)

Orr, S., and L. S. Lau. 1987. Trace organic (DBCP) transport simulation of Pearl Harbor aquifer, Oahu, Hawaii: Multiple mixing-cell model, Phase I. Technical Report No. 174, Water Resources Research Center, University of Hawai'i at Mānoa, Honolulu. 60 p.

———. 1988. Modeling of trace organic (DBCP) transport in Pearl Harbor aquifer, Oahu, Hawaii: Method of characteristics, Phase II. Technical Report No. 175, Water Resources Research Center, University of Hawai'i at Mānoa, Honolulu. 42 p.

*Orr, S., and J. Li. 1997. Modeling of virus transport in the subsurface, southern Oahu, Hawaii. A report submitted to the Water Resources Research Center, University of Hawai'i at Mānoa. 107 p.

Parham, J. E. 2002. Spatial models of Hawaiian streams and stream fish habitats. Ph.D. dissertation, Louisiana State University, Baton Rouge. 144 p.

Parks, N. M. 1998. Environmental quality. In *Atlas of Hawai'i*, 3rd ed., ed. S. P. Juvik and J. O. Juvik, 294–297. Honolulu: University of Hawai'i Press.

Patterson, S. H. 1971. Investigation of ferruginous bauxite and other mineral resources on Kauai and a reconnaissance of ferruginous bauxite deposits on Maui, Hawaii. U.S. Geological Survey Professional Paper 656, Washington, D.C.

Penman, H. L. 1948. Natural evaporation from open water, bare soil and grass. *Proceedings of the Royal Society of London, Series A* 193:120–146.
Phamwon, S., and Y.-S. Fok. 1977. Urban runoff digital computer model. *Journal of the Hydraulics Division, American Society of Civil Engineers* 103 (7): 723–735.
Philip, J. R. 1957. The theory of infiltration: 1. The infiltration equation and its solution. *Soil Science* 83 (5): 345–357.
Piechota, T. C., and J. A. Dracup. 1996. Drought and regional hydrologic variation in the United States: Association with the El Nino–Southern Oscillation. *Water Resources Research* 32 (5): 1359–1373.
Pilgrim, D. H., and I. Cordery. 1993. Flood runoff. In *Handbook of hydrology*, ed. D. R. Maidment, 9.1–9.42. New York: McGraw-Hill.
Ponce, V. M., and R. H. Hawkins. 1996. Runoff curve number: Has it reached maturity. *Journal of Hydrologic Engineering, American Society of Civil Engineers* 1 (1): 11–19. [Also see discussions, 1997, same journal, 2 (July): 145–148.]
Pontius, F. W. 1992. A current look at the federal drinking water regulations. *American Water Works Association Journal* 84 (3): 36–50.
———. 2003. Update on U.S. EPA's drinking water regulations. *American Water Works Association Journal* 95 (3): 57–68.
Pratt, L. W., and S. M. Gon III. 1998. Terrestrial ecosystem. In *Atlas of Hawai'i*, 3rd ed., ed. S. P. Juvik and J. O. Juvik, 121–129. Honolulu: University of Hawai'i Press.
Price, S. 1993. Sources of information on climate and weather in Hawaii. In *Prevailing trade winds: Weather and climate in Hawai'i*, ed. M. Sanderson, 111–142. Honolulu: University of Hawai'i Press.
Priestley, C. H. B., and R. J. Taylor. 1972. On the assessment of surface heat flux and evaporation using large-scale parameters. *Monthly Weather Review* 100:81–92.
Pringle, K. W., C. C. K. Liu, and R. E. Green. 1984. DBCP volatilization from soil and water: A laboratory study with two Hawaiian soils. Technical Report No. 157, Water Resources Research Center, University of Hawai'i at Mānoa, Honolulu. 95 p.
Pruess, K. 1991. TOUGH2—a general purpose numerical simulator for multiphase fluid and neat air flow, LBL-29400, Lawrence Berkeley Laboratory, Berkeley, California.
Raine, L. M., B. Z. Siegel, and G. M. McMurtry. 1995. Mercury accumulation in sediments of the Ala Wai Canal and in soils and stream sediments of the central Honolulu watershed. *Pacific Science* 49:511–525.
Ramage, C. S. 1971. *Monsoon meteorology.* New York: Academic Press.
Rao, P. S. C., A. G. Hornsby, and R. E. Jessup. 1985. Indices for ranking the potential for pesticide contamination of groundwater. *Soil and Crop Science Society of Florida Proceedings 44:1–8.*
Rasmussen, R. M., and P. K. Smolarkiewicz. 1993. On the dynamics of Hawaiian cloud bands. Part III. Local aspects. *Journal of Atmospheric Sciences* 50 (11): 1560–1572.
Rasmussen, R. M., P. K. Smolarkiewicz, and J. Warner. 1989. On the dynamics of Hawaiian cloud bands: Comparison of model results with observations and island climatology (Part II). *Journal of Atmospheric Sciences* 46 (11): 1589–1608.
Rasmusson, E. M., and J. M. Wallace. 1983. Meteorological aspects of the El Nino/Southern Oscillation. *Science* 222:1195–1202.
Rawls, W. J., L. R. Ahuja, D. L. Brakensiek, and A. Shirmahammadi. 1993. Infiltration and soil water movement. In *Handbook of hydrology*, ed. D. R. Maidment, 5.1–5.51. New York: McGraw-Hill.
Rawls, W. J., D. L. Brakensiek, and N. Miller. 1983. Green-Ampt infiltration parameters from soil data. *Journal of Hydraulic Engineering, American Society of Civil Engineers* 109 (1): 62–70.
Raymond, D. J., and K. Wilson. 1973. Development of a new rain intensity gauge. *Journal of Applied Meteorology* 13:180–181.
Reed, S. C., R. W. Crites, and E. J. Middlebrooks. 1995. *National systems for waste management and treatment.* New York: McGraw-Hill. 433 p.
Reilly, E. R., M. H. Frimpter, D. R. LeBlanc, and A. S. Goodman. 1987. Analysis of steady-state salt-water upconing with application to Truro well field, Cape Cod, Massachusetts. *Ground Water* 25 (2): 194–206.
Richards, L. A. 1931. Capillary conduction of liquids in porous media. *Physics* 1:318–333.
Riehl, H., T. C. Yeh, J. J. Malkus, and N. E. La Seur. 1951. The northeast trade of the Pacific Ocean. *Quarterly Journal of the Royal Meteorological Society* 77:598–626.
Ripperton, J. C., and E. Y. Hosaka. 1942. Vegetation zones of Hawaii. Bulletin No. 89, Hawai'i Agricultural Experiment Station, University of Hawai'i at Mānoa, Honolulu. 60 p.
*Roberts, P. J. W. 1996. Plume modeling. In Mamala Bay Study, Final Report, Vol. I. Honolulu. 45 p.
Roberts, P. J. W., W. H. Snyder, and D. J. Baumgartner. 1989. Ocean outfalls. *Journal of Hydrologic Engineering, American Society of Civil Engineers* 115 (1): 1–70.
Rojstaczer, S., and F. S. Riley. 1990. Response of the water level in a well to earth tides and atmospheric loading under unconfined conditions. *Water Resources Research* 26 (8): 1803–1817.
Roll, B. M., and R. S. Fujioka. 1997. Sources of fecal indicator

bacteria in a brackish, tropical stream and their impact on recreational water quality. *Water Science and Technology* 35 (11–12): 179–186.

Rose, C. W. 1993. Erosion and sedimentation. In *Hydrology and water management in the humid tropics*, eds. M. Bonell, M. M. Hufschmidt, and J. S. Gladwell, 301–343. Cambridge: Cambridge University Press.

Rose, C. W., J. R. Williams, G. C. Sander, and D. A. Barry. 1983. A mathematical model of soil erosion and deposition processes. *Soil Science Society of America Journal* 47:991–995.

Rosenfelt, D., D. B. Wolff, and D. Atlas. 1993. General probability-matched relations between radar reflectivity and rain rate. *Journal of Applied Meteorology* 32:50–72.

Russo, A. R. 1994. A survey of selected coral and fish assemblages near the Waianae Ocean Outfall, Oʻahu, Hawaiʻi, 1994. Project Report PR-95-04, Water Resources Research Center, University of Hawaiʻi at Mānoa, Honolulu. 23 p.

———. 1995. Survey of selected coral and fish assemblages near the Waianae ocean outfall, Oahu, Hawaii, 1994. In *Challenges of our changing global environment*, vol. 2, 896–906. Oceans '95 MTS/IEEE Conference Proceedings, October 9–12, San Diego, California.

Salas, J. D. 1993. Analysis and modeling of hydrologic time series. In *Handbook of hydrology*, ed. D. R. Maidment, 19.1–19.72. New York: McGraw-Hill.

Sanderson, M., ed. 1993. *Prevailing trade winds: Weather and climate in Hawaiʻi*. Honolulu: University of Hawaiʻi Press. 126 p.

Saner, R. J., and F. W. Pontius. 1991. Federal groundwater laws. *American Water Works Association Journal* 83 (3): 20–26.

Sato, H. H., W. Ikeda, and R. Paeth. 1973. Soil survey of the island of Hawaii, state of Hawaii. U.S. Department of Agriculture, Soil Conservation Service, Washington, D.C. 115 p.

Sawyer, C. N., P. L. McCarty, and G. F. Parkin. 1994. *Chemistry for environmental engineering*. New York: McGraw-Hill. 658 p.

Schaake, J. C., J. C. Geyer, and J. W. Knapp. 1967. Experimental examination of the Rational method. *Journal of the Hydraulics Division, American Society of Civil Engineers* 93 (HY6): 353–370.

Schmorak, S., and A. Mercado. 1969. Upconing of fresh water–sea water interface below pumping wells, field study. *Water Resources Research* 5 (6): 1290–1311.

Schneider, R. C., R. E. Green, J. D. Wolt, R. K. H. Loh, D. P. Schmitt, and B. S. Sipes. 1995. 1, 3-dichloropropene distribution in soil when applied by drip irrigation or injection in pineapple culture. *Pesticide Science* 43:97–105.

Schoklitsch, A. 1934. Der Geschietrieb und die Geschiebefracht. *Wasserkraft und Wasserwirtschaft* 19 (4): 37–43.

Schroeder, T. A. 1976. Meteorological aspects of the Oahu flood of 19 April 1974. Technical Report No. 97, Water Resources Research Center, University of Hawaiʻi at Mānoa, Honolulu. 62 p.

———. 1977. Meteorological analysis of an Oahu flood. *Monthly Weather Review* 105:458–468.

———. 1993a. Climate controls. In *Prevailing trade winds: Weather and climate in Hawaiʻi*, ed. M. Sanderson, 12–36. Honolulu: University of Hawaiʻi Press.

———. 1993b. Wind and solar energy. In *Prevailing trade winds: Weather and climate in Hawaiʻi*, ed. M. Sanderson, 82–85. Honolulu: University of Hawaiʻi Press.

Schroeder, T. A., B. J. Kilonsky, and B. N. Meisner. 1977a. Diurnal variation in rainfall and cloudiness. Technical Report No. 112, Water Resources Research Center, University of Hawaiʻi at Mānoa, Honolulu. 67 p.

Schroeder, T. A., B. J. Kilonksy, and C. S. Ramage. 1977b. Diurnal rainfall variability over the Hawaiian Islands. In *Collected reprints, Volume V: 1978–1981*, ed. F. N. Fujimura and A. W. L. Kam, 12–17. Water Resources Research Center, University of Hawaiʻi at Mānoa, Honolulu.

Schulze-Makuch, D., D. A. Carlson, D. S. Cherkauer, and P. Malik. 1999. Scale dependency of hydraulic conductivity in heterogeneous media. *Ground Water* 37 (6): 904–919.

Seckel, G. R. 1962. Atlas of the oceanographic climate of the Hawaiian islands region. Fishery Bulletin 193, U.S. Fish and Wildlife Service 61:371–427.

Serrano, S. E. 2001. Explicit solution to Green and Ampt infiltration equation. *Journal of Hydrologic Engineering, American Society of Civil Engineers* 6 (4): 336–340.

Seto, Y.-B., R. A. Duce, and A. H. Woodcock. 1969. Sodium-to-chlorine ratio in Hawaiian rains as a function of distance inland and elevation. *Journal of Geophysical Research* 74 (4): 1101–1103.

Shade, P. J. 1984. Hydrology and sediment transport, Moanalua Valley, Oahu, Hawaii. Water-Resources Investigations Report 84-4156, U.S. Geological Survey, Honolulu, Hawaiʻi. 54 p.

Shade, P. J., and W. D. Nichols. 1996. Water budget and the effects of land-use changes on ground-water recharge, Oahu, Hawaii. U.S. Geological Survey Professional Paper 1412-C. 38 p.

Shapiro, A. H. 1961. Basic equations of fluid flow. In *Handbook of fluid dynamics*, ed. V. Streeter, 1–19. New York: McGraw-Hill.

Shen, H. W., and P. Y. Julien. 1993. Erosion and sediment

transport. In *Handbook of hydrology*, ed. D. R. Maidment, 12.1–12.61. New York: McGraw-Hill.

Sherman, L. K. 1932. Streamflow from rainfall by the unit-graph method. *Engineering News Record* 108:501–505.

Shuttleworth, W. J. 1991. Evaporation models in hydrology. In *Land surface evaporation: Measurement and parameterization*, ed. T. J. Schumgge and J. C. Andre, 93–120. New York: Springer-Verlag.

———. 1993. Evaporation. In *Handbook of hydrology*, ed. D. R. Maidment, 4.1–4.53. New York: McGraw-Hill.

Simonson, R. W. 1962. Soil classification in the United States. *Science* 137:1027–1034.

Slichter, C. S. 1905. Field measurement of the rate of movement of underground waters. U.S. Geological Survey Water Supply Paper No. 140.

Smith, J. A., D. J. Seo, M. L. Baeck, and M. D. Hudlow. 1996. An intercomparison study of NEXRAD precipitation estimates. *Water Resources Research* 32 (7): 2035–2045.

Smith, L., and S. W. Wheatcraft. 1993. Groundwater flow. In *Handbook of hydrology*, ed. D. R. Maidment, 6.1–6.58. New York: McGraw-Hill.

Smith, S. V., W. J. Kimmerer, E. A. Laws, R. E. Brock, and T. W. Walsh. 1981. Kaneohe Bay sewage diversion experiment: Perspectives on ecosystem responses to nutritional perturbation. *Pacific Science* 35:279–402.

Smolarkiewicz, P. K., R. M. Rasmussen, and T. L. Clark. 1988. On the dynamics of Hawaiian cloud bands, island forcing (Part I). *Journal of Atmospheric Sciences* 45 (13): 1872–1905.

Soicher, N. E. 1996. A revised parallel column model for simulating solute transport in soils. Ph.D. dissertation, University of Hawai'i at Mānoa, Honolulu. 113 p.

Souza, W. R., and W. Meyer. 1995. Numerical simulation of regional changes in ground-water levels and in the freshwater-saltwater interface induced by increased pumpage at Barbers Point shaft, Oahu, Hawaii. Water-Resources Investigations Report 95-4206, U.S. Geological Survey, Honolulu, Hawai'i. 47 p.

Souza, W. R., and C. I. Voss. 1987. Analysis of an anisotropic coastal aquifer system using variable-density flow and solute transport simulation. *Journal of Hydrology* 92:17–41.

*———. 1989. Assessment of potable groundwater in a freshwater lens using variable-density-flow and solute transport simulation. *Proceedings*, National Water Well Association Conference on solving ground water problems with models. Indianapolis, Indiana. 20 p.

Sposito, G. 1986. The physics of soil water physics. *Water Resources Research* 22 (9): 83S–88S.

Stearns, H. T. 1940. Geology and ground-water resources of the islands of Lanai and Kahoolawe, Hawaii. Bulletin No. 6, Hawai'i Division of Hydrography, Honolulu. 177 p.

———. 1947. Geology and ground-water resources of the island of Niihau, Hawaii, 3–38. Bulletin No. 12, Hawai'i Division of Hydrography, Honolulu.

———. 1966. *Geology of the state of Hawaii*. Palo Alto, California: Pacific Books Publishers. 266 p.

Stearns, H. T., and G. A. Macdonald. 1942. Geology and ground-water resources of the island of Maui, Hawaii. Bulletin No. 7, Hawai'i Division of Hydrography, Honolulu. 344 p.

———. 1946. Geology and ground-water resources of the island of Hawaii. Bulletin No. 9, Hawai'i Division of Hydrography, Honolulu. 363 p.

———. 1947. Geology and ground-water resources of the island of Molokai, Hawaii. Bulletin No. 11, Hawai'i Division of Hydrography, Honolulu. 113 p.

Stearns, H. T., and K. N. Vaksvik. 1935. Geology and ground-water resources of the island of Oahu, Hawaii. Bulletin No. 1, Hawai'i Division of Hydrography, Honolulu. 479 p.

Stephens, D. B. 1996. *Vadose zone hydrology*. Boca Raton, Florida: Lewis Publishers. 339 p.

*Stevenson, M., J. O'Connor, and J. Aldrick. 1996. Pollutant source identification. In Mamala Bay Study, Final Report, Vol. I. Honolulu.

Stidd, C. K., and L. B. Leopold. 1951. The geographic distribution of average monthly rainfall, Hawaii. *Meteorological Monographs* 1 (3): 24–33.

Strahler, A. N. 1964. Quantitative geomorphology of drainage basins and channel networks. In *Handbook of applied hydrology*, ed. V. T. Chow, 4-1 to 4-76. New York: McGraw-Hill.

———. 1965. *Introduction to physical geography*. New York: Wiley.

Streeter, H. W., and E. B. Phelps. 1925. A study of the pollution and natural purification of the Ohio River. U.S. Public Health Bulletin No. 146 (reprinted by U.S. Department of Health, Education, and Welfare, 1958).

Sverdrup, H. V., M. W. Johnson, and R. H. Fleming. 1946. *The oceans: Their physics, chemistry and general biology*. New York: Prentice-Hall. 1,087 p.

Swartz, J. H. 1937. Resistivity studies of some salt water boundaries in the Hawaiian islands. In *Transactions*, American Geophysical Union, 387–393.

Takahashi, T. 1988. Long-lasting trade-wind rain showers in a 3-dimensional model. *Journal of Atmospheric Sciences* 45 (22): 3333–3353.

Takasaki, K. J. 1978. Summary appraisals of the nation's

groundwater resources: Hawaii region. U.S. Geological Survey Professional Paper 813-M. 29 p.

Takasaki, K. J., G. T. Hirashima, and E. R. Lubke. 1969. Water resources of windward Oahu, Hawaii. Water-Supply Paper No. 1894, U.S. Geological Survey, Washington, D.C.

Takasaki, K. J., and J. F. Mink. 1981. Evaluation of major dike-impounded ground-water reservoirs, island of Oahu. Open-File Report 81-1119, U.S. Geological Survey, Honolulu. 139 p.

Tchobanoglous, G., and E. D. Schroeder. 1985. *Water quality.* Reading, Massachusetts: Addison-Wesley Publishing Co. 768 p.

Tenorio, P. A., R. H. F. Young, N. C. Burbank Jr., and L. S. Lau. 1970. Identification of irrigation return water in the subsurface; Phase III Kahuku, Oahu and Kahului and Lahaina, Maui. Technical Report No. 44, Water Resources Research Center, University of Hawai'i at Mānoa, Honolulu. 53 p.

*Tetra Tech Inc. 1994. Effluent fate study, Lahaina Wastewater Reclamation Facility, Maui, Hawaii. Final report. Lafayette, California. 73 p.

Thomas, D. M. 1977. The isotopic profile of gases from the summit and flank of Kilauea Volcano. Ph.D. dissertation, University of Hawai'i at Mānoa, Honolulu.

Thomas, D. M., F. L. Paillet, and M. E. Conrad. 1996. Hydrogeology of the Hawaii Scientific Drilling Project borehole KP-1, 2. Groundwater geochemistry and regional flow patterns. *Journal of Geophysical Research* 100 (B5): 11683–11694.

Timbol, A. S., and J. A. Maciolek. 1978. Stream channel modification in Hawaii. Part A statewide inventory of streams, habitat factors and associated biota. FWS/OBS-78/16, U.S. Fish and Wildlife Service, National Stream Alteration Team, Columbia, Missouri.

Timbol, A. S., A. J. Sutter, and J. D. Parrish. 1980. Distribution, relative abundance, and stream environment of *Lentipes concolor* (Gill, 1960), and associated fauna in Hawaiian streams. WRRC Cooperative Report No. 5, Water Resources Research Center, University of Hawai'i at Mānoa, Honolulu. 64 p.

Todd, D. K. 1959. *Groundwater hydrology.* New York: John Wiley and Sons. 336 p.

———. 1980. *Groundwater hydrology,* 2nd ed. New York: John Wiley and Sons. 535 p.

*Towill Corporation, R. M. 1978. Feasibility study: Surface water impoundment/recharge, Pearl Harbor basin, Oahu, Hawaii. Report prepared for U.S. Army Corps of Engineers Pacific Ocean Division.

Tucker, G. E., and R. Slingerland. 1997. Drainage basin responses to climate change. *Water Resources Research* 33 (8): 2031–2047.

Turnbull, S. J., N. J. Meyers, R. A. Howe, J. D. Daniel, and L. B. Haines. 1994. The use of stable and radioactive isotopes to evaluate regional groundwater flow in central Oahu, Hawaii. Harding Lawson Asssociates, Denver, Colorado. 9 p.

Uehara, G. 1973. Soils. In *Atlas of Hawaii,* ed. R. W. Armstrong, 39–41. Honolulu: University Press of Hawai'i.

Underwood, M. R., W. Meyer, and W. R. Souza. 1995. Ground-water availability from the Hawi aquifer in the Kohala area, Hawaii. Water-Resources Investigations Report 95-4113, U.S. Geological Survey, Honolulu, Hawai'i. 57 p.

U.S. Agricultural Research Service. 1961. A universal equation for predicting rainfall-erosion losses. U.S. Department of Agriculture. USDA-ARS Special Report, 22–26. Washington, D.C.

U.S. Army Corps of Engineers. 1971. HEC-4 monthly streamflow simulation. Hydrologic Engineering Center, Davis, California.

———. 1972. Hydrology, Kaneohe-Kailua area flood control and allied purposes, Oahu, Hawaii. Design Memorandum No. 1, Pacific Ocean Division, Honolulu. 11 p.

U.S. Environmental Protection Agency. 1987. Quality criteria for water. EPA-440/5-86-001. Washington, D.C.

———. 1993. Remediation technologies screening matrix: Reference guide. EPA 542-B-93-005. Solid Waste and Emergency Response, Washington, D.C. 142 p.

U.S. Federal Council for Science and Technology. 1962. Scientific hydrology. Ad-hoc Panel on Hydrology, Washington, D.C. 37 p.

U.S. Natural Resources Conservation Service. 1993. *Engineering field handbook.* Washington, D.C.

———. 2001. Physical properties of the soil, Island of Hawaii. Honolulu, Hawai'i.

U.S. Soil Conservation Service. 1969. *National engineering handbook,* section 4, Hydrology. U.S. Department of Agriculture, Washington, D.C.

———. 1970. Hawaiian supplement to *National engineering handbook,* section 4, Hydrology. U.S. Department of Agriculture, Washington, D.C.

U.S. Water Resources Council. 1976. Guidelines for determining flood flow frequency. Bulletin No. 17. Washington, D.C.

U.S. Weather Bureau. 1962. Rainfall-frequency atlas of the Hawaiian Islands. Technical Paper No. 43. U.S. Department of Commerce, Washington, D.C. 60 p.

———. 1963. Probable maximum precipitation in the Hawaiian Islands. Hydrometeorological Report No. 39. Department of Commerce, Washington, D.C.

Valenzuela, H. 1994. Drip irrigation. Vegetable crop update 4(2), Hawai'i Cooperative Extension Service, College of Tropical Agriculture and Human Resources, University of Hawai'i at Mānoa, Honolulu. 5 p.

Van Bavel, C. H. M. 1966. Potential evaporation: The combination concept and its experimental verification. *Water Resources Research* 2 (3): 455–467.

van't Woudt, B. D. 1960. Water level control in evaporation pans. *Journal of Geophysical Research* 65 (12): 4031–4035.

———. 1963. A pan evaporimeter for rainy areas. Technical Bulletin 57, Hawai'i Agricultural Experiment Station, University of Hawai'i at Mānoa, Honolulu. 39 p.

Veihmeyer, F. J. 1964. Evapotranspiration. In *Handbook of applied hydrology*, ed. V. T. Chow, 11-1 to 11-38. New York: McGraw-Hill.

Visher, F. N., and J. F. Mink. 1964. Ground-water resources in southern Oahu, Hawaii. Water-Supply Paper 1778, U.S. Geological Survey, Washington, D.C. 133 p.

Vogel, R. M., and I. Wilson. 1996. Probability distribution of annual maximum, mean, and minimum streams in the United States. *Journal of Hydrologic Engineering, American Society of Civil Engineers* 1 (2): 69–76.

Voss, C. I. 1984. SUTRA: A finite-element model for saturated-unsaturated, fluid-density dependent ground-water flow with energy transport or chemically-reactive singe species solute transport. Water-Resources Investigations Report 84-4369, U.S. Geological Survey, Reston, Virginia. 409 p.

Voss, C. I., and W. R. Souza. 1987. Variable density flow and solute transport simulation of regional aquifers containing a narrow freshwater-saltwater transition zone. *Water Resources Research* 23 (10): 1851–1866.

Voss, C. I., and W. W. Wood. 1993. Synthesis of geochemical, isotopics and groundwater modeling analysis to explain regional flow on a coastal aquifer of southern Oahu, Hawaii. In *Proceedings*, Mathematical models and their applications to isotope studies in groundwater hydrology, 147–178. International Atomic Energy Agency, Vienna, Austria.

Wang, R.-Y., and I.-P. Wu. 1972. Characteristics of short-duration unit hydrograph. *Transactions*, American Society of Agricultural Engineers 15 (3): 452–456.

Wentworth, C. K. 1928. Principles of stream erosion in Hawaii. *Journal of Geology* 36 (5): 385–410.

———. 1938. Geology and ground-water resources of the Palolo-Waialae district. Honolulu Board of Water Supply, Honolulu, Hawai'i. 274 p.

———. 1942. Storage consequences of the Ghyben-Herzberg theory. *Transactions*, American Geophysical Union 23:683–693.

———. 1946. Laminar flow in the Honolulu aquifer. *Transactions*, American Geophysical Union 27:540–548.

———. 1951. Geology and ground-water resources of the Honolulu–Pearl Harbor area, Oahu, Hawaii. Honolulu Board of Water Supply, Honolulu, Hawai'i. 111 p.

Wentworth, C. K., and G. A. Macdonald. 1953. Structures and forms of basaltic rocks in Hawaii. Bulletin 994, U.S. Geological Survey. 98 p.

Wheatcraft, S. W., and F. L. Peterson. 1979. Numerical modeling of liquid waste injection into a two-phase fluid system. Technical Report No. 125, Water Resources Research Center, University of Hawai'i at Mānoa, Honolulu. 103 p.

Wheatcraft, S. W., F. L. Peterson, and D. L. Heutmaker. 1976. Waste injection into the Hawaiian Ghyben-Herzberg aquifer: A laboratory study using a sand-packed hydraulic model. Technical Report No. 96, Water Resources Research Center, University of Hawai'i at Mānoa, Honolulu, Hawai'i. 88 p.

Wilcox, C. 1996. *Sugar water: Hawaii's plantation ditches*. Honolulu: University of Hawai'i Press. 191 p.

Williams, J. A., and T.-C. Liu. 1975. Response to tides of coastal aquifers: Analog simulation vs. field observation. Technical Report No. 86, Water Resources Research Center, University of Hawai'i at Mānoa, Honolulu. 67 p.

Williams, J. A., and R. L. Soroos. 1973. Evaluation of methods of pumping test analyses for application to Hawaiian aquifers. Technical Report No. 70, Water Resources Research Center, University of Hawai'i at Mānoa, Honolulu. 159 p.

Willocks, S. P., R. H. Esaki, R. K. Yukumoto, and L. S. Lau. 1961. A field investigation of infiltration for some soils on Oahu, Hawaii. Department of Civil Engineering, University of Hawai'i at Mānoa, Honolulu. 30 p.

Wilson, J. T. 1963. A possible origin of the Hawaiian Islands. *Canadian Journal of Physics* 41:863–870.

Wilson Okamoto and Associates. 1983. Instream use study: Windward Oahu. Report R68, Division of Water and Land Development, Department of Land and Natural Resources, State of Hawai'i, Honolulu. 154 p.

Wishmeier, W. H., and D. D. Smith. 1978. Predicting rainfall erosion losses — A guide to conservation planning. Agriculture Handbook No. 537, U.S. Department of Agriculture, Washington, D.C.

Wolman, A. 1962. Water resources: A report to the Committee on Natural Resources, National Academy of Sciences — National Research Council. Publication 1000-B, Washington, D.C.

Wong, M. F. 1994. Estimation of magnitude and frequency of flood for streams on the island of Oahu, Hawaii. Water-

Resources Investigation Report 94-4052, U.S. Geological Survey, Honolulu. 37 p.
Wood, H. B. 1971. Land use effects on the hydrologic characteristics of some Hawaiian soils. *Journal of Soil and Water Conservation* 26 (4): 158–160.
———. 1977. Hydrologic differences between selected forested and agricultural soils in Hawaii. *Soil Science Society of America Journal* 41:132–136.
Wu, I.-P. 1967. Hydrological data and peak discharge determination of small Hawaiian watersheds: Island of Oahu. Technical Report No. 15, Water Resources Research Center, University of Hawai'i at Mānoa, Honolulu. 97 p.
———. 1969. Hydrograph study and peak discharge determination of Hawaiian small watersheds: Island of Oahu. Technical Report No. 30, Water Resources Research Center, University of Hawai'i at Mānoa, Honolulu. 31 p.
———. 1972. Consumptive use and irrigation scheduling of Hawaiian vegetable crops. In *International Conference on Tropical and Subtropical Agriculture*, Honolulu, Hawai'i, 247–251. Special Pububliсation SP-01-72, American Society of Agricultural Engineers, St. Joseph, Michigan.
Wyrtki, K. 1982. The Southern Oscillation, ocean-atmosphere interaction and El Nino. *Marine Technology Society Journal* 16:3–10.
Yamamoto, T. 1963. Soil moisture constants and physical properties of selected soils in Hawaii. U.S. Forest Service Research Paper PSW-P2, Pacific Southwest Forest and Range Experiment Station, Berkeley, California, U.S. Forest Service, U.S. Department of Agriculture. 10 p.
Yeh, J. T.-C. 1989. One-dimensional steady state infiltration in heterogeneous soils. *Water Resources Research* 25 (10): 2149–2158.
Yeh, T. C., J. E. Carson, and J. J. Marciano. 1951a. On the relation between the circumpolar westerly current and rainfall over the Hawaiian Islands. *Meteorological Monographs* 1 (3): 47–55.
Yeh, T. C., C. C. Wallen, and J. E. Carson. 1951b. A study of rainfall over Oahu. *Meteorological Monographs* 1 (3): 34–46.
Yen, B. C., and K. T. Lee. 1997. Unit hydrograph derivation for ungauged watersheds by stream-order laws. *Journal of Hydrologic Engineering, American Society of Civil Engineers* 2 (1): 1–9.
Yevjevich, V. 1963. Fluctuation of wet and dry years, part 1, research data assembly and mathematical models. Hydrology Paper no. 1, Colorado State University, Fort Collins, Colorado.
Yim, S. K., and G. L. Dugan. 1975. Quality and quantity of nonpoint pollution sources in rural surface water runoff on Oahu, Hawaii. Technical Report No. 93, Water Resources Research Center, University of Hawai'i at Mānoa, Honolulu. 60 p.
Yost, R. S., G. Uehara, and R. L. Fox. 1982. Geostatistical analysis of soil chemical properties of large areas: I. Semivariograms and II. Kriging. *Soil Science Society of America Journal* 46:1028–1037.
Young, R. H. F. 1978. Urban storm water runoff in Hawaii. In *Proceedings*, International symposium on urban storm water management, 129–134. University of Kentucky, Lexington, July 24–27, 1978.
Young, R. H. F., G. L. Dugan, L. S. Lau, and H. Yamauchi. 1975. Entrophication and fish toxicity potentials in a multiple-use subtropical reservoir. Technical Report No. 89, Water Resources Research Center, University of Hawai'i at Mānoa, Honolulu. 175 p.
*Yuen and Associates. 1992. State water resources protection plan. Prepared for State Water Resource Management Commission, Honolulu.

# INDEX

## ABOUT THE AUTHORS

**L. Stephen Lau** is professor emeritus of civil engineering at the University of Hawai'i at Mânoa, where he was the director of the Water Resources Research Center from 1971 to 1990. He was also a visiting professor at the University of California, Berkeley, and a Senior Fulbright Scholar at the University of Malaya in Kuala Lumpur. He has been a consultant to the World Health Organization, the World Bank, and the United Nations; he has also served on the Hawai'i State Water Commission and as a pesticide expert for the State's Office of Environmental Quality Control. In 1989 he received the George Warren Fuller Award from the American Water Works Association for his distinguished service in the field of water supply. He received his Ph.D. from the University of California, Berkeley.

**John F. Mink** was engaged in water resources investigations over the last 50 years, mostly in Hawai'i but also in many other Pacific islands including Guam, the Northern Mariana Islands, the Marshall Islands, Okinawa, and the Philippines. He also conducted studies in Sri Lanka, Haiti, Korea, Taiwan, Venezuela, and Egypt. After several years with the U.S. Geological Survey, he was a hydrologist-geologist for the Honolulu Board of Water Supply, where he authored the O'ahu Water Plan. He later entered private consulting and was vice president of Mink and Yuen, Inc., in Hawai'i. He had an M.S. degree in the geophysical sciences from the University of Chicago and studied environmental engineering at the Johns Hopkins University under a U.S. Public Health Services fellowship. He passed away on October 18, 2005, before the manuscript entered production.

Production Notes for LAU AND MINK | HYDROLOGY OF THE HAWAIIAN ISLANDS
Cover and Interior designed by April Leidig-Higgins in MinionPro, with display type in Meta
Composition by Copperline Book Services and ASCO Typesetting
Printing and binding by The Maple-Vail Book Manufacturing Group
Printed on 60# Finch White Opaque, 500 ppi